S0-CAH-301

ORGANOTIN COMPOUNDS

VOLUME I

ORGANOTIN COMPOUNDS

IN THREE VOLUMES

Edited by ALBERT K. SAWYER

Department of Chemistry
University of New Hampshire
Durham, New Hampshire

Volume I

MARCEL DEKKER, INC., New York 1971

MARCEL DEKKER, INC.
95 Madison Avenue, New York, New York 10016

LIBRARY OF CONGRESS CATALOG CARD NUMBER 71-142895

ISBN 0-8247-1595-0

PRINTED IN THE UNITED STATES OF AMERICA

DEDICATED TO HENRY G. KUIVILA

for whom I have the greatest respect, both as an individual and as a chemist, and who is responsible for my first contact with organotin chemistry.

PREFACE

There has been a marked increase of activity in organotin chemistry in recent years. This work was undertaken to provide up-to-date, comprehensive coverage of this field, prepared by individuals who are well informed in their specialized areas. As editor I have been most fortunate to have such well-qualified authors for the individual chapters.

It is hoped that this book will be of value not only to active workers in organotin chemistry and related areas, but also to new workers in providing the present state of knowledge in the field. In addition to chapters on the chemistry of compounds containing tin bonded to the main group elements, there are chapters on tin, other element bonds, and on such specialized topics as organotin polymers, applications, biological effects and analyses. Complete referencing has not been attempted although such referencing has been attempted for publications since the comprehensive review article by Ingham, Rosenberg, and Gilman in *Chemical Reviews*, October, 1960.

I am particularly indebted to Drs. William Considine and Gerald Reifenberg for critical reviews of several chapters. In addition I wish to thank Dr. Paul Jones for translating the chapter on "Organotin Compounds with Sn–S, Sn–Se, and Sn–Te Bonds," and Mr. Ingo Hartmann for translating the chapter on "Organotin Compounds with Sn–P, Sn–As, Sn–Sb, and Sn–Bi Bonds." I wish also to express appreciation for the help, advice, and encouragement given by many friends too numerous to mention, but particularly for the reading of individual chapters by Drs. Henry Kuivila, Paul Jones, and John Uebel.

Durham, New Hampshire A. K. SAWYER

FOREWORD

There can be little question of the usefulness of a book that brings together up-to-date material in an active area of organometallic chemistry. Professor Sawyer has done this with *Organotin Compounds*. Not only is there included his continuing, extensive studies but also the cooperative efforts of a group of eminent chemists distinguished for the breadth and depth of their researches in different aspects of this area.

Where does organotin chemistry stand relative to organometallic chemistry as a whole? This is a query that carries with it a relatively high degree of subjectivity. Organotin compounds were among the first organometallic species to be investigated. They have an increasing importance as synthetic agents. However, for a long period no organometallic group equaled the versatility of Grignard reagents as synthetic tools. What of industrial applications? On a tonnage basis, tetraethyllead and tetramethyllead (used primarily as antiknock agents) for many years have exceeded other organometallic groups. In this connection we wrote only about 35 years ago: "Undoubtedly the greatest value of organometallic compounds is their laboratory use for synthesis. It is doubtful that any other group of organic compounds combines at the same time an astonishingly high utility in the laboratory with an equally low usefulness in industry." However, the industrial picture of organometallic chemistry has changed markedly in the last two or three decades.

What shall be said of the use of organotin compounds in the investigation of bonding? Here studies with every type of organometallic compound are important with this most fundamental of all concepts. Some organometallic compounds currently lend themselves to a greater extent than others to the development and testing of new principles by a variety of kinetic and spectroscopic measurements and techniques. One that comes to mind with organotin compounds is the Mössbauer effect.

What of biological or physiological properties? One is aware of the expanding applications of organotin compounds as biocides. But how is one to compare tonnage use of some materials with the extraordinary importance of trace quantities of metal combinations in basically vital processes such as animal enzymatic transformations and photosynthesis? In an exercise of this kind, involved with some comparisons or correlations of similarities and differences, how shall one evaluate the relative impact of organometallic types in the effective interdisciplinary bridging of different branches of chemistry as well as currently disparate areas of science generally.

How much consideration should be given to the recent and current rate of growth of research interest, both academic and industrial? Here organotin chemistry has a highly distinguished record. Might this be ephemeral, or may there be great pauses or interruptions in activity? It is known that more than 100 years passed before the problem of dialkyltins was clarified, and that more than 40 years went by between the times when the structure of tetrakis-(triphenylstannyl)tin was suggested and confirmed. But interruptions in some developments are not atypical. With greater activity in the broad domain of organometallic chemistry, the prospects of fewer discontinuities will improve.

Whatever the prognosis may be, it is reasonably certain that the significant development of organotin chemistry will continue. This book will assist by providing a most helpful background; by suggesting new avenues of approach; by indicating useful correlations, particularly because of the special situation of tin in the periodic arrangement of the elements; and by setting up research targets for those who hunt intentionally or unknowingly.

HENRY GILMAN

Iowa State University
Ames, Iowa

CONTRIBUTORS TO THIS VOLUME

A. J. BLOODWORTH, University College, London, England

ALWYN G. DAVIES, University College, London, England

EUGENE J. KUPCHIK, Department of Chemistry, St. John's University, Jamaica, New York

J. G. A. LUIJTEN, Institute for Organic Chemistry TNO, Utrecht, The Netherlands

E. V. VAN DEN BERGHE, Rijksuniversiteit Gent, Laboratorium voor Algemene en Anorganische Chemie-B, Gent, Belgium

G. P. VAN DER KELEN, Rijksuniversiteit Gent, Laboratorium voor Algemene en Anorganische Chemie-B, Gent, Belgium

G. J. M. VAN DER KERK, Institute for Organic Chemistry TNO, Utrecht, The Netherlands

L. VERDONCK, Rijksuniversiteit Gent, Laboratorium voor Algemene en Anorganische Chemie-B, Gent, Belgium

Volume 3 contains the Author and Subject Indexes

CONTENTS

CONTENTS OF OTHER VOLUMES

VOLUME 2

VOLUME 3

ORGANOTIN COMPOUNDS

VOLUME I

I. INTRODUCTION

G. J. M. VAN DER KERK AND J. G. A. LUIJTEN

Institute for Organic Chemistry TNO
Utrecht, The Netherlands

In 1849 Frankland, when studying the reaction between tin and ethyl iodide, obtained a product which in 1853 he characterized as diethyltin diiodide. In 1852 Löwig described the action of ethyl iodide on a tin-sodium alloy. Remarkably enough he did not observe any formation of tetraethyltin but isolated, in addition to triethyltin iodide and hexaethylditin, a product which he indicated as "diethyltin."

These first recorded preparations of organotin compounds mark the beginnings of organotin chemistry. The early development of organotin chemistry—and of organometal chemistry in general—contributed significantly to the establishment of the concept of valence. In its turn, main group organometallic chemistry gained much from the orderliness created in organic chemistry by the general acceptance of valence theory. On the other hand, organometallic chemistry as a whole has suffered badly under the early dictatorship of the strict valence rules of organic chemistry, which offered no room for π-complexes and related types of coordination compounds.

The development of organotin chemistry during the first fifty years was rather slow because of the lack of really efficient and attractive methods for making well-defined compounds at will. This situation was changed profoundly when, around the turn of the century, the organomagnesium halides became available as general-purpose alkylating and arylating agents. The easily accessible and manageable Grignard reagents have contributed enormously to the further development of organometal chemistry since the high electropositivity of magnesium allows the transfer of hydrocarbon radicals from this metal to almost any other main group metal. Through these reagents a wide variety of compounds of the type R_4Sn became readily available for the first time.

Of decisive importance for the development of organotin chemistry to its present scope ha^c also been the discovery by Kocheshkov in 1929 of the so-called redistribution reaction between compounds R_4Sn and $SnCl_4$ (or

$SnBr_4$). Through this reaction, in dependence on reaction conditions, the various organotin compounds of a lower degree of organic substitution can be obtained in high yields.

As a consequence of these landmarks an upsurge of organotin chemistry, both scientifically and technically, has occurred since 1920 and in particular since the thirties. In fact, the present large-scale application of organotin compounds found its start in the latter period. In 1936 Yngve, working with Carbide and Carbon Chemical Corp., discovered the heat-stabilizing effect of organotins on polyvinyl chloride (PVC) and other chlorinated-hydrocarbon polymers. Originally the development of this application was slow, but since 1950 a great variety, in particular of dibutyltin and lately also of dioctyltin derivatives, have gained ever increasing industrial importance for this purpose.

Factually, around 1950 organotin chemistry was already quite extensive but its scope had scarcely widened beyond that pictured in Krause and Von Grosse's monumental treatise on organometallic chemistry of 1937 "Die Chemie der Metall-organischen Verbindungen." The vast majority of organotin compounds fall within the classes:

$$R_4Sn \qquad R_3SnX \qquad R_2SnX_2 \qquad RSnX_3 \qquad R_3Sn{-}SnR_3$$

With very few exceptions, all known compounds contained simple, non-functional alkyl and aryl groups R, and the nature of the anionic substituent X showed little variation (mainly halogen, O, S, OH, OR and OCOR).

Some functionally substituted phenyl groups have been introduced at tin quite early by Nesmeyanov and Kocheshkov (1930) via the corresponding mercury compounds. A more direct method, developed by the same authors in 1935, involved the decomposition of stannic halide/aryl diazonium complexes by powdered metals. In this way *p*-halo-, *p*-methoxy- and *p*-methoxycarbonylphenyl groups have been attached to tin. The advent of organolithium chemistry around 1930, exploited especially by Gilman, facilitated the accessability of functionally substituted aryltins but did not widen very much the range of functional groups. Even at present, little further progress has been made since no better methods have become available, and aryltin chemistry must still be regarded as relatively underdeveloped.

Until 1955, the possibilities of introducing functionally substituted alkyl groups at tin were even worse. This situation was changed drastically as a consequence of the exploitation by Noltes, starting in 1955, of the properties of tin-hydrogen bonds. The high reactivity of these bonds allowed addition reactions of the following type:

$$R_3SnH + CH_2{=}CH{-}R' \longrightarrow R_3Sn{-}CH_2CH_2R'$$

Variation of the nature of R′ led to the introduction at tin of a great variety of aliphatic groups carrying a functional substituent in the β-position or beyond. Because of the stability of the tin-carbon bond formed, several of these substituents can be modified afterwards by means of nucleophilic reagents. In this way it has been possible to introduce into organotin compounds aliphatic groups containing alcohol, ether, acetal, carboxylic, ester, amide, cyano, amino and carbonyl functions. Similar addition reactions with alkynes allowed the attachment to tin of substituted vinyl groups. The addition reactions of tin-hydrogen bonds—generally referred to as hydrostannations—have been extended to the use of organotin di- and trihydrides and very recently even to tin tetrahydride, and to dienes and diynes. As a result, interesting new types of tin-containing polymers and cyclic systems have become available as well.

A further useful extension of the addition reactions of tin-hydrogen bonds has been the hydrostannation of other types of double bonds, such as C=O, C=S, C=N, and N=N. In these cases tin-hetero atom bonds are formed exclusively and the method thus allows the introduction at tin of a great variety of groups not attached via carbon (indicated as X in the general formulae given above). Organotin hydrides have, as a result, also served in the development of this side of organotin chemistry.

The high reactivity of the tin-hydrogen bond has in still another way proved useful for the synthesis of new types of organotin compounds. The catalytic decomposition of tin-hydrogen bonds by amines, first observed at Utrecht, leads to the formation of tin-tin bonds. Hexaorganoditins and compounds of the formula $(R_2Sn)_n$, which latter have been shown by Kuivila and by Neumann to be polymers or cyclic oligomers, can thus be obtained. A more versatile and purposeful method for the synthesis of compounds with tin-tin bonds became available, however, when the so-called hydrostannolysis reactions were discovered at Utrecht. Especially suited to the purpose mentioned is the hydrostannolysis of the tin-nitrogen bond, generally represented by:

$$\geq Sn-H + \,>N-Sn\leq \;\longrightarrow\; \geq Sn-Sn\leq + \,>NH$$

The hydrostannolysis reaction has been used to prepare a great diversity of polytin compounds as well as compounds with bonds between tin and other main group or transition metals.

Exchange reactions of organotin hydrides with organic halides in which halogen bound to carbon is replaced by hydrogen have become much-studied both with respect to mechanism and as a synthetic tool in organic chemistry.

Detailed studies concerned with the kinetics and mechanism of reactions of organotin hydrides carried out in particular by the Utrecht group have shown that there is a close analogy with hydrocarbon chemistry in that these compounds may react by a nucleophilic, an electrophilic or a radical pathway.

Groups attached to tin via chalcogen atoms or via atoms of the group V elements have been introduced by many methods during the past ten years. Noteworthy is the synthesis of the first aminotin compounds reported by various workers in 1962. Compounds with phosphorus, arsenic, and antimony bound to tin have been described since and their chemistry as well as that of the organotin-chalcogen compounds is now well-developed.

The fruitful development resulting from the discovery of the high reactivity of the tin-hydrogen bond has recently been followed by similar developments, based on the reactivity of tin-oxygen and tin-nitrogen bonds. The investigations in question have been mainly carried out by Davies and Valade, and by Lappert, respectively. Even the tin-carbon bond, generally regarded as rather inert, in special cases enters into addition reactions. This was first demonstrated for phenylethynyl- and allyltin compounds, but quite recently also for cyanomethyl-, acetonyl-, alkoxycarbonylmethyl- and *N*,*N*-dialkylcarbamylmethyltin derivatives. The reaction involved is the addition of the particular tin-carbon bond across the carbon-oxygen double bond in aldehydes and ketones. The tin-oxygen compounds thus formed can easily be hydrolyzed, and among the hydrolysis products organic compounds occur which are not accessible by other routes. This is another example of the use of organotin compounds as synthetic tools in organic chemistry. Further examples are the use of the bulkiness of the tributyltin group in the preparation of a polymer with a *cis* configuration, *viz. cis*-polychloroprene, and the use of the same group to protect a carboxyl group during the establishment of a peptide linkage in pencillin synthesis.

It is impossible to cover in a short survey like this all modern developments in organotin chemistry. The work of Seyferth on dihalocarbene insertion reactions with organotin compounds of various types and on Diels-Alder reactions of alkenyl- and alkynyltin compounds, for example, which have resulted in the synthesis of several highly interesting structures, can only be mentioned here. Also the numerous investigations on ring systems in which tin atoms occur as hetero atoms, as well as the many mechanistic studies concerned with the cleavage of tin-carbon bonds (Eaborn, Nasielski) must be skipped.

It may nevertheless be gathered from the foregoing, rather arbitrary, choice of subjects that research in the organotin field has shown a tremendous growth during the past twenty years. This is also reflected in the number of papers which appear per annum. From some twenty in 1950 it has grown to over four hundred in 1968. The industrial importance of organotin compounds has similarly risen during this period, as is clear from a consideration of production figures. In 1950 not more than one hundred tons of organotin compounds were produced, as compared with over ten thousand in 1968.

The only method of production of the starting materials, the tin tetraalkyls and tetraaryls, at first was the Grignard procedure. This is still the only one applicable to the production of tetraphenyltin. For the manufacture of tetrabutyl- and tetraoctyltin, however, two new methods have been introduced in recent years, *viz.* the Wurtz procedure and the alkylation by alkylaluminum compounds. Still more recent is the industrial realization of the direct synthesis of dialkyltin dihalides from tin and alkyl halides.

The first practical application of organotin compounds has been, as indicated, in the stabilization of polyvinyl chloride (PVC). This has remained up till now also the most important application. Development in this area has shown a gradual shift from the early dibutyltin dilaurate, via dibutyltin maleate and dibutyltin derivatives of maleic half esters to the highly efficient thiotin compounds, dibutyltin derivatives of mercaptoacetic esters. In the past few years a rapid growth in importance is observed of the non-toxic dioctyltin stabilizers in PVC for food-packaging purposes.

The latter use was suggested by the present authors in 1955 on the basis of several years of research into the biocidal properties of organotin compounds. This research, started in 1950, was successful in itself in that it very soon revealed the high antifungal activity of certain triorganotin compounds. Parallel investigations in industry have led since 1957 to the use of triphenyltin acetate, and later also the hydroxide, as agricultural fungicides. Tributyltin compounds, in most cases the oxide, have subsequently gained importance as general industrial biocides, e.g., for antifouling, water treatment, and paint and wood preservation. Organotin compounds are, moreover, used in a number of smaller applications as diverse as the removal of intestinal worms from chickens and the cold curing of silicone rubber so that it may be said that tin has become a metal which is unmatched in the diversity of its organic applications.

Not yet touched upon is the vast amount of research into the physical chemistry and the toxicology of organotin compounds. The tin atom is particularly suited to physical methods of investigation because of its nuclear properties which enable the use of nuclear magnetic resonance and Mössbauer spectroscopical techniques. An interesting feature of tin in its organic compounds is the variety of coordination numbers it can have. The investigations of years into the pharmacology and toxicology of organotin compounds have been an invaluable aid in the development of their many present applications.

In conclusion, it may be stated that among all organometallic compounds organotin compounds belong to the best studied in every respect: chemistry, physical properties, applications, and biological effects. There are no signs which would indicate the approach of a change in this position.

2. ORGANOTIN HYDRIDES

EUGENE J. KUPCHIK

Department of Chemistry
St. John's University
Jamaica, New York

I. Introduction

During the past few years organotin chemistry has been one of the most active areas in the field of organometallic chemistry. Within this area, the organotin hydrides have received considerable attention. Much of this interest stems from their unique reducing properties and from their ability to add across multiple bonds leading to compounds which could not otherwise be easily obtained.

Organotin hydrides are derivatives of stannane, SnH_4, in which at least one of the hydrogens is replaced by an alkyl or aryl group. Stannane may also be called tin hydride. In naming the organic derivatives of stannane, the hydride nomenclature is commonly used. Thus, $(n\text{-}C_4H_9)_3SnH$ is tri-*n*-butyltin hydride, $(C_6H_5)_2SnH_2$ is diphenyltin dihydride, and $(CH_3)_2SnClH$ is dimethylchlorotin hydride.

Although stannane was described in 1919 (*180*), its organic derivatives received little attention until about 1956 (*74*). The discovery in 1947 that lithium aluminum hydride reduces organotin halides to organotin hydrides (*53*) paved the way for future studies. The chemistry of stannane itself has only recently been expanded (*190c*). Since 1956, organotin hydride chemistry has undergone a rapid development, and new results of great theoretical and practical significance have emerged. The literature on organotin chemistry, including organotin hydride chemistry, prior to 1959 (*74*, *236*) and between 1959 and 1963 (*138*) has been reviewed. The reactions of organotin hydrides with organic compounds, in which their use as reducing agents is emphasized, have been reviewed up to about 1964 by Kuivila (*92*). A more recent review by this same author has appeared (*93*). The reactions of organotin hydrides with unsaturated organic compounds have been reviewed up to about 1965 (*139*, *238*); this aspect of organotin hydride chemistry is included in a book (*133*), in a doctoral dissertation (*113*), and in a review article (*75a*). A doctoral dissertation which describes how organotin hydrides can be used to prepare tin-metal bonds has been published (*30*). Organotin hydride chemistry is included in a recent book which covers the entire field of organotin chemistry up to about 1967 (*140*). Organotin chemistry, including organotin hydride chemistry, is being reviewed on a yearly basis (*209*). An attempt has been made in this chapter to cover the literature up to 1969. Some later work is also mentioned.

II. Preparation and Properties of Organotin Hydrides

A. Organotin Hydrides of the Type R_nSnH_{4-n}

Prior to 1947 the only method known for the preparation of organotin hydrides involved the reaction of a triorganotinsodium compound with ammonium chloride or ammonium bromide in liquid ammonia. In this manner trimethyltin hydride (*89*), triphenyltin hydride (*24*, *259*), and dimethylethyltin hydride (*20*) were prepared. In 1947 a more convenient and general method was reported involving the reaction of lithium aluminum hydride with an organotin halide in the presence of diethyl ether or dioxane (*53*). In this manner methyltin trihydride, dimethyltin dihydride, and trimethyltin hydride were prepared. This method was subsequently used for the preparation of triphenyltin hydride (*64*, *259*) and, with slight modification of the isolation procedure, for the preparation of triethyltin hydride, tri-*n*-propyltin hydride, tri-*n*-butyltin hydride, triphenyltin hydride, di-*n*-propyltin dihydride, di-*n*-butyltin dihydride, diphenyltin dihydride, and *n*-butyltin trihydride (*241*). Prior to 1963 triphenyltin hydride was the only known member of the triaryltin hydride series. The following triaryltin hydrides have since been prepared by the lithium aluminum hydride method: tris(*p*-tolyl), tris(*p*-fluorophenyl), and trimesityltin hydrides (*132*), and tris(*m*-tolyl), tris(*o*-tolyl), and tris(*o*-biphenylyl)tin hydrides (*225*). Lithium aluminum hydride also reduces bis(tri-*n*-butyltin) oxide, which is commercially available, to give tri-*n*-butyltin hydride in 88% yield (*29*). Tri-*n*-butyltin hydride can also be obtained from the reaction of lithium aluminum hydride with hexa-*n*-butylditin, but the yield is low (*10*).

Good yields of organotin hydrides can be obtained from the reaction of an organotin chloride with diethyl- or diisobutylaluminum hydride at $-30°$ to 0°C (*150*). By this method triethyltin hydride, diethyltin dihydride, di-*n*-butyltin dihydride, diphenyltin dihydride, ethyltin trihydride, *n*-butyltin trihydride, and phenyltin trihydride were prepared. Ethoxytriethyltin is also reduced to the hydride with diethylaluminum hydride. The following tri-organotin deuterides were prepared by allowing the corresponding chlorides to react with diethylaluminum deuteride: triethyltin deuteride, triisobutyltin deuteride, and triphenyltin deuteride (*162*). Di-*n*-butylaluminum hydride converts diethylaminostannanes into hydrides according to Eq. (1) (*105*):

$$R_{4-n}Sn[N(C_2H_5)_2]_n + n(n\text{-}C_4H_9)_2AlH \longrightarrow R_{4-n}SnH_n + n(n\text{-}C_4H_9)_2AlN(C_2H_5)_2 \quad (1)$$

The yield of hydride is 99% when R is methyl and *n* is one.

Tri-*n*-propyltin hydride (65% yield), tri-*n*-butyltin hydride (60% yield), and triphenyltin hydride (65% yield) were obtained by reduction of the corresponding halides with amalgamated aluminum and water (*242*). This method, however, does not appear to be generally satisfactory, since triethyltin chloride gave only about a 30% yield of hydride, while di-*n*-propyltin dichloride could not be reduced to the dihydride at all.

Triphenyltin hydride can be obtained in 82% yield by the hydrolysis, with aqueous ammonium chloride, of a tetrahydrofuran solution of bis(triphenyltin)magnesium (*227*). This latter compound is prepared by allowing triphenyltin chloride to react with magnesium in tetrahydrofuran using ethyl bromide as initiator. The hydrolysis of triphenyltinlithium with 1 *M* hydrochloric acid also gives triphenyltin hydride (69% yield) (*228*). The hydrolysis of tri-*n*-butyltinlithium with water affords the hydride in 54% yield (*229*). Triethyltin hydride results from the hydrolysis of triethylstannylbis(dimethylamino)borane, which is prepared by the reaction of triethyltinlithium with bis(dimethylamino)boron chloride (*176*).

Tri-*n*-propyltin hydride and tri-*n*-butyltin hydride can be prepared by thermal decomposition of the corresponding formate under vacuum (*177*):

$$R_3SnOOCH \xrightarrow{160°-180°C} R_3SnH + CO_2 \tag{2}$$

In the case of tri-*n*-butyltin formate, which is obtained from the reaction of bis(tri-*n*-butyltin) oxide with formic acid, a 60% conversion to hydride occurred.

Excellent yields of organotin hydrides result from the reaction of organotin methoxides with diborane (*6*):

$$4\ R_3SnOCH_3 + B_2H_6 \longrightarrow 4\ R_3SnH + 2\ HB(OCH_3)_2 \tag{3}$$

$$2\ R_2Sn(OCH_3)_2 + B_2H_6 \longrightarrow 2\ R_2SnH_2 + 2\ HB(OCH_3)_2 \tag{4}$$

By this method were prepared triethyltin hydride, tri-*n*-butyltin hydride, triphenyltin hydride, and diethyltin dihydride. Diborane also converts diethylaminostannanes into tin hydrides in good yield (*105*):

$$2\ R_{4-n}Sn[N(C_2H_5)_2]_n + n\ B_2H_6 \longrightarrow 2\ R_{4-n}SnH_n + 2n\ H_2BN(C_2H_5)_2 \tag{5}$$

Sodium borohydride has been employed in the preparation of phenyltin trihydride from phenyltin trichloride (*5*). In this procedure the trichloride in aqueous potassium hydroxide is cooled to 0°C, sodium borohydride is added, and the mixture is treated with 6 *N* hydrochloric acid. The yield, after work-up, is 50%. This method has been used also to prepare methyltin trihydride from $K(CH_3SnO_2)$ (*56*). Organotin hydrides can be obtained from the reaction of excess sodium borohydride with organotin chlorides in ethylene glycol dimethyl ether (monoglyme) or diethylene glycol dimethyl ether (diglyme) (*14*, *15*). This method is quite convenient, and the yields are comparable to those obtained with lithium aluminum hydride.

Methyltin trihydride has been prepared in 63% yield by the following sequence of reactions (*7*):

$$SnH_4 + K \longrightarrow SnH_3K + \tfrac{1}{2}\ H_2 \tag{5a}$$

$$SnH_3K + CH_3I \longrightarrow CH_3SnH_3 + KI \tag{5b}$$

Organotin hydrides can be obtained in good yield by the reduction of organotin oxides or organotin alkoxides with the silicon-hydrogen bond of methylhydropolysiloxane, $(CH_3HSiO)_n$ (*76*). The organotin compound and the polysiloxane are mixed at room temperature, and the organotin hydride is removed by distillation. By this method tri-*n*-butyltin hydride (87% yield), di-*n*-butyltin dihydride (24% yield), and diethyltin dihydride (61% yield) were prepared. This reduction may also be accomplished with triphenylsilane or 1,1,3,3-tetraphenyldisiloxane, the latter compound being especially effective (*67*). For example, admixture of the disiloxane and bis(tri-*n*-butyltin) oxide (1:1 mole ratio) at room temperature produces tri-*n*-butyltin hydride in 65% yield. Admixture of the disiloxane and di-*n*-butyltin oxide (1:1 mole ratio) followed by heating under vacuum to 120°C results in a 78% yield of di-*n*-butyltin dihydride. Similar reactions have been reported by others (*10a*). The tri-*n*-butyltin hydride prepared by this method can be used, without isolation, to reduce organic halides and 4-methylcyclohexanone (*64a*).

Triorganotin hydrides result from the reaction of secondary phosphine oxides with bis(triorganotin) oxides (*75*).

The deuterium oxide hydrolysis of organotinsodium compounds, prepared by cleavage of compounds having tin–tin bonds with naphthalene–sodium, affords organotin deuterides (*90b*). The reaction of tri-*n*-butyltin hydride with a Grignard reagent followed by treatment of the reaction mixture with deuterium oxide affords tri-*n*-butyltin deuteride (*110a*).

A very convenient method for preparing diorganotin dihydrides involves the reaction of tri-*n*-butyltin hydride with diorganotin dichlorides (*198*):

$$2\,(n\text{-}C_4H_9)_3SnH + R_2SnCl_2 \longrightarrow R_2SnH_2 + 2\,(n\text{-}C_4H_9)_3SnCl \qquad (6)$$

In summary, the most convenient and, at the same time, most general methods for preparing organotin hydrides of the type R_nSnH_{4-n} (R = alkyl or aryl) appear to be those involving the reaction of an organotin halide with lithium aluminum hydride or sodium borohydride and the reaction of an organotin oxide with an organosilicon hydride.

The thermal stability of organotin hydrides of the type R_nSnH_{4-n} (R = alkyl or aryl) decreases as n decreases. Also, for a given n the alkyl compounds are far more stable than the aryl compounds. For a given n in the alkyl series the compound with the larger R tends to be the more stable. The course of the decomposition depends upon the conditions. Amines strongly catalyze the decomposition of triphenyltin hydride to hexaphenylditin and hydrogen (*224*, *243*). Primary amines are particularly effective; the rate of formation of hexaphenylditin decreases with increasing methyl substitution of the amine group in *p*-toluidine (*224*).

If diethylamine is added to an ether solution of diphenyltin dihydride, hydrogen and polymeric diphenyltin are formed (*101*). If the ether solution is diluted with methanol, a hydride, which appears to have the structure $H[(C_6H_5)_2Sn]_6H$, is formed (*101*, *145*, *147*). Decomposition of this hydride at 20°C in the presence of pyridine gives the cyclic hexamer $[(C_6H_5)_2Sn]_6$ (*145*, *147*) whose six-membered ring structure is supported by x-ray studies (*179*). The cyclic hexamer is also obtained when diphenyltin dihydride is decomposed at 20°C in the presence of pyridine, while decomposition in the presence of dimethylformamide gives the cyclic pentamer (*145*, *147*). Other diaryltin dihydrides also decompose to give cyclic hexamers (*148*). The decomposition of diethyltin dihydride in pyridine containing a trace of diethyltin dichloride produces the cyclic nonamer $[(C_2H_5)_2Sn]_9$ (*155*). The decomposition of dibenzyltin dihydride at 50°C in dimethylformamide produces a cyclic tetramer (*146*). The first heterocycle containing divalent tin in the ring was prepared by decomposition of an organotin dihydride (*94*):

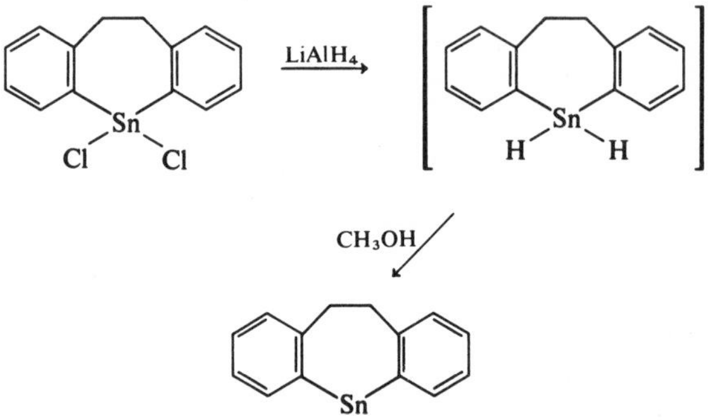

$$\xrightarrow{LiAlH_4} \qquad \xrightarrow{CH_3OH} \tag{7}$$

The decomposition of triphenyltin hydride in *n*-octane solution at 80°C is catalyzed by either oxygen or azobisisobutyronitrile (*45*). Thus, a free radical mechanism is indicated. The products of the azobisisobutyronitrile catalyzed decomposition in either refluxing *n*-hexane or benzene are hexaphenylditin, tetraphenyltin, and a trace of polymers. Photolysis of a *n*-octane solution of the hydride at 25°–50°C produces the same products. Dramatic evidence for the homolytic decomposition of the tin-hydrogen bond has been obtained by Schmidt and co-workers (*205*) who found that photolysis of triisobutyl- and triphenyltin hydride produced $R_3Sn\cdot$ radicals which could be collected on a cold finger and studied by the electron spin resonance technique. The esr signal of the triisobutyltin radical exhibited the greater line width and anisotropy, which indicates a greater localization of the unpaired electron on tin compared to the triphenyltin radical. The SnH_3 radical has been studied by the esr technique also (*76a*).

Because of the sensitivity of organotin hydrides to oxygen and to substances such as silicone stopcock grease, metallic tin, and aluminum halides (*150*), operations involving them should be carried out in clean glassware under an inert atmosphere (*92*).

Two compilations of many of the organotin hydrides prepared up to the present have been made (*92*, *140*). Some organotin hydrides of the type R_nSnH_{4-n}, for which proton magnetic resonance data are available, are listed in Appendix 1. Also given is the position of the SnH infrared absorption band. Several interesting trends are evident. For a given R in R_nSnH_{4-n}, ν(SnH) decreases with increasing *n*. For any R_nSnH_{4-n}, τ(SnH), which is a measure of the shielding of the proton bound to tin, is lower than for stannane itself. For a given R in R_nSnH_{4-n}, τ(SnH) decreases with increasing *n*. A theoretical interpretation of this trend in terms of the magnetic anisotropy of the SnC bond has been given (*49*). For R_3SnH, $\tau(SnH)$ tends to increase in the order $i\text{-}C_3H_7 < n\text{-}C_4H_9 \cong n\text{-}C_3H_7 \cong C_2H_5 < CH_3$. An explanation in terms of the hyperconjugation effect of the alkyl group has been suggested (*85*). For $(CH_3)_3SnH$ the following resonance structures may be written:

$$H{-}\overset{\displaystyle CH_3}{\underset{\displaystyle CH_3}{\overset{|}{\underset{|}{Sn}}}}{-}\overset{\displaystyle H}{\underset{\displaystyle H}{\overset{|}{\underset{|}{C}}}}{-}H \longleftrightarrow H{-}\overset{\displaystyle CH_3}{\underset{\displaystyle CH_3}{\overset{|\ominus}{\underset{|}{Sn}}}}{=}\overset{\displaystyle H^{\oplus}}{\underset{\displaystyle H}{\overset{}{\underset{|}{C}}}}{-}H$$

The second structure indicates that the electrons of the CH bonds are involved in bonding with vacant $5d$ orbitals on tin. The electron density of the proton bound directly to tin is thus increased resulting in a higher $\tau(SnH)$ value. The lower $\tau(SnH)$ value for triphenyltin hydride compared to trialkyltin hydrides may be due to a combination of the $-I$ effect and the ring current effect of the phenyl group. For the series $(CH_3)_nSnH_{4-n}$, the spin-spin coupling constants $J(^{119}SnH)$ and $J(^{117}SnH)$ decrease with increasing n. This trend has been explained by a combination of the theory of spin-spin coupling and the effects of the electronegativity of a substituent on the hybridization of a central atom (*56*). The ^{119}SnH and ^{117}SnH coupling constants are proportional to the *s*-character of the tin orbital directed toward hydrogen, assuming a dominant Fermi spin-spin interaction (*83, 190*). According to Bent (*11, 12*) the *s*-character of an atom tends to concentrate in orbitals directed toward the less electronegative groups. Since the methyl group is less electronegative than hydrogen (*226*), substitution of methyl for hydrogen in SnH_4 should cause a decrease in the amount of *s*-character of the tin orbital directed toward hydrogen. The coupling constants $J(^{119}SnH)$ and $J(^{117}SnH)$ should therefore decrease, as is found experimentally. The values of $J(^{119}SnH)$ and $J(^{117}SnH)$ for $(CH_3)_2(CF_2CF_2H)SnH$ are seen (Appendix 1) to be considerably higher than those for $(CH_3)_2SnH_2$. In this case replacement of a hydrogen on tin in $(CH_3)_2SnH_2$ by the more electronegative CF_2CF_2H group causes an increase in the *s*-character of the tin orbital directed toward the remaining hydrogen with a consequent increase in the values of the coupling constants. The *s*-character of the carbon orbital directed toward hydrogen in a methyl group is also increased since the carbon is now attached to a group which is more electronegative than hydrogen. Consequently, the $J(^{13}CH)$ value also increases. A comparison of the coupling constants $J(^{119}SnH)$ and $J(^{117}SnH)$ for $(CH_3)_3SnH$ and $(C_6H_5CH_2)_3SnH$ indicates that there is more *s*-character in the tin orbital directed toward hydrogen in the latter compound. It has been pointed out that the coupling constants $J(^{119}SnCH)$ and $J(^{117}SnCH)$ for these compounds cannot be interpreted by changes in *s*-electron density around the tin atom only (*244*). It has been suggested that there should be a linear relationship between these indirect tin-proton coupling constants and the percentage of *s*-character of the tin orbitals directed toward carbon (*73*). If this relationship holds for the hydrides under discussion, $J(\text{SnCH})$ for $(C_6H_5CH_2)_3SnH$ should be less than $J(\text{SnCH})$ for $(CH_3)_3SnH$. However, the opposite is true. Clearly, further research is indicated.

Nuclear magnetic double resonance experiments have been performed on some organotin hydrides (*48a*).

A good linear relationship has been found between $\nu(\text{SnH})$ and $J(^{119}\text{SnH})$ for alkyltin hydrides, R_nSnH_{4-n} (*134*):

$$\nu(\text{SnH}) = 946.2 - 707.1\,\sigma^* + (0.4947 + 0.3691\,\sigma^*)J(^{119}\text{SnH}) \qquad (8)$$

In Eq. (8) $\sigma^* = \sigma^*(\text{alkyl})\text{-}\sigma^*(\text{H})$ where $\sigma^*(\text{alkyl})$ is the polar substituent constant for R in a given R_nSnH_{4-n} and $\sigma^*(\text{H})$ is the polar substituent constant for hydrogen. Values for $\sigma^*(\text{alkyl})$ and $\sigma^*(\text{H})$ have been tabulated (*226*). A linear relationship between $\Sigma\sigma^*$ of alkyltin hydrides and the chemical shift of the hydrogen attached to tin has been found also (*104*). A correlation of tin-proton spin-spin coupling constants by pairwise interactions has been made (*245*). The proton magnetic resonance spectra of organotin hydrides have also been studied by Reeves (*190a*). The infrared spectra of CH_3SnH_3, CH_3SnD_3, and CD_3SnH_3 have been obtained, and a complete vibrational analysis has been made (*87*). The infrared and Raman spectra of $(CH_3)_3SnH$, $(C_2H_5)_3SnH$, $(n\text{-}C_4H_9)_3SnH$, and $(C_6H_5)_3SnH$ have been measured and the frequencies have been assigned (*90a*).

The Mössbauer spectra of stannane and some R_nSnH_{4-n} compounds have been measured (*71*). The isomer shifts (with respect to SnO_2) for the R_nSnH_{4-n} compounds fall into two groups: 1.24 ± 0.03 mm/sec for compounds with at least one CH_3—Sn bond and 1.42 ± 0.04 mm/sec for molecules not having a CH_3—Sn bond. As was observed for $(p\text{-}FC_6H_4)_3SnH$ (*72*), the R_nSnH_{4-n} spectra exhibit no quadrupole splitting. These results support a previous observation that quadrupole splitting is only observed when one of the atoms directly bonded to tin in a Sn(IV) compound has an unshared electron pair (*63*). It has been suggested that the quadrupole splitting is a result of overlap of the orbital containing the electron pair with an empty orbital on the tin atom (*63*, *71*). Some recent data seem to indicate that electronegativity effects alone can induce quadrupole splitting (*182*). It has been pointed out that organotin compounds in which the ligand symmetry around the tin atom is other than tetrahedral and in which no atom directly bonded to tin contains an unshared electron pair would be of interest for Mössbauer spectral studies (*70*).

A correlation of Mössbauer and nuclear magnetic resonance parameters for methyltin hydrides has been made (*135b*).

B. ORGANOTIN HYDRIDES WITH NEGATIVE SUBSTITUENTS

Di-*n*-butylacetoxytin hydride, the first example of an organotin compound containing a negative group on the tin bearing the hydrogen, was prepared by Sawyer and Kuivila (*199*). When equimolar amounts of di-*n*-butyltin dihydride and di-*n*-butyltin diacetate were mixed, the infrared spectrum showed a new SnH band at 1875 cm^{-1} in addition to the band of the dihydride at 1835 cm^{-1}. The relative intensities of the two bands varied suggesting the equilibrium:

$$(n\text{-}C_4H_9)_2SnH_2 + (n\text{-}C_4H_9)_2Sn(OCOCH_3)_2 \rightleftarrows 2(n\text{-}C_4H_9)_2SnH(OCOCH_3) \quad (9)$$

If the mixture is immediately cooled to $-70°C$, the di-*n*-butylacetoxytin hydride precipitates, and it can be recrystallized from diethyl ether at $-70°C$. The same equilibrium mixture, after evolution of hydrogen, is obtained when one mole of di-*n*-butyltin dihydride is allowed to react with one mole of acetic acid. When this equilibrium mixture is allowed to stand at room temperature, hydrogen is evolved and 1,1,2,2-tetra-*n*-butyl-1,2-diacetoxyditin is formed; the ditin compound reacts with acetic acid to give di-*n*-butyltin diacetate (*203*).

The reaction of di-*n*-butyltin diethoxide with an equimolar amount of methylhydropolysiloxane affords di-*n*-butylethoxytin hydride (*67*).

Di-*n*-butylchlorotin hydride, the first organohalotin hydride, was prepared by Sawyer and Kuivila (*201*) by allowing equimolar quantitites of di-*n*-buyltin dihydride and di-*n*-butyltin dichloride to react at room temperature:

$$(n\text{-}C_4H_9)_2SnH_2 + (n\text{-}C_4H_9)_2SnCl_2 \rightleftarrows 2\,(n\text{-}C_4H_9)_2SnClH \quad (10)$$

In contrast to the reaction represented by Eq. (9), the equilibrium shown in Eq. (10) lies far to the right at room temperature since the yield is quantitative. In this reaction the SnH band of the dihydride is immediately replaced by that of the chlorotin hydride at 1853 cm^{-1}. This reaction can be reversed, however, by distilling the reaction mixture under vacuum in which case the distillate affords di-*n*-butyltin dihydride, while the residue yields di-*n*-butyltin dichloride (95%). Di-*n*-butyltin dihydride reacts with other di-*n*-butyltin dihalides to give $(n\text{-}C_4H_9)_2SnXH$ ($X = F$, Br, or I) (*197*). These same di-*n*-butylhalotin hydrides may be obtained also from the reaction of tri-*n*-butyltin hydride with an equimolar amount of the appropriate di-*n*-butyltin dihalide (*196*):

$$(n\text{-}C_4H_9)_3SnH + (n\text{-}C_4H_9)_2SnX_2 \longrightarrow (n\text{-}C_4H_9)_3SnX + (n\text{-}C_4H_9)_2SnXH \quad (11)$$

Further reaction of the halotin hydride with tri-*n*-butyltin hydride affords di-*n*-butyltin dihydride. This latter reaction possibility was pointed out earlier by Neumann and Pedain (*156*) who also reported the preparation of R_2SnClH ($R = i\text{-}C_4H_9$, C_6H_5, or C_2H_5) and $(C_2H_5)_2SnXH$ ($X =$ Br or I). Tri-*n*-butyltin hydride also reacts with other R_2SnCl_2 compounds ($R = CH_3$,

C_2H_5, n-C_3H_7, i-C_4H_9, n-C_8H_{17}, cyclo-C_6H_{11}, or C_6H_5) to give the corresponding diorganochlorotin hydride (*198*). Addition of a second mole of tri-*n*-butyltin hydride converts the chlorotin hydrides to the corresponding dihydrides in each case. These reactions represent a convenient method for preparing diorganotin dihydrides.

Organohalotin hydrides are formed when organotin dihydrides are treated with halogen acids. Di-*n*-butylchlorotin hydride results from the reaction of di-*n*-butyltin dihydride with an equimolar amount of hydrochloric acid in dioxane at room temperature (*197*):

$$(n\text{-}C_4H_9)_2SnH_2 + HCl \longrightarrow (n\text{-}C_4H_9)_2SnClH + H_2 \qquad (12)$$

Ethylbromotin dihydride, a white crystalline compound, results when ethyltin trihydride is allowed to react with an equimolar amount of hydrogen bromide at $-78°C$ (*59, 60*). Reaction of triphenyltin hydride with hydrogen bromide at $-78°C$ gives diphenylbromotin hydride and benzene. In this case a phenyl group is eliminated from tin in preference to a hydrogen atom. Reaction of triphenyltin hydride with two moles of hydrogen bromide at $-78°C$ gives phenyldibromotin hydride. Cleavage of the third phenyl group with additional hydrogen bromide does not occur at $-78°C$.

Di-*n*-butylchlorotin hydride affords di-*n*-butyltin dichloride and hydrogen on treatment with hydrogen chloride (*197, 201*). Although di-*n*-butylhalotin hydrides are quite stable at room temperature, they are less stable than di-*n*-butyltin dihydride (*197*). As in the case of di-*n*-butylacetoxytin hydride, thermal decomposition affords hydrogen and a ditin compound:

$$2\,(n\text{-}C_4H_9)_2SnXH \longrightarrow H_2 + (n\text{-}C_4H_9)_2\underset{\displaystyle X}{\underset{|}{Sn}}\underset{\displaystyle X}{\underset{|}{Sn}}(n\text{-}C_4H_9)_2 \qquad (13)$$

The decomposition is catalyzed by pyridine and other amines (*156*). Ethylbromotin dihydride decomposes at room temperature to give hydrogen and $(C_2H_5SnBr)_n$ (*59, 60*). The phenyl derivative, on the other hand, affords benzene instead of hydrogen.

Proton magnetic resonance data and infrared data for some diorganohalotin hydrides are given in Table 1. A comparison of Appendix 1 and Table 1 reveals that for a given R in R_2SnH_2 replacement of H by any X ($X = Cl$, Br, or I) causes an increase in the SnH infrared absorption frequency. Also, for a given R in R_2SnXH the SnH absorption frequency tends to increase in the order $I < Br < Cl$. For a given R in R_2SnXH, τ(SnH) decreases as the *-I* effect of X increases. For a given X in R_2SnXH, the dimethylhalotin hydride exhibits the largest τ(SnH) value; this has been explained by hyperconjugation between the methyl groups and the tin atom (*84*). For a given R in R_2SnXH the coupling constants $J(^{117}SnH)$ and $J(^{119}SnH)$ increase in the order $H < I < Br < Cl$, which is to be expected, since the *s*-character of the

TABLE 1

PMR AND IR DATA FOR DIORGANOHALOTIN HYDRIDES

Compound	τ(SnH) (ppm)	$J(^{119}SnH)$ (cps)	$J(^{117}SnH)$ (cps)	ν(SnH) (cm^{-1})	Solvent	Ref.
$(CH_3)_2SnClH$	2.88	2228	2128	1877	Cyclohexane	(84)
	3.02			1877	Cyclohexane	(198)
$(CH_3)_2SnBrH$	3.31	2178	2082	1874	Cyclohexane	(84)
$(CH_3)_2SnIH$	3.92	2128	2032	1862	Cyclohexane	(84)
$(C_2H_5)_2SnClH$	2.67	2031	1940	1859	Cyclohexane	(84)
	2.55			1857	Neat	(198)
$(C_2H_5)_2SnBrH$	3.18				Cyclohexane	(84)
$(C_2H_5)_2SnIH$	3.79	1908	1823	1843	Cyclohexane	(84)
$(n\text{-}C_3H_7)_2SnClH$	2.68	2002	1914	1852	Cyclohexane	(84)
	2.41				Neat	(198)
$(n\text{-}C_3H_7)_2SnBrH$	3.15	1954	1862	1844	Cyclohexane	(84)
$(n\text{-}C_3H_7)_2SnIH$	3.63	1926	1836	1841	Cyclohexane	(84)
$(i\text{-}C_3H_7)_2SnClH$	2.70	1828	1745	1830	Cyclohexane	(84)
$(i\text{-}C_3H_7)_2SnBrH$	3.13	1786	1706	1830	Cyclohexane	(84)
$(i\text{-}C_3H_7)_2SnIH$	3.56	1736	1666	1830	Cyclohexane	(84)
$(n\text{-}C_4H_9)_2SnFH$	2.44			1875	Methanol	(197)
$(n\text{-}C_4H_9)_2SnClH$	2.58			1853	Neat	(197)
	2.80	2208	2108	1848	Tetrahydrofuran	(84)
	2.78	2145	2049	1852	Dioxane	(84)
	2.90	2113	2019	1835	Tetrahydro-thiophene	(84)
	2.69	2045	1956	1845	Diethyl ether	(84)
	2.66	2002	1911	1855	Cyclohexane	(84)
	2.69	1983	1890	1852	Neat	(84)
	2.72				Neat	(71)
$(n\text{-}C_4H_9)_2SnBrH$	2.91			1847	Neat	(197)
	3.13	1964	1875	1845	Cyclohexane	(84)
$(n\text{-}C_4H_9)_2SnIH$	3.92			1836	Neat	(197)
	3.85	1902	1817	1838	Cyclohexane	(84)
$(i\text{-}C_4H_9)_2SnClH$	2.46			1849	Neat	(198)
$(n\text{-}C_8H_{17})_2SnClH$	2.56			1845	Neat	(198)
$(cyclo\text{-}C_6H_{11})_2SnClH$	2.66			1831	Neat	(198)
$(C_6H_5)_2SnClH$	2.16			1879	Neat	(198)

tin orbital directed toward hydrogen should increase in this order. For a given X in R_2SnXH the coupling constants increase from isopropyl to methyl.

C. TETRAORGANODITIN DIHYDRIDES

The first tetraorganoditin dihydride, 1,1,2,2-tetra-*n*-butyl-1,2-dihydroditin, was obtained by Sawyer and Kuivila (*200*) from the reaction of the corresponding dichloride with lithium aluminum hydride:

$$(n\text{-}C_4H_9)_2\underset{\substack{|\\Cl}}{Sn}\underset{\substack{|\\Cl}}{Sn}(n\text{-}C_4H_9)_2 \xrightarrow{LiAlH_4} (n\text{-}C_4H_9)_2\underset{\substack{|\\H}}{Sn}\underset{\substack{|\\H}}{Sn}(n\text{-}C_4H_9)_2 \quad (14)$$

The required dichloride was prepared by allowing the corresponding diacetate to react with hydrogen chloride in diethyl ether. The SnH infrared absorption band of the ditin dihydride at 1795 cm^{-1} is considerably lower than that of di-*n*-butyltin dihydride. The structure of the ditin dihydride is supported by its conversion to the ditin diacetate with acetic acid and by quantitative reaction of the ditin diacetate with bromine to give di-*n*-butylacetoxybromotin. Furthermore, its pmr spectrum shows two sets of satellites resulting from coupling between tin and the directly bound proton [$J(^{119}SnH)$, 1671; $J(^{117}SnH)$, 1601] and from couplings acting over two bonds [$J(^{119}SnSnH)$, 1389; $J(^{117}SnSnH)$, 1453] (*235*). Lithium aluminum hydride reduction of tetraisobutylditin 1,2-dichloride, obtainable from the decomposition of diisobutylchlorotin hydride, affords the corresponding ditin dihydride in 45% yield [ν(SnH), 1793 cm^{-1}] (*222*). Tetraisobutylditin 1,2-dihydride reacts with $R_3SnN(C_2H_5)_2$ to give tetrastannanes (*222*) and with $(i\text{-}C_4H_9)_2Sn[N(C_2H_5)_2]_2$ to give the cyclostannane, $[(i\text{-}C_4H_9)_2Sn]_9$ (*157*).

The interesting inorganic hydrides, H_3SnMH_3 (M = Si or Ge), have been prepared by lithium aluminum hydride reduction of the corresponding hexachlorides or hexaacetates (*257*). The hexachlorides are obtained by treating the hexaacetates with hydrogen chloride; the hexaacetates are obtained by treating the hexaphenyl derivatives with acetic acid. These hexahydrides decompose above −80°C according to the equation:

$$H_3SnMH_3 \longrightarrow MH_4 + Sn + H_2 \qquad (M = Si \text{ or } Ge) \qquad (14a)$$

Ditin hexahydride can be obtained in low yield by the addition of aqueous potassium stannite and potassium borohydride to aqueous hydrochloric acid (*78*, *79*); it decomposes below room temperature.

D. Pentaorganoditin Monohydrides

Pentaorganoditin monohydrides can be prepared by allowing a dialkyl- or diaryltin dihydride to react with an equimolar amount of a trialkyl(*N*-phenylformamido)tin (*30*, *33*):

$$R_3SnN(C_6H_5)CH{=}O + R'_2SnH_2 \longrightarrow R_3SnSnR'_2H + C_6H_5NHCH{=}O \quad (14b)$$

The required trialkyl(*N*-phenylformamido)tin is prepared by allowing a trialkyltin hydride to react with an equimolar amount of phenyl isocyanate (see Sec. III. F.).

The pentaorganoditin monohydrides are useful for the synthesis of linear tri-, tetra-, penta-, and hexatin derivatives (*30*, *33*). For example, catalytic decomposition of a pentaorganoditin monohydride by pyridine or diethylamine affords a linear tetratin compound:

$$2\ R_3SnSnR'_2H \longrightarrow R_3SnSnR'_2SnR'_2SnR_3 + H_2 \quad (14c)$$

A linear pentatin compound can be prepared by the following sequence of reactions:

$$R_3SnSnR_2H + C_6H_5N{=}C{=}O \longrightarrow R_3SnSnR_2N(C_6H_5)CH{=}O \quad (14d)$$

$$2\ R_3SnSnR_2N(C_6H_5)CH{=}O + R'_2SnH_2 \longrightarrow R_3SnSnR_2SnR'_2SnR_2SnR_3 + 2\ C_6H_5NHCH{=}O \quad (14e)$$

III. Reactions of Organotin Hydrides with Organic Compounds

A. Reduction of Alkyl and Aryl Halides

1. *Alkyl Halides*

Organotin hydrides reduce alkyl halides:

$$(4\text{-}n)RX + R'_nSnH_{4-n} \longrightarrow (4\text{-}n)RH + R'_nSnX_{4-n} \quad (15)$$

In this equation R is an alkyl or substituted alkyl group. This reaction was discovered by Noltes and van der Kerk (*172*) who observed that triphenyltin hydride reacts with allyl bromide to give triphenyltin bromide (98% yield) and propene. Significantly, reduction occurs in preference to addition across the double bond (see Sec. III. C). The scope of this reaction has been thoroughly explored by Kuivila et al. (*98*, *99*). The order of reactivity for alkyl halides is $RI > RBr > RCl > RF$. *n*-Butyl and phenyltin hydrides follow the reactivity sequence $(C_6H_5)_2SnH_2 \cong n\text{-}C_4H_9SnH_3 > (C_6H_5)_3SnH \cong (n\text{-}C_4H_9)_2SnH_2 > (n\text{-}C_4H_9)_3SnH$. Catalysis by azobisisobutyronitrile, a well-known radical initiator, not only suggests a free radical mechanism but also broadens the scope of the reaction. For example, tri-*n*-butyltin hydride in toluene at 80°C reduces benzyl chloride and cyclohexyl chloride to the extents of 100% and 70% in the presence of 1.5 mole% of catalyst but only to the extents of 26% and 1% in the absence of catalyst. The vicinal dibromides, 1,2-dibromopropane and meso-stilbene dibromide, undergo elimination on reaction with tri-*n*-butyltin hydride:

$$RCHBrCHBrR' + 2\ (n\text{-}C_4H_9)_3SnH \longrightarrow RCH{=}CHR' + H_2 + 2\ (n\text{-}C_4H_9)_3SnBr \quad (16)$$

α-Haloketones are reduced to the unhalogenated ketones in high yields. An organotin halide may be used as a "hydride carrier" for the reduction of organic halides by lithium aluminum hydride (*98*):

$$LiAlH_4 + 4\ R_3SnX \longrightarrow LiAlX_4 + 4\ R_3SnH \quad (17)$$

$$R_3SnH + R'X \longrightarrow R'H + R_3SnX \quad (18)$$

This method is based upon the fact that lithium aluminum hydride reacts more rapidly with organotin halides than with organic halides.

There is much experimental evidence to support the radical chain mechanism proposed by Kuivila et al. (*99*, *137*):

$$\equiv SnH + Q\cdot \longrightarrow \equiv Sn\cdot + QH \quad (19)$$

$$\equiv Sn\cdot + RX \longrightarrow \equiv SnX + R\cdot \quad (20)$$

$$R\cdot + \equiv SnH \longrightarrow \equiv Sn\cdot + RH \quad (21)$$

In this sequence Q· is a free radical or molecule capable of abstracting a hydrogen atom from a triorganotin hydride. Q· may come from a variety of sources, including the reactants. The catalysis by azobisisobutyronitrile, mentioned above, strongly supports this mechanism. Further support is provided by the observation that the reaction is retarded by the radical inhibitor hydroquinone (*137*).

Many observations provide support for the formation of intermediate alkyl radicals. The reduction of optically active halides produces racemic hydrocarbons. The reaction of optically active α-phenylethyl chloride with triphenyltin deuteride gives racemic α-deuterioethylbenzene (*99*, *137*):

$$(+)\text{-}C_6H_5CHClCH_3 + (C_6H_5)_3SnD \longrightarrow (\pm)\text{-}C_6H_5CHDCH_3 + (C_6H_5)_3SnCl \quad (22)$$

The reaction of optically active 1-bromo-1-methyl-2,2-diphenylcyclopropane with trimethyltin hydride affords racemic 1-methyl-2,2-diphenylcyclopropane (*218*):

$$(-)\text{-}\underset{C_6H_5\;\;C_6H_5}{\triangle}(CH_3)(Br) + (CH_3)_3SnH \longrightarrow (\pm)\text{-}\underset{C_6H_5\;\;C_6H_5}{\triangle}(H)CH_3 + (CH_3)_3SnBr \quad (23)$$

These results eliminate an Sn2 mechanism and a four-center mechanism involving flank attack by the tin-hydrogen bond, since these would yield an optically active hydrocarbon. On the other hand, the reduction of (+)-1-bromo-1-methyl-2,2-diphenylcyclopropane with triphenyltin hydride occurs with net inversion (*4a*). In this case the bromine side of the radical may be blocked by triphenyltin bromide, so that hydrogen abstraction from triphenyltin hydride occurs predominantly from the other side.

The observation that bridgehead bromides can be reduced with triphenyltin hydride (*106*, *107*) tends to eliminate an Sn1 or Sn2 mechanism, since bridgehead halides are known to be very unreactive in these kinds of reactions. On the other hand, it is well known that bridgehead free radicals can be easily formed (*188*).

Propargylic bromide (*137*) and propargylic chlorides (*52*) afford a mixture of the corresponding acetylene and allene upon reaction with tri-*n*-butyltin hydride. These results are consistent with the formation of an intermediate hybrid propargyl radical:

$$[HC{\equiv}C\dot{C}R_2 \longleftrightarrow H\dot{C}{=}C{=}CR_2] \xrightarrow{(n\text{-}C_4H_9)_3SnH} HC{\equiv}CCR_2H + H_2C{=}C{=}CR_2 \quad (24)$$

In the case of the propargylic chlorides, the free radical nature of the reaction was further indicated by initiation and inhibition studies. The observation that separate reaction of α-methallyl chloride and γ-methallyl chloride with triphenyltin hydride gives the same mixture of butenes can also be explained by assuming an intermediate hybrid radical (*137*):

$$\begin{matrix} CH_3CH{=}CHCH_2Cl \\ \text{or} \\ CH_2{=}CHCHCl(CH_3) \end{matrix} \xrightarrow{(C_6H_5)_3SnH} [CH_3CH{=}CH\dot{C}H_2 \longleftrightarrow CH_3\dot{C}HCH{=}CH_2]$$

$$\xrightarrow{(C_6H_5)_3SnH} CH_3CH{=}CHCH_3 + CH_3CH_2CH{=}CH_2 \quad (25)$$

Allyl radicals produced by reaction of a *cis*- or *trans*-allyl chloride with triphenyltin hydride have been found to undergo facile *cis-trans* isomerization (*46*). The fact that reduction of either a norbornenyl halide or nortricyclyl halide with tri-*n*-butyltin hydride produces the same mixture of norbornene and nortricyclene is in accord with a mechanism involving common free radical intermediates (*255*):

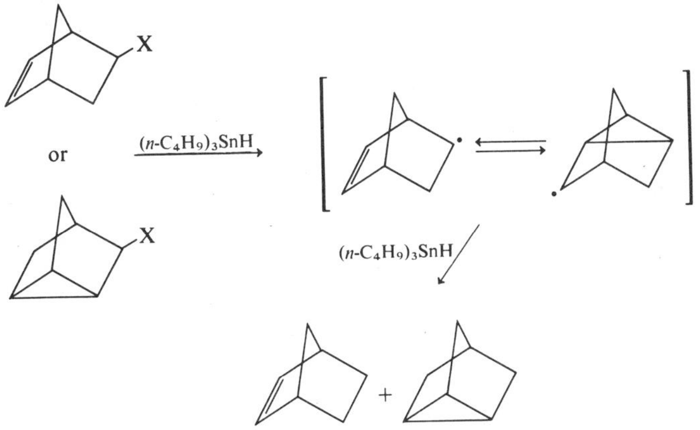

X
or $\xrightarrow{(n\text{-}C_4H_9)_3SnH}$
X

$(n\text{-}C_4H_9)_3SnH$

+

(26)

The chloride and bromide give the same results. Also, the same product composition is obtained irrespective of whether or not the reaction is catalyzed by azobisisobutyronitrile. Temperature has only a slight effect on the product composition. In these reactions the initially formed radical undergoes equilibration faster than it can abstract a hydrogen atom from tri-*n*-butyltin hydride. The ratio of nortricyclene to norbornene from the neat reduction of nortricyclyl bromide is higher when triphenyltin hydride is employed. In this case the initially formed nortricyclyl radical abstracts hydrogen from the more reactive triphenyltin hydride before equilibration is complete.

The following relative rates of reduction of cyclic bromides by tri-*n*-butyltin hydride at 45°C have a bearing on the question of anchimeric assistance in these reactions: cyclopentyl, 1.0; cyclopent-3-enyl, 1.73; cyclohexyl, 0.62; exo-bicyclo[2.2.1]hept-2-yl, 0.81; exo-norbornenyl, 1.40; endo-norbornenyl, 0.37; and nortricyclyl, 0.31. These results indicate that, compared to solvolysis reactions, there is no significant anchimeric assistance. For example, exo-norbornenyl bromide is seen to be only about 3.5 times more reactive than the endo isomer; the acetolysis of exo-norbornenyl brosylate, on the other hand, proceeds about 7000 times faster than that of the endo isomer (*258*). The reduction of *syn*- or *anti*-7-bromonorbornene with tri-*n*-butyltin deuteride has been reported to give *anti*-7-deuterionorbornene, possibly by way of a nonclassical free radical (*254*). However, the intervention of a nonclassical free radical in this case (*40a*, *192a*) and in others (*40a*, *76b*, *76c*) has been seriously questioned. The fact that cholesteryl chloride (**1**) yields only 5-cholestene (**2**) on reaction with triphenyltin hydride, whereas cyclocholestanyl chloride (**3**) affords a mixture of 3,5-cyclocholestane (**4**) and 5-cholestene has been interpreted in terms of the classical free radicals (**5**) and (**6**) (*40*):

(3) Cl $\xrightarrow{(C_6H_5)_3Sn\cdot}$ (5) $\xrightarrow{(C_6H_5)_3SnH}$ (4)

(5) ↓ (6)

(1) Cl $\xrightarrow{(C_6H_5)_3Sn\cdot}$ (6) $\xrightarrow{(C_6H_5)_3SnH}$ (2)

(26a)

Radical (**5**) rearranges partially to the more stable radical (**6**) before undergoing reaction with triphenyltin hydride.

The fact that *cis*-α-bromostilbene and *trans*-α-bromostilbene react separately with triphenyltin hydride at 26°–30°C to give the same mixture of *cis*- and *trans*-stilbene appears to be consistent with a mechanism involving the formation of intermediate vinyl-type radicals (*107*):

$$\begin{array}{c} C_6H_5\backslash C{=}C/C_6H_5 \\ H/ \quad \backslash Br \\ \text{or} \\ C_6H_5\backslash C{=}C/Br \\ H/ \quad \backslash C_6H_5 \end{array} \xrightarrow{(C_6H_5)_3SnH} \left[\begin{array}{c} C_6H_5\backslash C{=}\dot{C}/C_6H_5 \\ H/ \end{array} \rightleftarrows \begin{array}{c} C_6H_5\backslash C{=}\dot{C} \\ H/ \quad \backslash C_6H_5 \end{array} \right] \tag{27}$$

$$\xrightarrow{(C_6H_5)_3SnH} \begin{array}{c} C_6H_5\backslash C{=}C/C_6H_5 \\ H/ \quad \backslash H \end{array} + \begin{array}{c} C_6H_5\backslash C{=}C/H \\ H/ \quad \backslash C_6H_5 \end{array}$$

The *trans* isomer predominates (about 60% yield) which indicates that the equilibrium shown in Eq. (27) lies to the right. At 26°–42°C the equilibrium is further displaced to the right since the yield of *trans*-stilbene from *cis*-α-bromostilbene is about 98%. Another mechanistic possibility is a free radical addition of the triphenyltin hydride to the double bond followed by a four-center elimination of triphenyltin bromide. Such an addition-elimination mechanism has been demonstrated in some cases (*2*, *44*).

Still further support for the free radical chain mechanism shown in Eqs. (19)–(21) comes from the observation that reaction of *trans*- or *cis*-9-chlorodecalin with tri-*n*-butyltin hydride affords the same mixture of *trans*- and *cis*-decalin in which the *trans* isomer predominates (*65*).

Menapace and Kuivila (*137*) have obtained important structure-reactivity data which clearly eliminates an Sn1 or Sn2 mechanism and supports a radical mechanism. The reactivity of butyl bromides was found to be in the order tertiary > secondary > primary, which is the reverse of that observed in a typical Sn2 reaction. Although this is the same order as that found for Sn1 reactions, other results eliminate this reaction possibility. For example, the greater reactivity of propargyl bromide compared to allyl bromide is the reverse of that observed in the Sn1 reaction. On the other hand, the rate sequence found was similar to that obtained for the abstraction of halogen atoms from halides by methyl radicals (*51*, *57*). It has been suggested that the reactivity sequence I > Br > Cl, previously mentioned, and the reactivity order tertiary > secondary > primary is primarily due to changes in the carbon-halogen bond energy; also, resonance stabilization of the incipient free radical is important in determining the reactivities of the benzyl, allyl, and propargyl halides (*137*). The importance of polar factors in these reactions is indicated by the fact that electron-withdrawing substituents facilitate reaction. For example, *m*-trifluoromethylbenzyl chloride at 80°C reacts with tri-*n*-butyltin hydride 1.64 times faster than benzyl chloride; furthermore the most reactive halide was $BrCCl_3$. One possible explanation is that electron-withdrawing groups stabilize the following kind of transition state (*137*):

$$R{-}X \cdot Sn{\equiv} \longleftrightarrow \overset{\ominus}{R}{:}\cdot X\ \overset{\oplus}{Sn}{\equiv} \longleftrightarrow R\cdot{:}\overset{\ominus}{X}\ \overset{\oplus}{Sn}{\equiv} \longleftrightarrow R\cdot\ X{-}Sn{\equiv}$$

The rate constants for the chain propagation steps, Eqs. (20) and (21), and for two out of the three possible termination steps, Eqs. (27a)–(27c), have been evaluated (*22*, *23*):

$$R\cdot + R\cdot \longrightarrow \text{inactive products} \qquad (27a)$$

$$R\cdot + {\equiv}Sn\cdot \longrightarrow \text{inactive products} \qquad (27b)$$

$$\equiv Sn\cdot + {\equiv}Sn\cdot \longrightarrow \text{inactive products} \qquad (27c)$$

In general, with alkyl chlorides the reaction represented by Eq. (20) is rate-determining, whereas with alkyl bromides and methyl iodide the reaction represented by Eq. (21) is rate-determining. For the reaction of *t*-butyl bromide with either tri-*n*-butyltin hydride or triphenyltin hydride, termination occurs by reaction (27a), whereas for the *t*-butyl chloride—triphenyltin hydride reaction, termination occurs by reaction (27c). For the *t*-butyl chloride—tri-*n*-butyltin hydride reaction, reactions (20) and (21) are of comparable rate; the kinetics further indicate that all three possible termination reactions can occur.

The reaction of organic halides with organotin hydrides provides a convenient method for cleanly generating a variety of radicals. The azobisisobutyronitrile-catalyzed reduction of neophyl chloride with tri-*n*-butyltin hydride at 80°C gives *t*-butylbenzene as the sole product (*98*). In this case the neophyl radical abstracts a hydrogen atom from tri-*n*-butyltin hydride:

$$C_6H_5C(CH_3)_2CH_2\cdot + (n\text{-}C_4H_9)_3SnH \longrightarrow C_6H_5C(CH_3)_2CH_3 + (n\text{-}C_4H_9)_3Sn\cdot \qquad (28)$$

faster than it rearranges to the 2-methyl-1-phenyl-2-propyl radical. The 2,2,2-triphenylethyl radical, which apparently possesses a great tendency to rearrange since numerous attempts to generate it have resulted only in rearranged products, has been prepared and trapped before rearrangement by allowing 2,2,2,-triphenylethyl chloride to react with triphenyltin hydride (*81*):

$$(C_6H_5)_3Sn\cdot + (C_6H_5)_3CCH_2Cl \longrightarrow (C_6H_5)_3SnCl + (C_6H_5)_3CCH_2\cdot \qquad (29)$$

$$(C_6H_5)_3CCH_2\cdot + (C_6H_5)_3SnH \longrightarrow (C_6H_5)_3CCH_3 + (C_6H_5)_3Sn\cdot \qquad (30)$$

The yield of 1,1,1-triphenylethane from the organic chloride (0.27 mmole) and hydride (1.43 mmoles) in benzene (0.10 ml) at 68°C is $>90\%$. In some cases cyclization of the radical competes favorably with reaction of the uncyclized radical with the organotin hydride. The product obtained from the reduction of γ-chlorobutyrophenone with tri-*n*-butyltin hydride is composed of 20%

butyrophenone and 80% of the cyclization product, 2-phenyltetrahydrofuran (*137*):

$$C_6H_5\overset{O}{\overset{\|}{C}}CH_2CH_2CH_2\cdot \longrightarrow C_6H_5\dot{C}\langle O, CH_2, H_2C\text{—}CH_2\rangle \xrightarrow{(n\text{-}C_4H_9)_3SnH} C_6H_5\overset{H}{C}\langle O, CH_2, H_2C\text{—}CH_2\rangle \qquad (31)$$

The cyclization is favored by the formation of the 5-membered ring and by resonance stabilization of the resulting radical by the phenyl group and by the oxygen atom. Certain organic bromides possessing a 5,6-double bond react with tri-*n*-butyltin hydride to give a 5-membered ring compound as the major product (*252*):

$$H_2C(\overset{H}{C}{=}CH_2)\text{—}X\text{—}CH_2\dot{C}R_2 \longrightarrow \text{cyclo}[H_2C\text{—}\overset{CH_2\cdot}{C}H\text{—}CR_2\text{—}CH_2\text{—}X] \xrightarrow{(n\text{-}C_4H_9)_3SnH} \text{cyclo}[H_2C\text{—}\overset{CH_3}{C}H\text{—}CR_2\text{—}CH_2\text{—}X] \qquad (32)$$

$X = CH_2$, $R = H$ or CH_3
$X = O$, $R = H$

The question of why the 5-membered ring is formed in preference to the 6-membered ring remains to be answered. The cyclization : reduction ratio for the reaction of 1,3-diiodopropane with group IV hydrides has been determined (*82a*).

Organotin hydrides are useful reagents for the stepwise reduction of geminal polyhalides. Carbon tetrachloride can be reduced successively with triphenyltin hydride (*130*) or tri-*n*-butyltin hydride (*215*) to chloroform, methylene chloride, and methyl chloride. Similarly, benzotrichloride can be reduced successively with tri-*n*-butyltin hydride to benzal chloride, benzyl chloride, and toluene (*98*). The reaction of trialkylstannanes with carbon tetrachloride or chloroform to form trialkyltin chlorides has been reported by others also (*90a*). The partial reduction of bromoform to methylene bromide and of tribromofluoromethane to dibromofluoromethane with tri-*n*-butyltin hydride is possible (*215*). Tri-*n*-butyltin hydride or triphenyltin hydride reduces the C—Br bond of $C_6H_5HgCCl_2Br$(*207, 212*). The reduction of gem-dihalocyclopropanes to the monohalides can be conveniently accomplished with tri-*n*-butyltin hydride or triphenyltin hydride. The reduction of 7,7-dihalobicyclo [4.1.0] heptanes with tri-*n*-butyltin hydride affords a mixture of the *cis* and *trans* monohalides in which the *cis* isomer predominates (*215*) :

$$\text{7,7-}X_2\text{-bicyclo[4.1.0]heptane} + (n\text{-}C_4H_9)_3SnH \longrightarrow \underset{cis}{\text{7-X-bicyclo[4.1.0]heptane}} + \underset{trans}{\text{7-X-bicyclo[4.1.0]heptane}} + (n\text{-}C_4H_9)_3SnX \qquad (33)$$

The ratio of *cis* : *trans* is 2.5 : 1 when X is Br and 1.8 : 1 when X is Cl. It was suggested that the greater amount of the apparently less thermodynamically stable product may be due to steric factors. Sterically, the tri-*n*-butyltin radical would be expected to break the more exposed carbon-halogen bond. Furthermore, it may be easier for the bulky tri-*n*-butyltin hydride to transfer its hydrogen atom to the less hindered side of the intermediate cyclopropyl radical, which is the side having the cyclopropane hydrogens. A similar explanation has been used to account for the partial stereospecificity observed in the reduction of 1,1-diiodo-2,3-*cis*-dimethylcyclopropane with tri-*n*-butyltin hydride (*178*). In this case the ratio of 1-iodo-*cis,cis*-2,3-dimethylcyclopropane to 1-iodo-*trans,trans*-2,3-dimethylcyclopropane is 3 to 1. Steric factors also probably account for the stereospecific reduction of a mixture of *cis* and *trans* 7-chloro-7-phenyl [4.1.0] heptane to the *cis* hydrocarbon with triphenyltin hydride (*77*). The first synthesis of monofluorocyclopropanes by reduction of 1-chloro-1-fluorocyclopropanes with organotin hydrides was accomplished by Oliver et al. (*178a*). Stereospecificity has been observed in the reduction of gem-halofluorocyclopropanes with tri-*n*-butyltin hydride (*8*) :

Cl, F $\xrightarrow{(n\text{-}C_4H_9)_3SnH}$ H, F (34)

F, Cl $\xrightarrow{(n\text{-}C_4H_9)_3SnH}$ F, H (35)

These results are consistent with a mechanism involving flank attack by the Sn—H bond, but in view of the catalysis observed with di-*t*-butyl peroxide and azobisisobutyronitrile, the reaction most certainly involves free radicals. Apparently the intermediate fluorocyclopropyl radical abstracts hydrogen from tri-*n*-butyltin hydride faster than it inverts its configuration. It was suggested that a complex might be formed between fluorine and tin compounds which may restrict inversion. Several other examples of the mono-reduction of gem-dihalocyclopropanes with tri-*n*-butyltin hydride have appeared (*9a*, *41*, *136*, *189*, *246*). 2,2-Dibromoalkylidenecyclopropanes can be successively reduced to the monobromides and the hydrocarbons with tri-*n*-butyltin hydride (*189*):

Br_2, $(CH_3)_2$, CH_2 $\xrightarrow{(n\text{-}C_4H_9)_3SnH}$ Br, H, $(CH_3)_2$, CH_2 $\xrightarrow{(n\text{-}C_4H_9)_3SnH}$ H, H, $(CH_3)_2$, CH_2 (35a)

Significantly, reduction occurs rather than addition across the double bond (see Sec. III.C.). 1,1-Dichloro-2-vinylcyclopropane and trimethyltin hydride, on the other hand, react according to (*208*):

$$(CH_3)_3SnH + CH_2{=}CH{-}\underset{Cl_2}{\triangleleft} \xrightarrow{100^\circ C} (CH_3)_3SnCH_2CH{=}CHCH_2CCl_2H \quad (35b)$$

The following radical mechanism has been suggested for this reaction:

$$R_3Sn\cdot + CH_2{=}CH{-}\underset{Cl_2}{\triangleleft} \longrightarrow R_3SnCH_2\dot{C}H{-}\underset{Cl_2}{\triangleleft} \quad (35c)$$

$$R_3SnCH_2\dot{C}H{-}\underset{Cl_2}{\triangleleft} \longrightarrow R_3SnCH_2CH{=}CHCH_2CCl_2\cdot \quad (35d)$$

$$R_3SnCH_2CH{=}CHCH_2CCl_2\cdot + R_3SnH \longrightarrow R_3SnCH_2CH{=}CHCH_2CCl_2H + R_3Sn\cdot \quad (35e)$$

The reaction of tetrachlorocyclopropene with tri-*n*-butyltin hydride has been used to prepare 3,3-dichlorocyclopropene (*19*) and 3-chlorocyclopropene (*18*). Hydrolysis of the former compound affords cyclopropenone, while reaction of the latter compound with antimony pentachloride affords the cyclopropenyl cation. Reduction of tropylium bromide with triphenyltin hydride affords cycloheptatriene (isolated as the chloroplatinate) (*17*).

Tri-*n*-butyltin deuteride can be used to convert aliphatic halogen compounds into their corresponding deuterium derivatives (*251*). 9-Deuteriodecalin has been obtained from the reaction of 9-chlorodecalin with tri-*n*-butyltin deuteride (*65*).

2. *Aryl Halides*

Whereas the reduction of many alkyl halides with organotin hydrides is exothermic, fairly high temperatures are required for the satisfactory reduction of aryl halides. The reduction of aryl halides by triaryltin hydrides was independently discovered by Rothman and Becker (*191*, *192*), who obtained benzene and triphenyltin bromide from the reaction of bromobenzene with triphenyltin hydride, and by Noltes and van der Kerk (*172*), who reported triphenyltin bromide from this same reaction. The scope and mechanism of this reaction has been thoroughly studied by Becker et al. (*132*, *192*). The order of reactivity for halobenzenes and halonaphthalenes is $I > Br > Cl$, which is the opposite of that found in nucleophilic aromatic substitution reactions. For the reaction of bromobenzenes with triphenyltin hydride at 124°C electron-withdrawing groups in the *ortho* and *para* positions tend to accelerate the reaction, while electron-donating groups hinder the reaction.

At 154°C there is little or no selectivity since the yield of triphenyltin bromide is at least 90% in every case. 1-Iodonaphthalene was found to react faster than 2-iodonaphthalene. Also, 1-bromo-4-methylnaphthalene was more reactive than 1-bromo-2-methylnaphthalene. The addition of azobisisobutyronitrile or benzoyl peroxide increases the extent of reduction of 4-bromobiphenyl. The extent of reduction was found to increase also with the dielectric constant of the solvent. Complicating the mechanistic picture somewhat is the catalysis observed with triphenylboron, a Lewis acid, and the lack of inhibition of the reduction of 4-bromobiphenyl by hydroquinone. As was pointed out by Pang and Becker (*181*), the catalytic effect of triphenylboron does not necessarily rule out a free radical mechanism, since trialkylborons are known to polymerize vinyl monomers, probably by a radical mechanism. The above data are probably best explained by a radical mechanism involving the same type of polar-radical transition state shown above for the alkyl halide reductions. The reactivities of the halonaphthalenes, mentioned above, have been explained by steric effects (*132*).

B. Reaction with Acid Halides

An acid chloride or acid bromide yields a mixture of aldehyde, ester, and organotin halide on reaction with an organotin hydride. The scope and mechanism of this reaction with tri-*n*-butyltin hydride has been thoroughly studied by Kuivila and Walsh (*103*, *253*). The relative yields of ester and aldehyde depend upon the structure of the acid chloride and upon whether or not a solvent is employed. Some of the results are given in Table 2. The importance of a steric factor is immediately apparent since the yield of ester decreases and the yield of aldehyde increases as the alkyl group becomes more bulky. With the exception of tri-*n*-butyltin chloride, the presence of a solvent increases the yield of aldehyde and decreases the yield of ester. The two acid bromides studied gave a much higher yield of aldehyde compared to the corresponding chlorides. A test of the generality of this result with other acid bromides would be of interest. Other studies indicate that the relative yields of ester and aldehyde depend also upon the structure of the hydride employed (*108*, *109*, *253a*). The free radical mechanism for aldehyde formation :

$$\equiv SnH + Q\cdot \longrightarrow \equiv Sn\cdot + QH \tag{36}$$

$$\equiv Sn\cdot + RCOCl \longrightarrow R\dot{C}{=}O + \equiv SnCl \tag{37}$$

$$R\dot{C}{=}O + \equiv SnH \longrightarrow RCHO + \equiv Sn\cdot \tag{38}$$

proposed by Kuivila and Walsh (*103*) is supported by several observations. Azobisisobutyronitrile catalyzes the reaction of ethyl chloroformate with tri-*n*-butyltin hydride. The reaction of tri-*n*-butyltin hydride with triphenylacetyl chloride in xylene at 100°C affords carbon monoxide, triphenylmethane, and triphenylacetaldehyde. These products are consistent with the formation

TABLE 2

REACTION OF ACID HALIDES WITH TRI-*n*-BUTYLTIN HYDRIDE AT THE AMBIENT TEMPERATURE (*103*)

R	% Yield (neat)		Solvent	% Yield (solution)	
	RCHO	RCO_2CH_2R		RCHO	RCO_2CH_2R
RCOCl					
CH_3	5	95			
C_2H_5	0	87	2,3-Dimethylbutane	75	25
			Methyl acetate	90	10
			Tri-*n*-butyltin chloride	27	73
n-C_4H_9	19	81	2,3-Dimethylbutane	91	9
i-C_3H_7	36	64	2,3-Dimethylbutane	52	48
t-C_4H_9	56	33			
$C_6H_5CH_2$			Toluene	55	45
C_6H_5	65	35			
$(C_6H_5)_3C$			*m*-Xylene	90	0
RCOBr					
C_2H_5	79	21	2,3-Dimethylbutane	60	40
C_6H_5	99	0			

of an intermediate acyl radical which readily loses carbon monoxide to form the more stable triphenylmethyl radical:

$$(C_6H_5)_3C{-}\dot{C}{=}O \longrightarrow CO + (C_6H_5)_3C\cdot \tag{39}$$

Toluene (39% yield) is obtained from the reaction of tri-*n*-butyltin hydride with benzyl chloroformate. In this case the intermediate benzyloxycarbonyl radical loses carbon dioxide to form the more stable benzyl radical. The relative reactivities of substituted benzoyl chlorides were determined by allowing each aroyl chloride to compete with benzyl bromide or 2-bromooctane for a deficiency of tri-*n*-butyltin hydride. The results are given in Table 3. Electron-withdrawing groups are seen to accelerate the reaction, which eliminates a mechanism involving acylium ions. A polar-radical transition state similar to the one suggested for the reaction of alkyl and aryl halides with organotin hydrides is indicated:

$$\mathrm{Ar\overset{\overset{O}{\|}}{C}Cl\cdot Sn{\equiv}} \longleftrightarrow \mathrm{Ar\overset{\overset{O}{\|}}{C}\!:^{\ominus}\ \cdot Cl\ Sn^{\oplus}{\equiv}} \longleftrightarrow \mathrm{Ar\overset{\overset{O}{\|}}{C}\cdot\ :Cl^{\ominus}\ Sn^{\oplus}{\equiv}} \longleftrightarrow \mathrm{Ar\overset{\overset{O}{\|}}{C}\cdot\ ClSn{\equiv}}$$

The Hammett ρ value for the reaction was found to be 2.6 in *m*-xylene, 2.7 in *o*-dichlorobenzene, and 4.0 in methyl acetate. A more detailed study of this solvent effect would be of interest.

Four possible mechanisms for the formation of ester from the aldehyde formed by Eqs. (36)–(38) were considered by Walsh and Kuivila (*253*):

TABLE 3

RELATIVE REACTIVITIES OF SUBSTITUTED BENZOYL CHLORIDES TOWARD TRI-*n*-BUTYLTIN HYDRIDE IN *m*-XYLENE AT 25°C (*103*)

Substituent	Relative rate
p-CN	61.40
p-CF_3	38.10
m-CF_3	9.50
m-Cl	8.20
m-F	2.34
p-Cl	1.27
H	1.00
m-CH_3	0.937
p-CH_3	0.290
p-OCH_3	0.166

(a) reaction of aldehyde with acid chloride to form an α′-chloroester which is reduced by hydride to ester; (b) reduction of aldehyde to alcohol by hydride followed by reaction of the alcohol with acid chloride; (c) reaction of the aldehyde with hydride to form an organotin alkoxide which reacts with acid chloride to give ester; and (d) reaction of aldehyde with acyl radicals to give an α-acyloxy radical which reacts with hydride to form ester. Mechanisms (a) and (b) were shown to be insignificant under ordinary conditions; there were indications that (c) occurs to a minor extent, which leaves (d) as the major pathway:

$$RCHO + R\dot{C}{=}O \longrightarrow RC(=O)O\dot{C}HR \qquad (40)$$

$$RC(=O)O\dot{C}HR + {\equiv}SnH \longrightarrow RC(=O)OCH_2R + {\equiv}Sn\cdot \qquad (41)$$

The relative reactivities of substituted benzaldehydes toward the propionyl radical are given in Table 4. Electron-withdrawing groups are seen to increase reactivity while electron-donating groups decrease reactivity. The ρ value was found to be 0.43. The following polar-radical transition state in which the aldehyde is acting as a nucleophile was suggested:

$$\left[(R)(H)C \overset{\delta\ominus}{\cdots} O \cdots C^{\delta\oplus}(R)(=O) \right]^{\bullet}$$

With the exception of triptoyl chloride, the principal reduction product of the reaction of several aliphatic, bridgehead, and aromatic acid chlorides

TABLE 4

RELATIVE REACTIVITIES OF SUBSTITUTED BENZALDEHYDES TOWARD THE PROPIONYL RADICAL IN 2,3-DIMETHYLBUTANE AT 25°C (*253*)

Substituent	Relative rate
m-Cl	1.57
p-Cl	1.26
p-CH_3	1.25
H	1.00
m-CH_3	0.94
p-OCH_3	0.468

with triphenyltin hydride with or without solvent was found to be the corresponding ester; a minor reduction product (less than 16%) was the corresponding aldehyde (*108*, *109*). The results are summarized in Table 5. The reaction of benzoyl chloride with triphenyltin hydride was found to be accelerated by azobisisobutyronitrile and retarded by each of three well-known radical inhibitors: hydroquinone, *trans*-stilbene, and galvinoxyl (*108*). This result is consistent with the free radical mechanism proposed for the tri-*n*-butyltin hydride reduction of acid halides, Eqs. (36)–(38) and Eqs. (40) and (41). The high yield of aldehyde obtained from triptoyl chloride is very likely due to steric factors which prevent the occurrence of the reaction represented by Eq. (40). Various *p*-substituted benzoyl chlorides were allowed to react with triphenyltin hydride in the absence of solvent for 6 h, and the extent of reaction was estimated from the amount of triphenyltin chloride isolated (*108*). There was an indication that electron-releasing groups facilitate the reaction, while electron-withdrawing groups retard the reaction. This result is opposite to that observed with tri-*n*-butyltin hydride (Table 3). More quantitative studies of the triphenyltin hydride reduction of substituted benzoyl chlorides are needed.

The reaction of succinyl dichloride with tri-*n*-butyltin hydride either in a 1 : 1 or 1 : 2 mole ratio affords γ-chloro-γ-butyrolactone (*91*). This product is consistent with the formation of an intermediate acyl radical which reacts according to Eq. (42):

$$\begin{array}{c} CH_2\overset{O}{\overset{\|}{C}}\cdot \\ | \\ CH_2COCl \end{array} \longrightarrow \begin{array}{c} CH_2\overset{O}{\overset{\|}{C}}\diagdown \\ | \quad\quad O \\ CH_2\underset{Cl}{\underset{|}{C}}\cdot\diagup \end{array} \xrightarrow{(n\text{-}C_4H_9)_3SnH} \begin{array}{c} CH_2\overset{O}{\overset{\|}{C}}\diagdown \\ | \quad\quad O \\ CH_2\underset{Cl}{\underset{|}{C}}\diagup\diagdown H \end{array} \quad (42)$$

TABLE 5

REACTION OF ACID CHLORIDES WITH TRIPHENYLTIN HYDRIDE (*109*)

R in RCOCl	% Yield: Ester	% Yield: Aldehyde
C_2H_5[a]	50	14(15)[b]
$C_6H_5CH_2$[a]	90	7(13)[b]
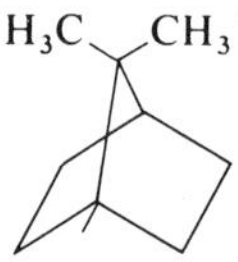 H_3C CH_3[a]	77	8(13)[b]
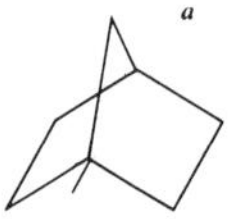 [a]	80	11
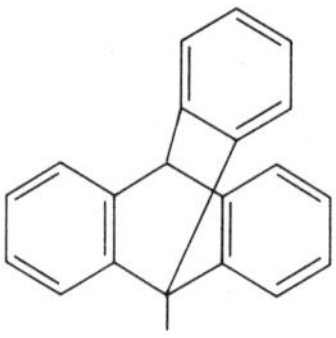	[c]	62[d]
C_6H_5[a]	87(70)[b](72)[d]	0
p-$CH_3C_6H_4$[a]	74	[c]
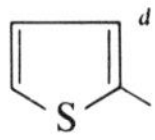 S[d]	29	1

[a] Neat, ambient, 24 h.
[b] Diethyl ether, reflux, 24 h.
[c] Not determined.
[d] Benzene, reflux, 24 h.

The product does not react with the second mole of hydride. The reaction of phthalyl dichloride with tri-*n*-butyltin hydride in a 1 : 2 mole ratio affords phthalide (*91*). In this case reduction of the intermediate 3-chlorophthalide with the second mole of hydride is favored by the formation of a stable benzyl-type radical:

$$\text{(3-chlorophthalide)} \xrightarrow{(n\text{-}C_4H_9)_3Sn\cdot} \text{(phthalidyl radical, }\cdot C\text{-H)} \xrightarrow{(n\text{-}C_4H_9)_3SnH} \text{phthalide} \quad (43)$$

The reaction of a mixture of aldehyde or ketone, acid chloride, and organotin hydride results in reductive acylation of the aldehyde or ketone (*80*, *82*, *103*):

$$RCOCl + \equiv Sn\cdot \longrightarrow R\overset{O}{\overset{\|}{C}}\cdot + \equiv SnCl \quad (44)$$

$$R\overset{O}{\overset{\|}{C}}\cdot + R'\overset{O}{\overset{\|}{C}}R'' \longrightarrow R\overset{O}{\overset{\|}{C}}O\dot{C}R'R'' \quad (45)$$

$$R\overset{O}{\overset{\|}{C}}O\dot{C}R'R'' + \equiv SnH \longrightarrow R\overset{O}{\overset{\|}{C}}OC(H)R'R'' + \equiv Sn\cdot \quad (46)$$

The quantitative reductive acylation of some ketones has been accomplished with an acid chloride and triphenyltin hydride (Table 6) (*80*, *82*).

TABLE 6

REDUCTIVE ACYLATION OF SOME KETONES WITH ACID CHLORIDES AND TRIPHENYLTIN HYDRIDE IN BENZENE AT 25°C (*82*)

R in RCOCl	Ketone	Product
CH_3	$C_6H_5COCH_3$	$C_6H_5CH(OCOCH_3)CH_3$
C_2H_5	$C_6H_5COCH_3$	$C_6H_5CH(OCOC_2H_5)CH_3$
CH_3	$C_6H_5COC_2H_5$	$C_6H_5CH(OCOCH_3)C_2H_5$
CH_3	$C_6H_5COCH(CH_3)_2$	$C_6H_5CH(OCOCH_3)CH(CH_3)_2$
C_6H_5[a]	$C_6H_5COCH_3$	$C_6H_5CH(OCOC_6H_5)CH_3$
CH_3	$C_2H_5COC_2H_5$	$C_2H_5CH(OCOCH_3)C_2H_5$
$CH_3O\overset{O}{\overset{\|}{C}}CH_2CH_2$	$C_6H_5COCH_3$	$C_6H_5CH(OCOCH_2CH_2\overset{O}{\overset{\|}{C}}OCH_3)CH_3$

[a] At 122°C.

C. Addition to Alkenes

1. *Scope*

Triorganotin hydrides undergo noncatalyzed addition to terminal alkenes at 60°–100°C (*237, 239, 240*):

$$(C_6H_5)_3SnH + CH_2{=}CHX \longrightarrow (C_6H_5)_3SnCH_2CH_2X \qquad (47)$$

X = CN, CH_2CN, CO_2CH_3, CO_2H, $CONH_2$, $CH(OC_2H_5)_2$, CH_2OH, OC_6H_5, $OCOCH_3$, C_6H_5, n-C_6H_{13}, CH_2OCOCH_3, $CH_2OCH_2CH_2CN$, $CH_2NHCOCH_3$, $CH_2CH_2COCH_3$, $CH_2Sn(C_6H_5)_3$,

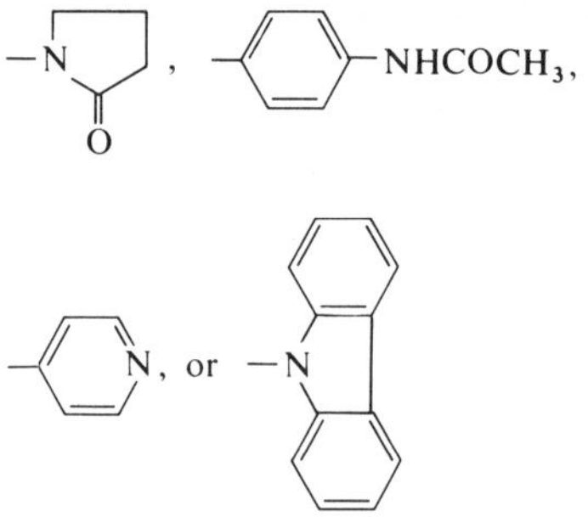

—N(pyrrolidinone), —C_6H_4—$NHCOCH_3$,

—C_5H_4N (pyridyl), or —N (carbazolyl)

$$(C_6H_5)_3SnH + CH_3CH{=}CHCN \longrightarrow (C_6H_5)_3SnCH(CH_3)CH_2CN \qquad (48)$$

$$(C_6H_5)_3SnH + CH_2{=}C(CH_3)X \longrightarrow (C_6H_5)_3SnCH_2CH(CH_3)X \qquad (49)$$

X = CO_2CH_3 or CH_2OH

$$(n\text{-}C_3H_7)_3SnH + CH_2{=}CHX \longrightarrow (n\text{-}C_3H_7)_3SnCH_2CH_2X \qquad (50)$$

X = CN, CO_2CH_3, $CONH_2$, $CH_2CO_2C_2H_5$, CH_2CN, C_6H_5, or —C_5H_4N (pyridyl)

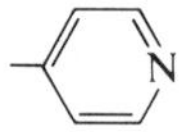

$$(n\text{-}C_3H_7)_3SnH + CH_3CH{=}CHCO_2C_2H_5 \longrightarrow (n\text{-}C_3H_7)_3SnCH(CH_3)CH_2CO_2C_2H_5 \qquad (51)$$

$$(n\text{-}C_3H_7)_3SnH + NCCH{=}CHC_6H_5 \longrightarrow (n\text{-}C_3H_7)_3SnCH(CN)CH_2C_6H_5 \qquad (52)$$

$$(n\text{-}C_4H_9)_3SnH + CH_2{=}CHX \longrightarrow (n\text{-}C_4H_9)_3SnCH_2CH_2X \qquad (53)$$

X = CN or CO_2CH_3

This reaction, which is referred to as hydrostannation, is of considerable importance since it enables the convenient synthesis of a variety of functionally substituted organotin compounds. The use of dialkenes affords bis-triphenylstannyl compounds, for example (*174, 237*):

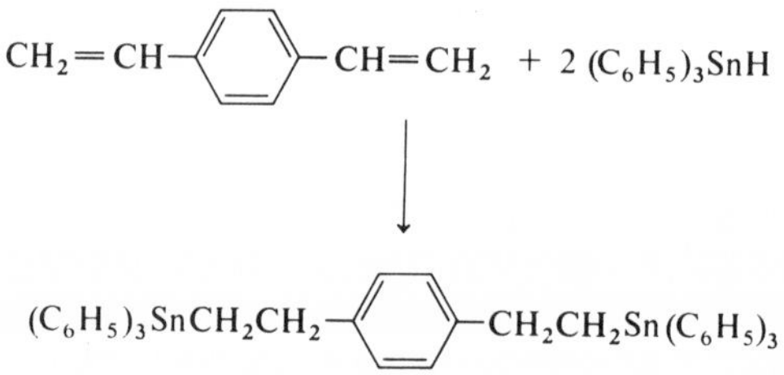

$$CH_2{=}CH\text{—}C_6H_4\text{—}CH{=}CH_2 + 2\,(C_6H_5)_3SnH \longrightarrow (C_6H_5)_3SnCH_2CH_2\text{—}C_6H_4\text{—}CH_2CH_2Sn(C_6H_5)_3 \qquad (54)$$

Hydrostannation has been used to prepare organotin-substituted carboxylic esters and carbonamides (*67a*, *179a*, *179b*). The reactions of phenyltin hydrides with acrylonitrile afford cyanoethyltin compounds (*190b*).

The location of the tin atom at the terminal carbon has been established in the case of the adducts obtained with acrylonitrile, styrene, and 1-octene (*240*). The triphenyltin hydride-styrene adduct was identical with the product of the reaction of triphenyltin chloride with 2-phenylethylmagnesium bromide. Similarly, the triphenyltin hydride-1-octene adduct was identical with the product of the reaction of triphenyltin chloride with 1-octylmagnesium bromide. The triphenyltin hydride-acrylonitrile adduct must be triphenyl-(2-cyanoethyl)tin since it was not identical with triphenyl(1-cyanoethyl)tin which was prepared by an unambiguous route:

$$(C_6H_5)_3SnO\overset{\overset{\displaystyle O}{\|}}{C}CH(CH_3)CN \xrightarrow{\text{heat}} CO_2 + (C_6H_5)_3SnCH(CH_3)CN \qquad (55)$$

The structures of the other adducts have been assigned on the basis of analogy with these examples. More recently, the use of gas chromatography has revealed that both possible adducts are formed in the thermal reaction of organotin hydrides with acrylonitrile (*126*).

Triphenyltin hydride is more reactive than trialkyltin hydrides. For example, the trialkyltin hydrides do not react with 1-octene, whereas triphenyltin hydride reacts with this alkene to give a 72% yield of adduct. Addition of triphenyltin hydride to 1-octene does not occur in the presence of free radical catalysts such as ultraviolet light, benzoyl peroxide, or phenylazotriphenylmethane; instead, tetraphenyltin is formed (*61*). Dialkylchlorotin hydrides also are more reactive than trialkyltin hydrides. For example, diisobutylchlorotin hydride reacts with 1-octene exothermally at 20°–45°C to give a 91% yield of adduct (*156*).

The use of free radical catalysts such as azobisisobutyronitrile (AIBN) broadens the scope of the hydrostannation reaction, especially for trialkyltin hydrides (*153*, *154*):

$$(C_2H_5)_3SnH + CH_2{=}CHX \xrightarrow{\text{AIBN}} (C_2H_5)_3SnCH_2CH_2X \qquad (56)$$

$X = (CH_2)_7CO_2C_2H_5$, CH_2OH, $(CH_2)_5CH_3$, $OCOCH_3$, CH_2OCOCH_3, OC_2H_5, OC_4H_9-*n*, $(CH_2)_7CH_3$, C_6H_5, CO_2CH_3, $CONH_2$, CN, $CH_2OCH_2CH(OH)CH_2NH_2$, $CH_2CH{=}CH_2$,

cyclohexenyl, $-C_6H_4-CH{=}CH_2$, or $-N$(carbazolyl)

$$(C_2H_5)_3SnH + CH_2{=}C(CH_3)X \xrightarrow{\text{AIBN}} (C_2H_5)_3SnCH_2CH(CH_3)X \quad (57)$$
$$X = C_6H_5,\ CO_2CH_3,\ \text{or}\ CONH_2$$

$$(i\text{-}C_4H_9)_3SnH + CH_2{=}CHX \xrightarrow{\text{AIBN}} (i\text{-}C_4H_9)_3SnCH_2CH_2X \quad (58)$$
$$X = (CH_2)_5CH_3,\ CH_2OCOCH_3,\ \text{or}\ OC_4H_9\text{-}n$$

$$(i\text{-}C_4H_9)_3SnH + CH_2{=}C(CH_3)CO_2CH_3 \xrightarrow{\text{AIBN}} (i\text{-}C_4H_9)_3SnCH_2CH(CH_3)CO_2CH_3 \quad (59)$$

The use of this catalyst extends X in Eq. (53) to include $OC_4H_9\text{-}i$, CH_2OH, and $CH_2CH{=}CH_2$. The addition of triethyltin hydride to B-trivinyl-*N*-triphenylborazine occurs only in the presence of azobisisobutyronitrile (*213*):

$$(C_2H_5)_3SnH + [CH_2{=}CHBNC_6H_5]_3 \xrightarrow{\text{AIBN}} [(C_2H_5)_3SnCH_2CH_2BNC_6H_5]_3 \quad (60)$$

No catalyst is required with triphenyltin hydride. Triethyltin hydride also adds to other vinylboranes (*17a*).

The addition of trialkyltin hydrides to terminal alkenes is catalyzed also by aluminum trialkyls, AlR_3, in which R corresponds to the alkene (*151*, *152*). For example, the reaction of triethyltin hydride with 1-octene at 60°–85°C in the presence of 1–5 mole% of tri-*n*-octylaluminum gives a 90% yield of adduct:

$$(C_2H_5)_3SnH + CH_2{=}CHC_6H_{13}\text{-}n \xrightarrow{Al(n\text{-}C_8H_{17})_3} (C_2H_5)_3SnC_8H_{17}\text{-}n \quad (61)$$

In this reaction the catalyst probably reacts with the triethyltin hydride to give the adduct in Eq. (61) and di-*n*-octylaluminum hydride which adds to the alkene to produce more of the catalyst:

$$(C_2H_5)_3SnH + Al(n\text{-}C_8H_{17})_3 \longrightarrow (C_2H_5)_3SnC_8H_{17}\text{-}n + HAl(n\text{-}C_8H_{17})_2 \quad (62)$$

$$HAl(n\text{-}C_8H_{17})_2 + CH_2{=}CHC_6H_{13}\text{-}n \longrightarrow Al(n\text{-}C_8H_{17})_3 \quad (63)$$

The net reaction is addition of the triethyltin hydride to the 1-octene, Eq. (61). Reactions of the type represented by Eq. (62) probably occur by way of a four-center transition state (*206*).

Trimethyltin hydride slowly forms an adduct with either *cis*- or *trans*-2-butene upon irradiation (*102*). Prior to this significant result, internal alkenes, with the exception of bicyclo[2,2,1]hepta-2,5-diene (*92*) and those whose double bond is activated by a conjugated electron-withdrawing group, Eqs. (48), (51), and (52), had been reported (*154*) not to undergo hydrostannation. More recently, a number of other internal alkenes were shown to undergo photocatalyzed addition of trimethyltin hydride (*219*).

The presence of other functional groups can interfere with the hydrostannation reaction. For example, allyl bromide (*172*), α- and γ-methallyl chloride (*137*), norbornenyl halides (*255*), and *cis*- and *trans*-α-bromostilbene (*107*) undergo reduction to the corresponding alkenes rather than hydrostannation. Allylamine cannot be hydrostannated with triphenyltin hydride because the amino group catalyzes the decomposition of the hydride (*224*, *240*, *243*). Although triphenyltin hydride adds to acrylic acid to give β-triphenylstannylpropionic acid, which subsequently loses benzene intramolecularly, tri-*n*-propyltin hydride reacts with the carboxyl group to give hydrogen and tri-*n*-propylstannyl acrylate (*240*). Triphenyltin hydride reduces the carbonyl group of methyl vinyl ketone and phenyl vinyl ketone instead of adding across the olefinic linkage (*172*, *239*):

$$2\,(C_6H_5)_3SnH + CH_2{=}CHCOR \longrightarrow (C_6H_5)_3SnSn(C_6H_5)_3 + CH_2{=}CHCH(OH)R \quad (64)$$

On the other hand, methyl γ-butenyl ketone undergoes hydrostannation instead of reduction of the carbonyl group (*239*). Reduction of the functional group rather than hydrostannation of the olefinic linkage has been reported to occur with divinyl sulfone, divinyl sulfoxide, and *p*-nitrostyrene (*172*). Triphenyltin hydride reacts normally with triphenylvinyltin at 70°C to give 1,2-bis(triphenylstannyl)ethane (*69*). With triphenylperfluorovinyltin, however, the reaction takes a different course (*210*):

$$(C_6H_5)_3SnH + (C_6H_5)_3SnCF{=}CF_2 \longrightarrow CF_2{=}CFH + (C_6H_5)_3SnSn(C_6H_5)_3 \quad (65)$$

Extensive reduction of the vinyl group occurs when trimethylperfluorovinyltin is allowed to react with trimethyltin hydride under ultraviolet irradiation at 25°C (*13*).

$$(CH_3)_3SnH + (CH_3)_3SnCF{=}CF_2 \longrightarrow (CH_3)_3SnCF{=}CFH\ (cis{:}trans = 3{:}1) + (CH_3)_3SnCH{=}CF_2 + (CH_3)_3SnC_2H_2F + (CH_3)_3SnF \quad (66)$$

Although trimethyltin hydride and triethyltin hydride form adducts with perfluorocyclobutene at 20°C, the adducts slowly eliminate the corresponding trialkyltin fluoride at 20°C (*44*):

$$\begin{array}{c} F_2C-CF_2 \\ | \quad\; | \\ F-C-C-H \\ | \quad\; | \\ R_3Sn \quad F \end{array} \longrightarrow \begin{array}{c} F_2C-CF_2 \\ | \quad\; | \\ FC=CH \end{array} + R_3SnF \quad (67)$$

Triethyltin hydride reduces 1,2-dichlorotetrafluorocyclobutene and 1,2-dichlorohexafluorocyclopentene at 100°C (*44*):

$$(C_2H_5)_3SnH + \underset{Cl\quad\;\; Cl}{\overset{(CF_2)_n}{C=C}} \longrightarrow \underset{H\quad\;\; Cl}{\overset{(CF_2)_n}{C=C}} + (C_2H_5)_3SnCl \quad (68)$$

$(n = 2 \text{ or } 3)$

Whether the mechanism involves direct reduction of the carbon-chlorine bond or addition-elimination has not been established, but the latter possibility is attractive in view of the result shown by Eq. (67). Under ultraviolet irradiation, trimethyltin hydride and trimethylperfluorovinylsilane form adducts by a free radical mechanism. These adducts undergo β-fluorine elimination to give organotin fluorides and fluorovinylsilanes (*2*). Organotin hydrides also form addition products with perfluorovinylgermanium compounds (*3*).

Diphenyl- and dialkyltin dihydrides can be added to terminal alkenes which have conjugated electron-withdrawing groups (*154*, *174*, *239*), for example (*154*):

$$(C_2H_5)_2SnH_2 + 2\,CH_2{=}CHCO_2CH_3 \xrightarrow{\text{AIBN}} (C_2H_5)_2Sn(CH_2CH_2CO_2CH_3)_2 \quad (69)$$

An alkene : hydride ratio of 3 : 1 is frequently employed, and azobisisobutyronitrile is used to catalyze the dialkyltin dihydride additions. The reaction of methyl methacrylate with di-*n*-propyltin dihydride in a 1 : 1 mole ratio affords a *bis*-adduct rather than the expected monohydride, which indicates that the monohydride is more reactive than the starting dihydride. It has been suggested that the increased reactivity of the monohydride may be due to coordination between the carbonyl oxygen and the tin atom (*92*). Alkyltin trihydrides form *tris*-adducts with alkenes (*154*, *239*).

Although dimethyltin dihydride is unreactive toward either *cis*- or *trans*-2-butene, equimolar quantities of this hydride and ethylene react under ultraviolet irradiation to give dimethylethyltin hydride (37% yield) and dimethyldiethyltin (31% yield) (*26*). With a 2 : 1 mole ratio of ethylene : hydride the yield of *bis*-adduct is considerably increased. Tetrafluorethylene is more reactive than ethylene toward dimethyltin dihydride. Equimolar amounts of tetrafluoroethylene and dimethyltin dihydride react in the dark at 25°C to give the monohydride in 52% yield and the *bis*-adduct in 9% yield (*25*). The *bis*-adduct was the only product obtained from the reaction of di-*n*-butyltin dihydride with tetrafluoroethylene (*90*). Dimethyltin dihydride reacts with trifluoroethylene, 1,1-difluoroethylene, and trifluorobromoethylene under ultraviolet irradiation to form unstable mono-adducts which decompose to alkene and unstable $(CH_3)_2SnHF$ or $(CH_3)_2SnHBr$. The products obtained using trifluoroethylene are explained by (*25*):

$$(CH_3)_2SnH_2 + C_2F_3H \longrightarrow (CH_3)_2SnH(C_2F_3H_2) \quad (70)$$

$$(CH_3)_2SnH(C_2F_3H_2) \longrightarrow [(CH_3)_2SnHF] + CFH{=}CFH \quad (71)$$

$$2\,[(CH_3)_2SnHF] \longrightarrow (CH_3)_2SnF_2 + (CH_3)_2SnH_2 \quad (72)$$

$$2\,(CH_3)_2SnH_2 \longrightarrow 2\,H_2 + Sn + (CH_3)_4Sn \quad (73)$$

Diphenyltin dihydride reacts abnormally with vinyltriphenylmetal compounds (*69*):

$$(C_6H_5)_2SnH_2 + 2\,CH_2{=}CHM(C_6H_5)_3 \longrightarrow (C_6H_5)_3SnCH_2CH_2M(C_6H_5)_3 \quad (74)$$
$M = Si, Ge, or Sn$

In the reaction of dimethyltin dihydride with dimethylbis(perfluorvinyl)tin, vinyl-hydrogen exchange occurs at 60°C in the dark to give dimethylperfluorovinyltin hydride. Under ultraviolet irradiation, dimethyltin difluoride and partially fluorinated vinyltin compounds are formed (*13*).

The reaction between di-*n*-butyltin dihydride and allyl alcohol affords hydrogen, di-*n*-butylbis(allyloxy)tin, and 2,2′-di-*n*-butyl-1-oxa-2-stannacyclopentane (*111*). Whether this latter compound is formed by pathway A or pathway B has not been definitely established:

$$(n\text{-}C_4H_9)_2SnH_2 + HOCH_2CH{=}CH_2 \xrightarrow{(A)} (n\text{-}C_4H_9)_2Sn(H)OCH_2CH{=}CH_2 + H_2$$

$$(n\text{-}C_4H_9)_2SnH_2 + HOCH_2CH{=}CH_2 \xrightarrow{(B)} (n\text{-}C_4H_9)_2Sn(H)CH_2CH_2CH_2OH \xrightarrow{-H_2} \text{2,2-di-}n\text{-butyl-1-oxa-2-stannacyclopentane}$$

$(n\text{-}C_4H_9)_2Sn(H)OCH_2CH{=}CH_2 \longrightarrow$ 2,2-di-*n*-butyl-1-oxa-2-stannacyclopentane (H_2C–CH_2, O, CH_2, Sn, n-C_4H_9, C_4H_9-n)

The reaction of diorganotin dihydrides with dialkenes can be used for the preparation of heterocyclic organotin compounds (*68*, *129*), for example (*129*):

$$(C_2H_5)_2SnH_2 + o\text{-}C_6H_4(CH{=}CH_2)_2 \longrightarrow \text{3,3-diethylbenzostannepine (benzo-fused seven-membered ring with } Sn(C_2H_5)_2) \quad (75)$$

or polymers containing tin in the backbone of the chain (*1*, *128*, *173–175*). The hydrostannation of α,ω-dialkenes to give polymers is catalyzed by diisobutylaluminum hydride (*160*).

Both 1,2- and 1,4-monoaddition products are obtained from the azobisisobutyronitrile catalyzed reactions of organotin hydrides with conjugated dienes (*54*, *163*, *165*). In most of the cases studied, the 1,4-adduct predominates. The results obtained with trimethyltin hydride are shown in Table 7.

Organotin hydrides react with alkylidenemalonitriles by polar 1,4-addition (*137a*, *167*):

$$RHC{=}C(CN)_2 + R'_3SnH \longrightarrow RCH_2C(CN){=}C{=}NSnR'_3 \quad (76)$$

$R = CH_3, C_2H_5O, C_6H_5, p\text{-}(CH_3)_2NC_6H_4,$
$p\text{-}CH_3OC_6H_4, p\text{-}ClC_6H_4, p\text{-}O_2NC_6H_4,$ or 2-furyl
$R' = C_2H_5$ or $n\text{-}C_4H_9$

The compounds $R_2C{=}C(CN)_2$ ($R{=}CH_3$ or C_6H_5) react analogously. Alkylidenecyanoacetates and alkylidenemalonic esters also react by 1,4-

TABLE 7

AZOBISISOBUTYRONITRILE-CATALYZED ADDITION OF TRIMETHYLTIN HYDRIDE TO DIENES (*54*)

Diene	%Yield	%1,4 Adduct	%1,2 Adduct
$CH_2{=}CHCH{=}CH_2$[a]	60	*cis* $CH_3CH{=}CHCH_2Sn(CH_3)_3$ (55.4) *trans* $CH_3CH{=}CHCH_2Sn(CH_3)_3$ (37.5)	$CH_2{=}CHCH_2CH_2Sn(CH_3)_3$ (7.1)
$CH_2{=}C(CH_3)CH{=}CH_2$[a]	66	$(CH_3)_2C{=}CHCH_2Sn(CH_3)_3$ (45.5) H_3C, $CH_2Sn(CH_3)_3$, C=C, H, CH_3 (43.6)	$CH_2{=}CHCH(CH_3)CH_2Sn(CH_3)_3$ (5.4) $CH_2{=}C(CH_3)CH_2CH_2Sn(CH_3)_3$ (5.5)
$CH_2{=}C(CH_3){-}C(CH_3){=}CH_2$[a]	66	$(CH_3)_2C{=}C(CH_3)CH_2Sn(CH_3)_3$ (77)	$CH_2{=}C(CH_3)CH(CH_3)CH_2Sn(CH_3)_3$ (23)
[a]	67	H $Sn(CH_3)_3$ (57)(89)[b]	H $Sn(CH_3)_3$ (43)(11)[b]
	55	H $Sn(CH_3)_3$ (66)	H $Sn(CH_3)_3$ (34)
[c]	28	H $Sn(CH_3)_3$ (87.5)	H $(CH_3)_3Sn$ (12.5)

[a] Twofold excess of diene at 100°C.
[b] 1:1 diene to hydride at 175°C without catalyst.
[c] 100% excess diene at 80°C.

addition but sometimes 1,2- addition of the C=C bond occurs also. The 1,4- addition products undergo numerous reactions to yield compounds which could not be easily obtained otherwise (*137a*, *220*):

$$RCH_2C(CN){=}C{=}NSnR'_3 \xrightarrow{H_2C{=}CHCH_2Br} RCH_2C(CN)_2CH_2CH{=}CH_2 \quad (77)$$

$$\xrightarrow{C_6H_5CH_2Cl} RCH_2C(CN)_2CH_2C_6H_5 \quad (78)$$

$$\xrightarrow{Br_2} RCH_2C(CN)_2Br \quad (79)$$

$$\xrightarrow{C_2H_5OH} RCH_2C(CN)_2H \quad (80)$$

$$\xrightarrow{CH_3C(=O)Cl} RCH_2C(CN)_2COCH_3 \quad (81)$$

Trimethyltin hydride can be added to allenes, as shown in Table 8 (*100*).

TABLE 8

AZOBISISOBUTYRONITRILE-CATALYZED ADDITION OF TRIMETHYLTIN HYDRIDE TO ALLENES AT 100°C (*100*)

Allene	%Yield	%Product
$H_2C{=}C{=}CH_2$[a]	67	$CH_2{=}C(CH_3)Sn(CH_3)_3$ (45.2) $CH_2{=}CHCH_2Sn(CH_3)_3$ (54.8)
$CH_3CH{=}C{=}CH_2$[b]	72.6	$CH_2{=}C(C_2H_5)Sn(CH_3)_3$ (31.5) $(H_3C)(H)C{=}C(CH_3)(Sn(CH_3)_3)$ (38.6) $(H_3C)(H)C{=}C(Sn(CH_3)_3)(CH_3)$ (16.3) *trans* $CH_3CH{=}CHCH_2Sn(CH_3)_3$ (10.0) *cis* $CH_3CH{=}CHCH_2Sn(CH_3)_3$ (3.5)
$(CH_3)_2C{=}C{=}CH_2$[c]	72	$CH_2{=}C(i\text{-}C_3H_7)Sn(CH_3)_3$ (26.8) $(CH_3)_2C{=}C(CH_3)Sn(CH_3)_3$ (73.2)

[a] 100% excess allene.
[b] 46% excess allene.
[c] 14% excess allene.

TABLE 8 (*continued*)

Allene	% Yield	% Product	
$CH_3CH{=}C{=}CHCH_3$[d]	65	$(H_3C)(H)C{=}C(Sn(CH_3)_3)(C_2H_5)$	(53.9)
		$(H_3C)(H)C{=}C(C_2H_5)(Sn(CH_3)_3)$	(46)
$(CH_3)_2C{=}C{=}CHCH_3$[e]	82.5	$(H_3C)(H)C{=}C(Sn(CH_3)_3)(C_3H_7\text{-}i)$	(42.2)
		$(H_3C)(H)C{=}C(C_3H_7\text{-}i)(Sn(CH_3)_3)$	(12.3)
		$(CH_3)_2C{=}C(C_2H_5)Sn(CH_3)_3$	(45.5)

[d] 7% excess allene.
[e] 10% excess allene.

2. *Mechanism*

The observed catalysis by radical initiators such as azobisisobutyronitrile (*45*, *164*) and ultraviolet irradiation (*9*, *25*, *26*, *45*) and retardation by the radical inhibitor galvinoxyl (*164*) provide evidence that the hydrostannation of alkenes can proceed by a radical chain mechanism. The following propagation steps have been proposed (*92*):

$$R_3Sn\cdot + \;\rangle C{=}C\langle \;\underset{k_{-1}}{\overset{k_1}{\rightleftharpoons}}\; R_3Sn\overset{|}{\underset{|}{C}}-\overset{|}{\underset{|}{C}}\cdot \tag{82}$$

$$R_3Sn\overset{|}{\underset{|}{C}}-\overset{|}{\underset{|}{C}}\cdot + R_3SnH \xrightarrow{k_2} R_3Sn\overset{|}{\underset{|}{C}}-\overset{|}{\underset{|}{C}}H + R_3Sn\cdot \tag{83}$$

This scheme explains the direction of addition of organotin hydrides to terminal alkenes. The organotin radical adds to the terminal carbon because the resulting carbon free radical can then be stabilized by the substituent attached to it. The energetics of reactions (82) and (83) have been discussed (*54*, *92*,

100). As a consequence of the instability of the SnH bond compared to the CH bond, reaction (83) is very probably exothermic and irreversible. Bond energy considerations indicate that reaction (82) may be endothermic or exothermic depending on the stability of the carbon free radical.

Although only a small amount of hydride was consumed when an excess of either *cis*- or *trans*-1-deuterio-1-hexene was irradiated in the presence of tri-*n*-butyltin hydride at 10°C, the recovered starting alkene was completely isomerized (*102*). Similarly, the thermal reaction of triphenyltin hydride at 70°C resulted in incomplete addition but complete isomerization of the starting *cis*- or *trans*-1-deuterio-1-hexene. These results indicate that reaction (82) is reversible with k_{-1} being much greater than k_2. The direction of addition of the organotin hydride is therefore thermodynamically controlled. In the case of the reaction of tri-*n*-butyltin hydride with either *cis*- or *trans*-β-deuteriostyrene the unreacted alkene was little isomerized which indicates a smaller value for k_{-1}. The lower k_{-1} value is a result of the greater stability of the benzyl-type radical compared to the *sec*-alkyl radical. The addition of trimethyltin hydride to *cis*- or *trans*-2-butene occurs rather slowly which indicates a relatively high value of k_{-1}. It has been pointed out that the high k_{-1} value for many internal alkenes explains why azobisisobutyronitrile is a relatively ineffective catalyst for the hydrostannation reaction in these cases (*219*). The high reversibility of reaction (82) enables the N=N bond of the catalyst to compete with the double bond of the alkene for the organotin hydride. The addition of organotin hydrides to the N=N bond is known (*169*). The reaction of *cis*- or *trans*-piperylene with the trimethyltin radical has been shown to be only partially reversible (*141*), which suggests that kinetic control may be important with conjugated dienes.

As previously mentioned, both possible adducts (α and β) are formed in the thermal reaction of organotin hydrides with acrylonitrile (*126*):

$$\underset{\alpha}{R_3SnCH(CH_3)CN} \qquad \underset{\beta}{R_3SnCH_2CH_2CN}$$

In the azobisisobutyronitrile-catalyzed reaction, on the other hand, only the β product is formed. The rate of formation of the α product was found to increase with increasing polarity of the solvent, whereas the effect of solvent on the rate of formation of the β product was negligible. The suggestion was made that the α product is formed by an ionic mechanism whereas the β product is formed by a free radical mechanism.

The fact that the *cis* alkene predominates in the 1,4- addition of trimethyltin hydride to dienes (Table 7) has been explained by assuming an equilibrium between open-chain radicals and a more stable bridged-radical in which the unpaired electron is delocalized by a 5*d* orbital on tin (*54*):

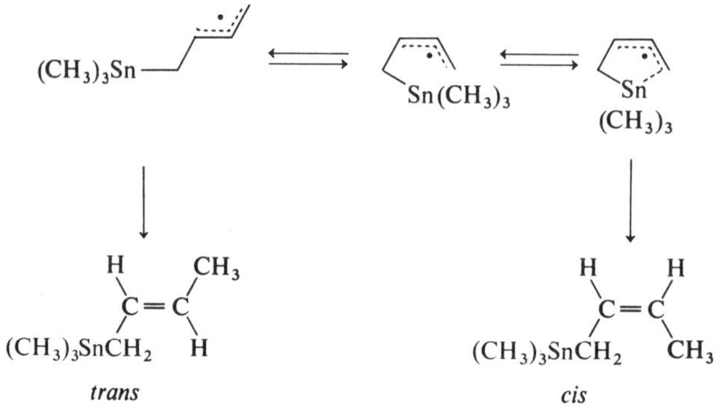

$$(CH_3)_3Sn\text{-allyl radical} \rightleftharpoons \text{bridged radical } Sn(CH_3)_3 \rightleftharpoons \text{bridged radical } Sn(CH_3)_3 \tag{84}$$

$$\downarrow \qquad\qquad\qquad\qquad \downarrow$$

$$\underset{trans}{(CH_3)_3SnCH_2(H)C{=}C(H)CH_3} \qquad \underset{cis}{(CH_3)_3SnCH_2(H)C{=}C(H)CH_3}$$

An alternate suggestion was that the bridged radical might be formed from a π-complex between the trimethyltin radical and butadiene.

The direction of addition and the geometry of the products observed in the reaction of trimethyltin hydride with allenes (Table 8) have been discussed in terms of a free radical mechanism (*100*). In the case of allene and 1,2-butadiene, 45.2% and 86.4%, respectively, of the products result from attack of the trimethyltin radical at the central carbon of the allene system. The remaining products results from attack at the terminal carbon. The hybrid allyl radical formed upon addition of the trimethyltin radical to the central carbon of 1,2-butadiene abstracts hydrogen predominantly at the less substituted carbon. The products from the last three allenes in Table 8 result exclusively from attack of the trimethyltin radical at the central carbon. Evidently the methyl groups in these allenes sterically hinder attack of the tin radical at the methyl substituted terminal carbons and stabilize the hybrid allyl radical formed by attack of the tin radical at the central carbon. It should be noted that attack of the trimethyltin radical at the terminal carbon of 1,2-butadiene results in more of the *trans* alkene than the *cis* alkene. It was pointed out that this result would be expected if the following equilibrium operates:

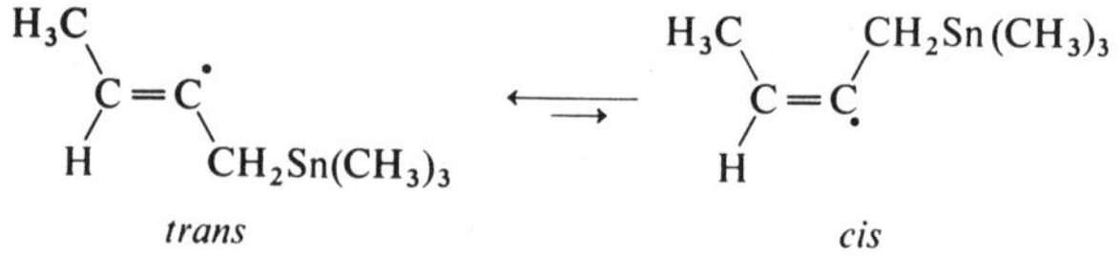

$$\underset{trans}{H_3C(H)C{=}\dot{C}CH_2Sn(CH_3)_3} \rightleftharpoons \underset{cis}{H_3C(H)C{=}\dot{C}CH_2Sn(CH_3)_3} \tag{85}$$

The *trans* radical, which would give a *trans* alkene on hydrogen abstraction, would be expected to predominate since it should be more stable for steric reasons. The isomer distribution resulting from attack of the trimethyltin radical at the central carbon of 1,2-butadiene was explained in terms of the two possible isomeric allyl radicals (**7**) and (**8**). Apparently (**7**), which would yield a *cis* alkene on hydrogen abstraction, predominates because the non-

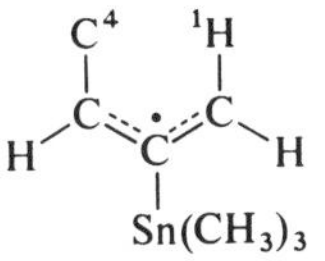

$$\underset{(7)}{\begin{array}{c} C^4 \quad {}^1H \\ H-C\cdots\dot{C}\cdots C-H \\ | \\ Sn(CH_3)_3 \end{array}} \qquad \underset{(8)}{\begin{array}{c} H \quad {}^1H \\ {}^4C-C\cdots\dot{C}\cdots C-H \\ | \\ Sn(CH_3)_3 \end{array}}$$

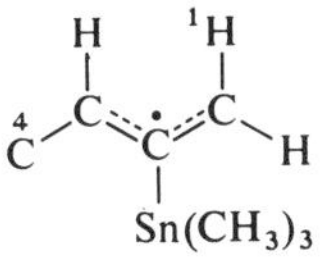

bonded interaction between C^4 and H^1 in (**7**) is less than that between C^4 and the trimethyltin group in (**8**).

D. Addition to Alkynes

Equimolar quantities of triphenyltin hydride and phenylacetylene react exothermally to give a quantitative yield of triphenyl-β-styryltin (*239*). The triphenyltin hydride adds across the triple bond in a *trans* manner to give a *cis* adduct which then rearranges to the more stable *trans* adduct (*62*):

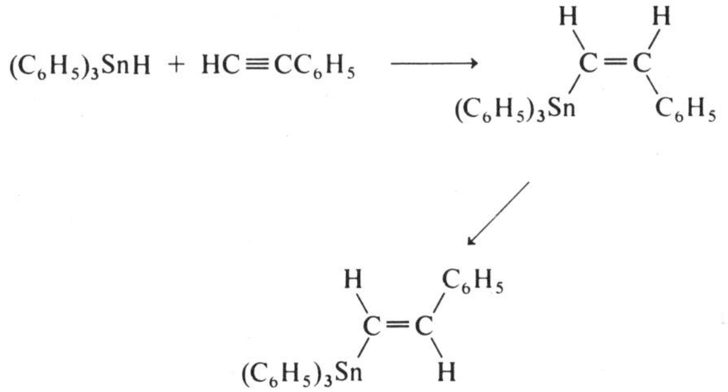

$$(C_6H_5)_3SnH + HC{\equiv}CC_6H_5 \longrightarrow \begin{array}{c} H \quad H \\ C{=}C \\ (C_6H_5)_3Sn \quad C_6H_5 \end{array} \longrightarrow \begin{array}{c} H \quad C_6H_5 \\ C{=}C \\ (C_6H_5)_3Sn \quad H \end{array} \tag{86}$$

Alkynes are more easily hydrostannated than alkenes. For example, although tri-*n*-propyltin hydride fails to react with 1-octene and allyl alcohol in the absence of a catalyst, this hydride readily adds to 1-hexyne and propargyl alcohol (*239*). The addition of triethyltin hydride across a C≡C bond which is conjugated with a C=C bond has been reported (*135, 135a*). Triethyltin hydride reacts with diacetylene and $C_2H_5C{\equiv}C{-}C{\equiv}CH$ at 60°–80°C to give $(C_2H_5)_3SnCH{=}CH{-}C{\equiv}CH$ and $(C_2H_5)_3SnCH{=}CH{-}C{\equiv}CC_2H_5$, respectively (*260*). Since the 1 : 1 adduct formed by reaction of an organotin hydride with an alkyne still contains a double bond, it can add a second molecule of organotin hydride. For example, the reaction of phenylacetylene with triphenyltin hydride (1 : 2 mole ratio) yields 1-phenyl-2,2-bis(triphenylstannyl)ethane (*45a, 127a, 239*):

$$2\,(C_6H_5)_3SnH + HC{\equiv}CC_6H_5 \longrightarrow [(C_6H_5)_3Sn]_2CHCH_2C_6H_5 \tag{87}$$

The second molecule of the hydride probably adds to the initially formed vinyltin compound by a free radical mechanism (*45a, 127a*).

The reaction of trimethyltin hydride with methyl ethynecarboxylate for 18 h at 55°C and subsequently for 4 h at 80°C affords five products (*124*):

$$(CH_3)_3SnH + HC{\equiv}CCOOCH_3 \longrightarrow \underset{\textbf{(9)}}{(H)_2C{=}C(COOCH_3)Sn(CH_3)_3}$$

$$+ \underset{\textbf{(10)}}{(CH_3)_3Sn(H)C{=}C(H)COOCH_3} + \underset{\textbf{(11)}}{(CH_3)_3Sn(H)C{=}C(COOCH_3)H} \qquad (88)$$

$$+ (CH_3)_3SnCH_2CH_2COOCH_3 + (CH_3)_3SnC{\equiv}CCOOCH_3$$

The nonterminal (**9**), *cis* (**10**), and *trans* (**11**) adducts constitute 47.5% of the mixture and are present in the ratio 77 : 18 : 5. The distribution of isomeric alkenes obtained with other monosubstituted ethynes are given in Table 9. It can be seen that when the ethyne contains a strongly electron-withdrawing group (CN, COOR), the major product is the nonterminal adduct. On the other hand, when the ethyne contains a weakly electron-withdrawing group (C_6H_5, CH_2OH) or an electron-releasing group (CH_3, n-C_4H_9, OC_2H_5), the major product is the *cis-trans* mixture.

Three 1 : 1 adducts are formed in the hydrostannation of ethyl 1-propynecarboxylate, the *trans* isomers (with respect to R_3Sn and H) predominating (*119, 123*):

$$R_3SnH + CH_3C{\equiv}CCOOC_2H_5 \longrightarrow$$

$$\underset{\textbf{(12)}}{(H_3C)(H)C{=}C(SnR_3)(COOC_2H_5)} + \underset{\textbf{(13)}}{(H)(H_3C)C{=}C(SnR_3)(COOC_2H_5)} + \underset{\textbf{(14)}}{(H_3C)(R_3Sn)C{=}C(H)(COOC_2H_5)} \qquad (89)$$

R = CH_3, ratio of (**12**):(**13**):(**14**) = 59:9:32
R = n-C_4H_9, ratio of (**12**):(**13**):(**14**) = 50:13:37

The *trans* isomer is the major product (*trans* : *cis* 9 : 1) formed in the hydrostannation of diethyl ethynedicarboxylate (*123*). Only the *trans* adduct is formed in the hydrostannation of hexafluoro-2-butyne (*43*) and dicyanoethyne (*119*).

Important information bearing on the mechanism of the hydrostannation

TABLE 9

REACTION OF TRIORGANOTIN HYDRIDES WITH MONOSUBSTITUTED ALKYNES

$$R_3SnH + HC{\equiv}CR' \longrightarrow H_2C{=}C(R')SnR_3\text{(nonterminal adduct)} + R_3SnCH{=}CHR'\text{(cis and trans)}$$

R	R′	Reaction conditions	Percentage yield of 1:1 adducts	Ratio of nonterminal adduct: *cis*: *trans*	Ref.
CH_3	$COOCH_3$	23h, 55°C	40	81:15:4	(*124*)
C_2H_5	$COOCH_3$	23h, 55°C	57	58:24:18	(*124*)
n-C_3H_7	$COOCH_3$	5.5h, 55°C	72	31:36:33	(*124*)
CH_3	$COOC_2H_5$	18h, 55°C 4h, 80°C	41	67:25:8	(*124*)
C_2H_5	CH_2OH	6h, 100°C	>70	25:45:30	(*123*)
C_6H_5	CH_2OH	3h, 60°C 16h, 20°C	80	5:45:50	(*119*)
CH_3	CN	1h, 20°C	90	100:0:0	(*123*)
C_2H_5	CN	1h, 20°C	100	100:0:0	(*123*)
n-C_3H_7	CN	0.75h, 20°C	90	100:0:0	(*123*)
CH_3	CF_3	24h, 20°C	43	22:13:65	(*43*)
CH_3	CH_3	18h, 100°C	30	4:28:68	(*214*)
CH_3	n-C_4H_9	7h, 60°C	85	2:29:69	(*119*)
C_2H_5	n-C_4H_9	6h, 50°C	55	3:51:46	(*119*)
		12h, 50°C 16h, 20°C	85	3:31:66	(*119*)
C_6H_5	n-C_4H_9	3h, 50°C	>80	0:85:15	(*119*)
CH_3	C_6H_5	66h, 50°C	75	6:66:28	(*119*)
C_2H_5	C_6H_5	7h, 60°C	75	0:70:30	(*119*)
C_6H_5	C_6H_5	2h, 20°C 1h, 60°C	90	0:28:72	(*119*)
C_2H_5	OC_2H_5	1.5h, 50°C	100	4:91:5	(*119*)
C_6H_5	OC_2H_5	1h, 20°C	>90	0:97:3	(*119*)
C_2H_5	SC_4H_9-n	3h, 45°C	70	0:77:23	(*119*)

of alkynes has been obtained by Leusink et al. (*114*, *115*, *116*, *122*). The addition of triethyltin deuteride to either methyl ethynecarboxylate or cyanoethyne affords mainly the *trans* (with respect to R_3Sn and D) nonterminal adduct, a result which is consistent with *trans* addition across the triple bond. The rate of formation of the nonterminal adduct from the reaction of triethyltin hydride with cyanoethyne increases with the polarity of the solvent and is unaffected by the radial inhibitor galvinoxyl. The rate of formation of the nonterminal adduct from the reaction of trimethyltin hydride with methyl ethynecarboxylate, Eq. (88), is unaffected by the radical inhibitor galvinoxyl or the radical initiator azobisisobutyronitrile. On the other hand, the rate increases with increasing polarity of the solvent. With diethyl ethynedicarboyxlate the rate increases in the order $(C_6H_5)_3SnH < (CH_3)_3SnH < (C_2H_5)_3SnH < (n\text{-}C_4H_9)_3SnH$. A similar sequence was found for cyanoethyne. The rate of reaction was found to be first order in both hydride and alkyne. The decrease in rate observed upon replacing the hydrogen of the tin hydride with deuterium indicates that the hydrogen is transferred in the rate-determining step of the reaction. Electron-withdrawing groups attached to the triple bond accelerate the reaction as illustrated by the following order of increasing rates: $HC{\equiv}CCOOCH_3 < C_2H_5OOCC{\equiv}CCOOC_2H_5 < HC{\equiv}CCN < NCC{\equiv}CCN$. The following mechanism involving nucleophilic *trans*-addition of the organotin hydride has been suggested for the formation of the nonterminal adduct (*115*):

$$R_3SnH + R'C{\equiv}CR'' \xrightarrow{\text{slow}} \left[\begin{array}{c} R' \\ \;\; C{\equiv}C^{\delta\ominus} \;\; R'' \\ H \cdots Sn R_3^{\delta\oplus} \end{array} \right] \longrightarrow R_3Sn^{\oplus} + \begin{array}{c} R' \qquad \\ C{=}\overset{\ominus}{C} \\ H \qquad R'' \end{array} \xrightarrow{\text{fast}} \begin{array}{c} R' \qquad SnR_3 \\ C{=}C \\ H \qquad R'' \end{array} \quad (89a)$$

The slow step is consistent with the isotope effect and the kinetics. The polar transition state explains the solvent effect. The incipient negative charge on carbon is best stabilized when R″ is a strongly electron-withdrawing group, while the incipient positive charge on tin is stabilized by electron-releasing groups. Only a small solvent effect was observed in the hydrostannation of diethyl ethynedicarboxylate. It was suggested that the transition state in this case is stabilized by intramolecular coordination between carbonyl oxygen and tin (*115*). In the above mechanism the hydrogen of the organotin hydride is transferred as a hydride ion. Organotin hydrides are also good hydride donors to carbonium ions (*21a*), isocyanates and isothiocyanates (*120*, *125*), and strongly electrophilic carbonyl compounds (*118*).

In the reaction of trimethyltin hydride with methyl ethynecarboxylate, Eq. (88), a catalytic amount of azobisisobutyronitrile increases the yield of *cis* β-adduct (**10**), while at the same time some *trans* β-adduct (**11**) is formed (*122*). Alkynes containing only an electron-releasing group, or both an electron-releasing group and an electron-withdrawing group, undergo hydrostannation in the absence of added catalysts to also give both *cis* and *trans* β-adducts (*114*). In these reactions the *cis* adduct is formed by a free radical *trans*-addition, and the *trans* adduct is formed by a free radical isomerization of the *cis* adduct (*114*, *116*).

The reaction of diorganotin dihydrides with acetylenes can be used to prepare tin-containing polymers and tin heterocycles. Di-*n*-propyltin dihydride reacts with phenylacetylene (1:1 mole ratio) to give a monoadduct which affords, upon heating, a linear polymer containing tin in the main chain (*175*).

$$n\ \mathrm{HSn(C_3H_7\text{-}}n)_2\mathrm{CH{=}CHC_6H_5} \xrightarrow{130^\circ C} \left[\mathrm{Sn(C_3H_7\text{-}}n)_2\mathrm{CH_2CH(C_6H_5)}\right]_n \tag{90}$$

The polymer is a viscous oil with an average degree of polymerization of about 11. The reaction of diphenyltin dihydride with diacetylene (*111b*) and the reactions of diorganotin dihydrides with α,ω-diynes afford linear polymers containing tin in the main chain (*175*). Polymers also have been obtained from the reaction of *p*-phenylene bis(dimethyltin hydride) with alkynes and dialkynes (*128*). The interesting tin heterocycle (**15**) has been obtained from

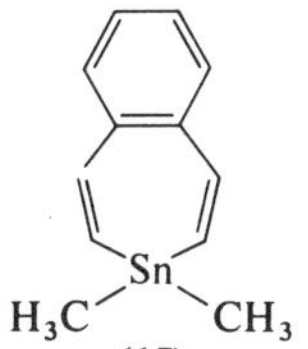

Sn, H_3C, CH_3

(15)

the reaction of dimethyltin dihydride with *o*-diethynylbenzene (*129*). Reaction of (**15**) with phenylboron dichloride affords dimethyltin dichloride and the boron heterocycle (**16**), the seven-membered ring of which appears to be aromatic as predicted by Hückel's rule (*121*).

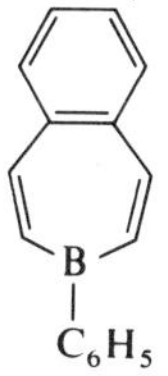

B, C_6H_5

(16)

The hydrostannation of alkynes has been used to prepare organotin-substituted vinyl ethers (*85a*), an unsaturated alcohol, and unsaturated acetals (*245a*). Dimetallic ethylene derivatives can be prepared by the hydrostannation of triorganometalacetylenes (*137b*, *c*, *d*, *e*,).

E. Reactions with Aldehydes and Ketones

Simple aldehydes and ketones are reduced to alcohols in good yield upon reaction with diphenyltin dihydride or di-*n*-butyltin dihydride (*95*, *96*):

$$R_2SnH_2 + R'_2CO \longrightarrow R_2Sn + R'_2CHOH \tag{91}$$

The unique feature of this reaction is that no hydrolysis step is necessary as in the case of reductions with lithium aluminum hydride. The alcohol is easily separated from the polymeric diorganotin which is also formed. Some simple carbonyl compounds which have been reduced with diphenyltin dihydride in diethyl ether at room temperature (the yield is given in parentheses) include cyclohexanone (82%), benzophenone (59%), and benzaldehyde (62%). Di-*n*-butyltin dihydride reduces benzophenone to the alcohol in 85% yield. Diphenyltin dihydride reduces the nitro group in preference to the carbonyl group of *m*-nitrobenzaldehyde, *p*-nitrobenzaldehyde, and *m*-nitroacetophenone. Diphenyltin dihydride has been found to reduce preferentially one of the keto groups of some steroids which contain two keto groups (*92*). It has been used also to reduce the ketone group of cycloheximide to an alcohol group (*77a*).

Tri-*n*-butyltin hydride and triphenyltin hydride reduce simple aldehydes and ketones (*96*):

$$2\,R_3SnH + R'_2CO \longrightarrow R_3SnSnR_3 + R'_2CHOH \qquad (92)$$

Triphenyltin hydride reduces methyl and phenyl 2-thienyl ketones in this manner (*256*). Reductions with phenyltin trihydride, which decomposes at room temperature, are slow and incomplete. An excess of *n*-butyltin trihydride, on the other hand, satisfactorily reduces 4-methylcyclohexanone and 4-*t*-butylcyclohexanone (*96*).

The organotin hydrides exhibit the following order of decreasing reactivity toward a given aldehyde or ketone:

$$(C_6H_5)_2SnH_2 > (n\text{-}C_4H_9)_2SnH_2 > n\text{-}C_4H_9SnH_3 > (C_6H_5)_3SnH > (n\text{-}C_4H_9)_3SnH$$

A given hydride reduces different aldehydes and ketones at different rates, which means that a dissociation of the hydride cannot be the rate-determining step of the reaction (*96*).

Some sterochemical information on the reduction of ketones has been obtained (*96*). The reduction of benzil with diphenyltin dihydride gives exclusively meso-hydrobenzoin. The organotin hydride reduction of 4-*t*-butylcyclohexanone and 4-methylcyclohexanone gives predominantly the *trans* alcohol. Furthermore, the proportion of *trans* alcohol is essentially independent of the kind of organotin hydride used. Compared to the organotin hydride reductions, lithium aluminum hydride reduction results in a slightly higher proportion of the *trans* isomer, while sodium borohydride reduction gives the same proportion of *trans* isomer.

Initially, the only reaction observed between α,β-unsaturated aldehydes and ketones and organotin hydrides was selective reduction of the carbonyl group (*95*, *96*, *172*). Thus, triphenyltin hydride and methyl vinyl ketone were reported to give hexaphenylditin and methyl vinyl carbinol (*172*). Hexaphenylditin was obtained also from the reaction of triphenyltin hydride with phenyl vinyl ketone (*239*). The following α,β-unsaturated aldehydes and ketones gave the corresponding α,β-unsaturated alcohols on reaction with diphenyltin dihydride in diethyl ether at room temperature (the yield is given in parentheses): methyl vinyl ketone (59%), cinnamaldehyde (75%), crotonaldehyde (59%), mesityl oxide (60%), and chalcone (benzalacetophenone) (75%) (*95, 96*).

Only the saturated ketones and hexaphenylditin were obtained from the reaction of triphenyltin hydride with mesityl oxide (*183, 185*), phorone, dihydrophorone, methyl propenyl ketone (5 h at 150°C) and methyl vinyl ketone (5 h at 70°C) (*183*). With tri-*n*-butyltin hydride, on the other hand, the saturated ketone is a minor product; the major product is an enol ether. Similar results have been obtained (Table 10) by Leusink and Noltes (*127*). It

TABLE 10

REACTION OF ORGANOTIN HYDRIDES WITH α,β-UNSATURATED KETONES (*127*)

$$R_3SnH + \;>C{=}C(-)-C(=O)- \;\longrightarrow$$

$$\underset{\text{(I)}}{>C(H)-C{=}C(OSnR_3)-} \;+\; \underset{\text{(II)}}{>C(H)-C(H)-C(=O)-} \;+\; \underset{\text{(III)}}{>C(R_3Sn)-C(H)-C(=O)-}$$

Ketone	Tin hydride	Mole ratio ketone: hydride	Reaction conditions	%Yield	Mole ratio I:II:III
Benzalacetophenone	$(n\text{-}C_4H_9)_3SnH$	1:1	5h, 70°C	90	90:10:—
(chalcone)	$(C_6H_5)_3SnH$	1:2	3h, 55°C	90	40:60:—
	$(C_6H_5)_2SnH_2$	1:1	3h, 55°C	55	0:100:—
Phenyl vinyl ketone	$(n\text{-}C_4H_9)_3SnH$	1:1	6h, 55°C	60	95:5:—
	$(C_6H_5)_3SnH$	1:2	4h, 55°C	85	0:100:—
	$(C_6H_2)_2SnH_2$	3:4	2h, 55°C	25	0:100:—
Methyl vinyl ketone	$(C_6H_5)_3SnH$	1:2	3h, 55°C	80	—:35:65
	$(C_6H_5)_2SnH_2$	1:1	1h, 55°C	40	—:100:0

was shown that the enol ethers (I) are easily cleaved by triphenyltin hydride to give the saturated ketones (II); on the other hand, only small amounts of (II) are obtained from the reaction of (I) with tri-*n*-butyltin hydride. It was reasonably suggested that both tri-*n*-butyltin hydride and triphenyltin hydride add in a 1,4-manner across the α,β-unsaturated system to give the enol ether (I). In the case of tri-*n*-butyltin hydride the reaction stops at the enol ether stage, but in the case of triphenyltin hydride further reaction occurs to give the enol form of the saturated ketone, which then tautomerizes to the more stable keto form (II). Other α,β-unsaturated ketones have been found to react exclusively by 1,4-addition to give enol ethers (*167*):

$$RHC{=}C(CN)\overset{\overset{\displaystyle O}{\|}}{C}C_6H_5 + R'_3SnH \longrightarrow RCH_2C(CN){=}\overset{\overset{\displaystyle OSnR'_3}{|}}{C}C_6H_5 \qquad (93)$$

$R = p\text{-}CH_3OC_6H_4$, C_6H_5, $p\text{-}O_2NC_6H_4$, or furyl
$R' = C_2H_5$ or $n\text{-}C_4H_9$

Organotin hydrides add in a 1,2-manner to α,β-unsaturated esters and nitriles. The amount of saturated ester or nitrile formed then depends upon the amount of α-adduct (tin attached to the α-carbon) formed, since an α-carbon–tin bond can be converted to a carbon–hydrogen bond by reaction with the organotin hydride (*182a*, *182b*).

Triphenyltin hydride has been reported to add across the carbon-carbon double bond of methyl γ-butenyl ketone to give (5-ketohexyl)triphenyltin (*239*). More recently, triorganotin hydrides have been found to react with γ-ethylenic ketones to give addition across the carbon-carbon double bond, reduction of the ketone to the unsaturated alcohol, or addition across the carbonyl group to form an organotin alkoxide (Table 11) (*184*).

A free radical mechanism has been proposed for the organotin hydride reduction of the carbonyl group of simple aldehydes and ketones without catalyst (*92*) and under ultraviolet irradiation (*186*). The steps represented by Eqs. (94) and (95) have been suggested (*92*):

$$\equiv SnH + \;>C{=}O \longrightarrow \;\equiv SnO\overset{|}{\underset{|}{C}}H \qquad (94)$$

$$\equiv SnO\overset{|}{\underset{|}{C}}H + \equiv SnH \longrightarrow \;\equiv SnSn\equiv + HO\overset{|}{\underset{|}{C}}H \qquad (95)$$

Reaction (94) can be effected at moderate temperatures in the presence of azobisisobutyronitrile (*142*, *143*, *153*) or ultraviolet irradiation (*21*, *186*, *234*) which suggests a free radical mechanism. Reaction (94) is catalyzed also by zinc chloride (*142*, *143*) which suggests that a polar mechanism may be operating in this case. Reaction (94) has been observed with hexafluoroacetone (*42*) and with other strongly electrophilic carbonyl compounds (*117*). The rate of the reaction of triethyltin hydride with 2,2,2-trifluoroacetophenone follows second-order kinetics, increases with the polarity of the solvent, and is unaffected by the radical inhibitor galvinoxyl; the rate with other

TABLE 11

REACTION OF TRIORGANOTIN HYDRIDES WITH γ-ETHYLENIC KETONES (*184*)

$$RR'C{=}CH(CH_2)_2C(=O)CH_3 + R''_3SnH \longrightarrow$$

$$\underset{\text{(I)}}{RR'C(SnR''_3){-}CH_2(CH_2)_2C(=O)CH_3} + \underset{\text{(II)}}{RR'C{=}CH(CH_2)_2CH(OH)CH_3} + \underset{\text{(III)}}{RR'C{=}CH(CH_2)_2CH(OSnR''_3)CH_3}$$

R	R′	R″	Reaction conditions	%Yield I	%Yield II	%Yield III
H	H	C_6H_5	9h, 90°C	64	0	0
			9h, 120°C	49	7	0
			5h, 150°C	4	28	0
			20h, −25°C, uv	91	2	0
		n-C_4H_9	9h, 70°C, uv	61	8	0
H	CH_3	C_6H_5	9h, 90°C	0	12	0
			5h, 150°C	0	19	0
			20h, −20°C, uv	0	26	0
		n-C_4H_9	9h, 70°C, uv	0	16	25
CH_3	CH_3	C_6H_5	9h, 90°C	0	25	0
			5h, 150°C	0	29	0
		n-C_4H_9	9h, 70°C, uv	0	13	22

triorganotin hydrides increases with the electron-supplying ability of the groups attached to tin (*118*). An ionic mechanism involving nucleophilic attack of the hydride hydrogen on carbonyl carbon was suggested (*118*):

$$\equiv SnH + \rangle C{=}O \xrightarrow{\text{slow}} \left[\equiv \overset{\delta\oplus}{Sn} \cdots H \cdots \overset{|}{\underset{|}{C}} \cdots \overset{\delta\ominus}{O} \right] \longrightarrow \equiv Sn^{\oplus} + H\overset{|}{\underset{|}{C}}O^{\ominus} \xrightarrow{\text{fast}} H\overset{|}{\underset{|}{C}}OSn\equiv \qquad (95a)$$

Reaction (95) has been observed (*194*) and shown to occur by an electrophilic attack of the organotin hydride hydrogen on oxygen (*35*, *38*). It was pointed out that this mechanism explains why organotin alkoxides can be isolated from the reaction of ketones with trialkyltin hydrides but not with triphenyltin hydride. The hydrogen of triphenyltin hydride is more electrophilic than that of a trialkyltin hydride, and therefore reaction (95) occurs more readily in the former case. This explanation has been used to account for the results in Table 10 (*127*).

F. Additions to Unsaturated Nitrogen Compounds

Trialkyltin hydrides react exothermally in the absence of catalysts with aromatic isocyanates (1:1 mole ratio) to form 1:1 adducts (**17**) containing a tin-nitrogen bond (*170*, *171*). The 1:1 adducts (**17**) readily react with water to

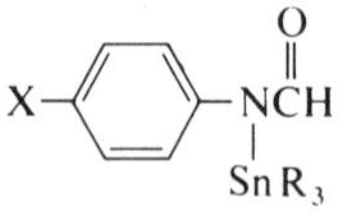

$$X-C_6H_4-N(SnR_3)-C(=O)H$$

$R = C_2H_5$, X = H, Cl, or NO_2
$R = n\text{-}C_4H_9$, X = H

(17)

afford *N*-arylformamides. Hexaphenylditin and the corresponding *N*-arylformamides are obtained from the reaction of phenyl isocyanate or α-naphthyl isocyanate with triphenyltin hydride (1:2 mole ratio) (*131*). The *N*-arylformamides probably result from hydrogenolysis of the initially formed 1:1 adducts by the second mole of triphenyltin hydride. This hydrogenolysis step has been separately realized with triethyl(*N*-phenylformamido)tin (*169*):

$$C_6H_5N(Sn(C_2H_5)_3)CH{=}O + (C_6H_5)_3SnH \longrightarrow (C_2H_5)_3SnSn(C_6H_5)_3 + C_6H_5NHCH{=}O \quad (96)$$

Triethyltin hydride and *n*-hexyl isocyanate do not react exothermally; nevertheless, the reaction is complete after 1 h at 90°C (*171*). The major product is the 1:1 adduct (**18**) containing a carbon-tin bond. The 1:1 adduct

$$n\text{-}C_6H_{13}NHC(=O)Sn(C_2H_5)_3 \quad \textbf{(18)} \qquad C_6H_5N{=}CHSSn(C_2H_5)_3 \quad \textbf{(19)}$$

(**19**) containing a tin-sulfur bond results from the reaction of triethyltin hydride with phenyl isothiocyanate (*170*, *171*).

Information bearing on the mechanism of the reaction of organotin hydrides with isocyanates has been obtained (*120*, *125*). The rate of addition increases with the polarity of the solvent and is not affected by catalytic amounts of the radical inhibitor galvinoxyl or of the radical initiator azobisisobutyronitrile. In cyclohexane-butyronitrile (3 : 1) the order of reactivity of triorganotin hydrides toward phenyl isocyanate is $(n\text{-}C_4H_9)_3SnH > (CH_3)_3SnH > (C_6H_5)_3SnH$. Phenyl isocyanate is more reactive than *n*-hexyl isocyanate. The reaction of triethyltin hydride with phenyl isocyanate appears to be first order in both hydride and isocyanate. The use of triethyltin deuteride reveals a negligible isotope effect: $k_H/k_D \approx 1$. The same effects of structure, solvent, and lack of activity of azobisisobutyronitrile and galvinoxyl have been found in the addition of triorganotin hydrides to isothiocyanates (*120*). The following polar mechanism involving nucleophilic attack of the organotin hydride hydrogen on carbon has been suggested for these reactions (*120*):

$$R_3SnH + R'N{=}C{=}X \xrightarrow{\text{slow}} \overbrace{R'N{\cdots}C{\cdots}X}^{\delta\ominus}\ (\text{C}\cdots H\cdots {}^{\delta\oplus}SnR_3)$$

$$\begin{matrix} R'N{=}\underset{|}{C}{-}X^{\ominus} \\ H \end{matrix} \longleftrightarrow \begin{matrix} R'\overset{\ominus}{N}{-}\underset{|}{C}{=}X \\ H \end{matrix} \quad + \quad {}^{\oplus}SnR_3 \xrightarrow{\text{fast}} \begin{cases} \xrightarrow{X=O} R'N(SnR_3)CH{=}O \\ \xrightarrow{X=S} R'N{=}CH(SSnR_3) \end{cases} \tag{96a}$$

The slow step is consistent with the kinetics of the reaction and with the solvent effect. Tri-*n*-butyltin hydride exhibits the highest reactivity since the *n*-butyl group can best inductively stabilize the incipient positive charge on tin in the transition state. The higher reactivity of phenyl isocyanate compared to *n*-hexyl isocyanate is probably due to the ability of the phenyl group to stabilize the incipient negative charge on nitrogen in the transition state both inductively and by a resonance effect. The experimental k_H/k_D ratio is consistent with the calculated values for the above mechanism which lie between 3.7 and 0.4, depending upon the extent of transfer of hydrogen from tin to carbon in the transition state (*120*).

Azobisisobutyronitrile or anhydrous zinc chloride catalyzes the addition of trialkyltin hydrides to azomethines (*144*):

$$R{-}C_6H_4{-}N{=}CH{-}C_6H_5 + R'_3SnH \longrightarrow R{-}C_6H_4{-}N(SnR'_3){-}CH_2{-}C_6H_5 \tag{97}$$

$R = H$, $R' = C_2H_5$ or $n\text{-}C_4H_9$
$R = CH_3$, $R' = C_2H_5$

The resulting adduct affords the corresponding secondary amine upon reaction with water.

Azoxybenzene results from the reaction of nitrosobenzene with triethyltin hydride (2 : 1 mole ratio) (*144*):

$$2C_6H_5NO + (C_2H_5)_3SnH \longrightarrow (C_2H_5)_3SnOH + C_6H_5\underset{\substack{\downarrow \\ O}}{N}{=}NC_6H_5$$

$$C_6H_5N\begin{matrix} \diagup OSn(C_2H_5)_3 \\ \diagdown H \end{matrix} \quad \textbf{(20)} \qquad (98)$$

The adduct (**20**) is undoubtedly an intermediate in this reaction since it can be isolated from the reaction of nitrosobenzene with triethyltin hydride (1 : 1 mole ratio) and affords azoxybenzene upon reaction with nitrosobenzene. The reaction of (**20**) with sulfuric acid gives phenylhydroxylamine. Upon standing, a mixture of triethyltin hydride and nitrosobenzene (1 : 1 mole ratio) affords azobenzene, which probably results from the reaction of the initially formed azoxybenzene with triethyltin hydride.

Triethyltin hydride adds to dicyclohexylcarbodiimide in the presence of azobisisobutyronitrile at 70°C, the tin becoming attached to the nitrogen (*144*).

The addition of triorganotin hydrides to azobenzene and diethyl azodicarboxylate (1 : 1 mole ratio) has been accomplished without catalysts (*169*). The resulting 1 : 1 adducts undergo hydrogenolysis upon reaction with a second mole of hydride, triphenyltin hydride being much more effective than triethyltin hydride:

$$C_6H_5N{=}NC_6H_5 + (C_6H_5)_3SnH \longrightarrow \begin{matrix} (C_6H_5)_3Sn & & H \\ & N{-}N & \\ C_6H_5 & & C_6H_5 \end{matrix} \qquad (99)$$

$$\xrightarrow{(C_6H_5)_3SnH} (C_6H_5)_3SnSn(C_6H_5)_3 + C_6H_5NHNHC_6H_5$$

G. Reactions with Acids

Upon reaction with halogen acids, triorganotin hydrides afford hydrogen and a triorganotin halide (*89*). The heats of formation of tri-*n*-propyltin hydride (−32.1 kcal/mole) and tri-*n*-butyltin hydride (−48.6 kcal/mole) were determined from measurements of the heats of reaction of these hydrides with gaseous hydrogen chloride at 25°C (*223*). Triphenyltin hydride reacts exothermally with nitric acid to produce nitrobenzene and diphenyltin oxide (*216*). Triphenyltin nitrate (*216*, *217*, *232*) is probably an intermediate in this reaction (*216*). A kinetic study of the acid- and base-catalyzed methanolysis of tri-*n*-butyltin hydride has been reported (*97*). In the acid-catalyzed reaction the rate-determining step appears to involve an electrophilic attack on the tin-hydrogen bond by molecular acid. In the base-catalyzed reaction the rate-determining step is preceded by the formation of a pentacoordinated tri-*n*-butyltin intermediate. Diorganotin dihydrides afford diorganohalotin hydrides upon reaction with halogen acids in a 1 : 1 mole ratio (*197*).

Triphenyltin hydride reacts with carboxylic acids to give the corresponding triphenyltin esters (I) and/or *sym*-tetraphenyldiacyloxyditins (II) (Table 12).

TABLE 12

REACTIONS OF TRIPHENYLTIN HYDRIDE WITH CARBOXYLIC ACIDS

$$(C_6H_5)_3SnH + RCOOH \longrightarrow \underset{\text{I}}{RCOOSn(C_6H_5)_3} + \underset{\text{II}}{\underset{RCOO\quad OCOR}{(C_6H_5)_2Sn{-}Sn(C_6H_5)_2}}$$

Carboxylic acid	Moles of hydride / mole of acid	% Yield I	% Yield II	Ref.
Acetic	0.5	17		(*256*)
	2	85		(*256*)
Propionic	2	90		(*256*)
1-Norbornanecarboxylic	1	78		(*109*)
1-Apocamphanecarboxylic	1	78		(*109*)
	2	80		(*109*)
1-Triptycenecarboxylic	1	84		(*109*)
	2	88		(*109*)
Benzoic	1		20	(*256*)
	2		46	(*256*)
Thiophene-2-carboxylic	1	31	21	(*256*)
	2		42	(*256*)
Furan-2-carboxylic	1		17	(*256*)
	2	31	35	(*256*)
Pyrrole-2-carboxylic	1	38		(*109*)
	2	47		(*109*)
Pyridine-2-carboxylic (picolinic)	2	[a]		(*109*)
Ferrocenecarboxylic	1	28		(*109*)
	2		58	(*109*)

[a] Diphenyltin dipicolinate was isolated in 90% yield.

An exception is pyridine-2-carboxylic acid (picolinic acid) which affords diphenyltin dipicolinate. This product may be accounted for by assuming the initial formation of triphenyltin picolinate. Coordination between nitrogen and tin in this compound may facilitate cleavage of a phenyl group by picolinic acid to give the observed diester. The mechanism of formation of the ditin compounds remains to be ascertained.

The acetolysis of tri-*n*-butyltin hydride in dimethyl sulfoxide and dimethyl sulfoxide-diglyme is assisted by halide ion (rate ~ [hydride][acetic acid][X^-]) (*47*). The following mechanism has been suggested:

$$(n\text{-}C_4H_9)_3SnH + X^{\ominus} \underset{}{\overset{\text{fast}}{\rightleftarrows}} (n\text{-}C_4H_9)_3Sn\begin{matrix}H \\ X\end{matrix} \xrightarrow[\text{slow}]{CH_3COOH} (n\text{-}C_4H_9)_3SnOOCCH_3 + H_2 + X^{\ominus} \quad (100)$$

Diphenyltin dihydride reacts with carboxylic acids to give *sym*-tetraphenyldiacyloxyditins (*202*). An exception is pyridine-2-carboxylic acid (picolinic acid) which affords diphenyltin dipicolinate (*109*). Di-*n*-butyltin dihydride reacts with carboxylic acids to give either the *sym*-tetra-*n*-butyldiacyloxyditin or the di-*n*-butyltin dicarboxylate, depending upon the mole ratio of acid : hydride (*199*, *202*). When the acid : hydride ratio is two, the simple dicarboxylate is formed; when the acid : hydride ratio is one or less than one, the ditin compound is formed. An exception is succinic acid which affords di-*n*-butyltin succinate irrespective of the acid : hydride ratio. It has been shown (*203*) that the following equilibrium occurs when one mole of acid is added to one mole of hydride:

$$2\ R_2Sn\begin{matrix}H \\ OCOR'\end{matrix}\ (\mathbf{21}) \rightleftarrows R_2SnH_2 + R_2Sn(OCOR')_2 \quad (101)$$

Upon standing, the intermediate, (**21**), decomposes to give hydrogen and the ditin compound. (**21**) reacts with a second mole of acid to produce hydrogen and the simple dicarboxylate. In the case of diphenyltin dihydride, decomposition to the ditin occurs before reaction with a second mole of acid can occur. As already mentioned, pyridine-2-carboxylic acid affords the simple dicarboxylate upon reaction with diphenyltin dihydride. In this case, coordination between tin and nitrogen in (**21**) may assist the reaction of the tin-hydrogen bond with the second mole of acid so that it effectively competes with the decomposition reaction.

H. Reduction of Esters

Organic esters are reduced by trialkyltin hydrides (*86*):

$$ROC(=O)R' + R''_3SnH \longrightarrow RH + R''_3SnOC(=O)R' \qquad (101a)$$

This reaction requires more vigorous conditions than is required for the reduction of alkyl halides. Examples of the reduction of benzoates are given in Table 13. The catalytic effect of free radical initiators indicates a free radical

TABLE 13

Reduction of Benzoates with Tri-*n*-butyltin Hydride[a] (*86*)

R in C_6H_5COOR	Catalyst	Reaction time, h	% Reduction[b]
t-C_4H_9[c]	uv	40	48
$C_6H_5CH_2$[c]	uv	10	70
$(C_6H_5)_2CH$[c,d]	AIBN[e]	24	78
$CH_2{=}CHCH_2$[c]	uv	7	30
n-C_4H_9	uv	50	10
sec-C_4H_9	uv	44	29
t-C_4H_9	uv	30	61
cyclo-C_6H_{11}	$(n\text{-}C_4H_9)_2O_2$[f]	27	20
C_6H_5	uv	6.5	33
$C_6H_5CH_2$	uv	3.5	77
$C_6H_5CHCH_3$	$(n\text{-}C_4H_9)_2O_2$	7.5	86
$(C_6H_5)_2CH$	$(n\text{-}C_4H_9)_2O_2$	6	89
$(C_6H_5)_2CH$	uv	3	80
$(C_6H_5)_3C$[d]	uv	2	77
$(C_6H_5)_3C$	$(n\text{-}C_4H_9)_2O_2$	5	74
$CH_2{=}CHCH_2$	uv	4	39
2-Cyclohexenyl	uv	4.5	74
$C_6H_5CH{=}CHCH_2$	uv	2.5	75

[a] Unless otherwise stated, equivalent amounts of the hydride and benzoate were allowed to react at 130°C without solvent.
[b] Based on $(n\text{-}C_4H_9)_3SnOCOC_6H_5$ obtained.
[c] Reaction temperature was 80°C.
[d] *t*-Butylbenzene was used as solvent.
[e] 20 mole% azobisisobutyronitrile.
[f] 3-5 mole%.

mechanism. Further support tor a free radical mechanism is provided by the observation that reduction of optically active α-phenethyl benzoate with tri-*n*-butyltin deuteride in benzene at 80°C with azobisisobutyronitrile as initiator affords racemic α-deuteriophenylethane.

IV. Reactions of Organotin Hydrides with Tin-Element Bonds

The tin-hydrogen bond can react with tin-element bonds (*33–36*, *38*, *157*, *159*, *161*, *169*, *194*, *210*, *221*, *222*):

$$\equiv SnH + XSn\equiv \longrightarrow \equiv SnSn\equiv + HX \qquad (102)$$

$$X = N\!\!<, \; P\!\!<, \; As\!\!<, \; O-, \text{ or } CF{=}CF_2$$

This reaction, for which the name "hydrostannolysis" has been proposed (*38*), has been found to be well-suited for the synthesis of polytin compounds.

The mechanism of the hydrostannolysis of the tin-oxygen bond has been investigated (*35*, *38*). The rate of the reaction of triethyltin-oxygen derivatives is unaffected by galvinoxyl or azobisisobutyronitrile, which eliminates a free radical mechanism. The rate increases with the polarity of the solvent, which suggests a polar transition state. Electron-supplying substitutents on oxygen have an accelerating effect. Thus, triethyltin *t*-butoxide reacts faster than triethyltin phenoxide. Furthermore, in the case of *para*-substituted phenoxides the rate increases with the electron-supplying ability of the *para* substituent. Electron-supplying groups attached to tin, on the other hand, have a retarding effect. Thus, the rate of reaction of triethyltin phenoxide increases in the order $(C_2H_5)_3SnH < (p\text{-}CH_3OC_6H_4)_3SnH < (C_6H_5)_3SnH$. A mechanism involving electrophilic attack on oxygen by the organotin hydride hydrogen has been proposed:

$$R_3SnH + R'_3SnOR'' \xrightarrow{\text{slow}} \left[\begin{array}{c} R'_3-\overset{\delta\oplus}{O}-R'' \\ \vdots \\ H \\ \vdots \\ R_3Sn^{\delta\ominus} \end{array} \right] \qquad (103)$$

$$\longrightarrow R_3Sn^{\ominus} + R'_3Sn\overset{\oplus}{\underset{H}{O}}R'' \xrightarrow{\text{fast}} R_3SnSnR'_3 + R''OH$$

Polar solvents increase the rate by stabilizing the polar transition state more than the ground state. Electron-withdrawing groups on tin increase the rate by stabilizing the incipient negative charge on tin in the transition state. An electron-supplying group on oxygen can best stabilize the incipient positive charge on oxygen in the transition state.

The same solvent effect, substituent effects, and lack of activity of azobisisobutyronitrile or galvinoxyl was observed for the hydrostannolysis of the tin-nitrogen bond (*34*, *38*). Furthermore, the hydrostannolysis of triethyl-(*N*-phenylformamido)tin by triphenyltin hydride was first order in both reactants. A study of the hydrogen isotope effect indicated that the triphenyltin hydride hydrogen is transferred in the rate-determining step of the reaction. Thus, the mechanism is analogous to that represented by Eq. (103).

The rate of hydrostannolysis of $(C_2H_5)_3SnM(C_6H_5)_2$ (M = N, P, or As) with triphenyltin hydride in butyronitrile decreases in the order: M = N > P > As (*38*). This trend has been attributed to increased d_π-p_π bonding in the heavier atoms which reduces the availability of the unshared electrons on M (*38*). The increased π bonding of the heavier atoms to tin is presumably due to more effective overlap of the atomic orbitals involved (*48*).

The hydrostannolysis of trimethyltin chloride, triethyltin cyanide, and tri-*n*-butyltin acetate can be accomplished with triphenyltin hydride if triethylamine is present to combine with the acid formed (*161*).

In addition to the hydrostannolysis reaction represented by Eq. (102), an exchange reaction can occur between the tin-hydrogen bond and the tin-element bond:

$$\equiv SnH + XSn'\equiv \rightleftharpoons \equiv SnX + HSn'\equiv \qquad (104)$$

In Eq. (104) X can be halogen (*84, 156, 195, 196, 197, 198, 201*), $OCOCH_3$ (*199, 203*), D (*162*), $OSnR_3$ (*161*), OR (*35, 161*), SR (*35*), $SSnR_3$ (*161*), CN (*161*), PR_2 (*161*), and AsR_2 (*38*). With organotin-nitrogen compounds the hydrostannolysis reaction, Eq. (102), tends to predominate (*38*). With organotin-thioalkoxides only the exchange reaction, Eq. (104), occurs (*35, 38*). Both exchange and hydrostannolysis occur in the reaction of triphenyltin hydride with triethyltin diphenylphosphide, triethyltin diphenylarside, or triorganotin alkoxides (*38*).

The exchange reaction can be conveniently studied using pmr spectroscopy since the chemical shift of the organotin hydride hydrogen depends upon the other groups attached to tin. The results of such studies are summarized in Table 14. An inspection of Table 14 reveals that the hydrogen tends to attach itself to the tin having the phenyl groups. This tendency is reversed, however, when X is $(n\text{-}C_4H_9)_3SnO$— or $(n\text{-}C_4H_9)_3SnS$—. The extent of exchange does not vary appreciably with the polarity of the solvent. In the case of the tin-nitrogen compounds the extent of exchange is only slightly greater in the more polar solvent, whereas the reverse is true for the tin-phosphorus compounds.

Further interesting results on halide-hydride exchanges have been reported by Sawyer (*195, 204*). An equimolar mixture of di-*n*-butylchlorotin hydride and di-*n*-butylbromotin hydride results from each of the following reactions: tri-*n*-butyltin hydride with di-*n*-butyltin bromide chloride, di-*n*-butyltin dihydride with di-*n*-butyltin bromide chloride, di-*n*-butylbromotin hydride with di-*n*-butyltin dichloride, and di-*n*-butylchlorotin hydride with di-*n*-butyltin dibromide. Interestingly, the pmr spectrum of the mixture contained a single broad peak at 7.23 ppm suggesting the occurrence of a rapid halogen exchange between the two halide-hydrides.

TABLE 14

EXCHANGE REACTIONS OF TRIORGANOTIN HYDRIDES

$$R_3SnH + R'_3SnX \rightleftharpoons R_3SnX + R'_3SnH$$

R	R′	X	Solvent	R_3SnH, %	$R_3'SnH$, %	Ref.
CH_3	C_2H_5	$N(C_6H_5)COH$	Cyclohexane	50	50	*(38)*
			Butyronitrile	47	53	
C_6H_5	C_2H_5	$N(C_6H_5)COH$	Cyclohexane	[a]	0	*(38)*
CH_3	C_2H_5	$N(C_2H_5)_2$	Cyclohexane	[a]	0	*(38)*
			Butyronitrile	77	23	
CH_3	C_2H_5	$N(C_6H_5)_2$	Cyclohexane	100	0	*(38)*
			Butyronitrile	90	10	
CH_3	C_2H_5	$P(C_6H_5)_2$	Cyclohexane	40	60	*(38)*
			Butyronitrile	53	47	
C_6H_5	C_2H_5	$P(C_6H_5)_2$	Cyclohexane	[a]	40	*(38)*
			Butyronitrile	[a]	34	
CH_3	C_2H_5	OCH_3	Cyclohexane	40	60	*(38)*
			Butyronitrile	33	67	
C_6H_5	n-C_4H_9	OCH_3	Neat	100	0	*(195)*
CH_3	C_2H_5	OC_6H_5	Cyclohexane	50	50	*(38)*
C_6H_5	C_2H_5	OC_6H_5	Cyclohexane	[a]	>20	*(38)*
			Butyronitrile	[a]	>20	
C_6H_5	n-C_4H_9	$OCOCH_3$	Cyclohexane	90	10	*(195)*
C_6H_5	n-C_4H_9	$OCOC_6H_5$	Neat	98	2	*(195)*
C_6H_5	n-C_4H_9	$OSn(n$-$C_4H_9)_3$	Neat	0	100	*(195)*
C_6H_5	n-C_4H_9	$SSn(n$-$C_4H_9)_3$	Neat	0	100	*(195)*
C_6H_5	CH_3	Cl	Neat	94	6	*(195)*
C_6H_5	C_2H_5	Cl	Neat	93	7	*(195)*
C_6H_5	n-C_3H_7	Cl	Neat	91	9	*(195)*
C_6H_5	n-C_4H_9	Cl	Neat	85	15	*(195)*
C_6H_5	n-C_4H_9	Br	Neat	87	13	*(195)*

[a] Not determined.

The extent of exchange between triphenyltin hydride and diphenyltin dihalides:

$$(C_6H_5)_3SnH + (C_6H_5)_2SnX_2 \rightleftharpoons (C_6H_5)_3SnX + (C_6H_5)_2Sn\begin{smallmatrix}H\\X\end{smallmatrix} \tag{105}$$

and between diphenyltin dihydride and diphenyltin dihalides:

$$(C_6H_5)_2SnH_2 + (C_6H_5)_2SnX_2 \rightleftharpoons 2\,(C_6H_5)_2Sn\begin{smallmatrix}H\\X\end{smallmatrix} \tag{106}$$

increases in the order $I < Br < Cl$. From the percentage of halide-hydride and triphenyltin hydride present at equilibrium in reaction (105), the preference of hydrogen to remain with a certain phenyltin moiety was found to decrease in the order $(C_6H_5)_2SnCl— > (C_6H_5)_2SnBr— \approx (C_6H_5)_3Sn— > (C_6H_5)_2SnI—$. The preference for hydrogen to remain on a certain phenyltin moiety in reaction (106) was found to be $(C_6H_5)_2SnCl— > (C_6H_5)_2SnBr— > (C_6H_5)_2SnH— > (C_6H_5)_2SnI—$. In neither reaction was there evidence for the formation of diphenyliodotin hydride. In the case of the reactions of triphenyltin hydride with dialkyltin dichlorides the hydrogen showed a strong tendency to remain on the triphenyltin group rather than the dialkylchlorotin group.

In exchanges where the organic group in the dihyride is different from that in the dihalide, there is a high percentage of a mixture of halide-hydrides at equilibrium, except when the phenyl group is involved. When the phenyl group is present, only about 50% of halide-hydrides is found, and the only dihydride found is diphenyltin dihydride.

A study of the exchanges of tri-*n*-butyltin hydride, di-*n*-butyltin dihydride, and di-*n*-butylchlorotin hydride with *n*-butyltin trichloride revealed the following order of decreasing tendency for the hydrogen to remain on a certain *n*-butyltin moiety: $n\text{-}C_4H_9SnCl_2— > n\text{-}C_4H_9SnH_2— > (n\text{-}C_4H_9)_2SnCl— > (n\text{-}C_4H_9)_2SnH— > (n\text{-}C_4H_9)_3Sn—$ (*195*, *196*).

Triphenyltin hydride and triethyltin hydride undergo exchange reactions with triorganolead imidazoles and triorganolead acetates (*31*). The reaction can be driven to completion by allowing the triphenyl or triethyllead hydride formed to add across the triple bond of an added acetylenic compound. Triethyltin hydride undergoes an exchange reaction with bis(triethylgermanium) telluride (*15a*, *247*).

Hydrostannolysis of the germanium-nitrogen bond has been accomplished (*32*).

V. Miscellaneous Reactions of Organotin Hydrides

Besides carbon-halogen bonds, other element-halogen bonds are reduced by organotin hydrides. The reaction between triphenyltin hydride and *N*-bromosuccinimide in refluxing benzene affords triphenyltin bromide (86%) and succinimide (95%) (*110*). Triphenyltin hydride reduces diphenylchlorophosphine to diphenylphosphine (*211*). High yields of triphenyltin chloride result from the reactions of triphenyltin hydride with phosphorous trichloride, antimony trichloride, and mercuric chloride (*16*). Bis(dimethylamino)borane and 1,3,2-benzodioxaborole can be conveniently prepared by reduction of the corresponding boron-chlorine compound with tri-*n*-butyltin hydride (*168*). The reaction proceeds faster in chlorobenzene than in the nonpolar *n*-hexane, and is unaffected by azobisisobutyronitrile or galvinoxyl, which rules out a free radical mechanism. The reaction of triethyltin hydride with $BrGe(CH_2CO_2CH_3)_3$ affords $HGe(CH_2CO_2CH_3)_3$ in 70% yield (*9b*).

The hydrostannolysis of carbon-metal bonds has been accomplished. Ethane, zinc, tetraethyltin, and hexaethylditin result from the reaction of diethylzinc with triethyltin hydride (*250*). Organomercury compounds undergo similar reactions (*249*). Compounds containing the zinc-tin or cadmium-tin bond can be prepared by allowing triphenyltin hydride to react with diethylzinc or dimethylcadmium. The reaction must either be carried out in a solvating solvent (diethyl ether, tetrahydrofuran, or dimethoxyethane), or a preformed coordination complex of the organo-zinc or -cadmium compound with *N,N,N′N′*-tetramethylethylenediamine or 2,2′bipyridyl must be used (*230*):

$$R_2M(\leftarrow D)_2 + 2\,(C_6H_5)_3SnH \longrightarrow [(C_6H_5)_3Sn]_2M(\leftarrow D)_2 + 2\,RH \quad (M = Zn \text{ or } Cd) \tag{107}$$

The reaction of triphenyltin hydride with ethylmagnesium bromide-triethylamine complex results in hydrostannolysis of the carbon-magnesium bond (*37*):

$$(C_6H_5)_3SnH + C_2H_5MgBr\cdot N(C_2H_5)_3 \longrightarrow (C_6H_5)_3SnMgBr\cdot N(C_2H_5)_3 + C_2H_6 \tag{107a}$$

Compounds containing tin-titanium and tin-zirconium bonds can be prepared by hydrostannolysis reactions (*39*); an example is the preparation of tetrakis(triphenylstannyl)titanium:

$$4\,(C_6H_5)_3SnH + Ti[N(CH_3)_2]_4 \longrightarrow [(C_6H_5)_3Sn]_4Ti + 4\,(CH_3)_2NH \tag{107b}$$

Bis(triphenylstannyl)mercury can be prepared by a hydrostannolysis reaction (*50*):

$$2\,(C_6H_5)_3SnH + Hg[N(SiMe_3)_2]_2 \longrightarrow Hg[Sn(C_6H_5)_3]_2 + 2\,(Me_3Si)_2NH \tag{107c}$$

Hydrostannolysis of ethyl-lead bonds occurs in the reaction of tetraethyllead with triethyltin hydride at 130°C (*248a*). The reactions of triethyltin hydride with $(C_2H_5)_2Te$, $C_2H_5TeSi(C_2H_5)_3$ (*246a*), $C_2H_5TeGe(C_2H_5)_3$(*246c*), and $[(C_2H_5)_3Ge]_2Te$ (*15a, 247*) afford compounds with Sn-Te bonds. Tris(triethylstannyl)antimony and -bismuth have been prepared by the reaction of triethyltin hydride with triethylantimony or -bismuth (*250a*). Trimethyltin hydride reacts with iridium(I) complexes to give compounds with tin-iridium bonds (*111a*). The reaction of triethyltin hydride with the product of the reaction of tris(triethylgermanyl)antimony with lithium gives antimony, lithium hydride, and $(C_2H_5)_3SnGe(C_2H_5)_3$(*248b*).

The reaction between tri-*n*-butyltin hydride and benzoyl azide affords benzamide and bis(tri-*n*-butyltin) oxide via the hydrolytically unstable *N*-(tri-*n*-butylstannyl)benzamide, which may be formed by a radical mechanism, since the reaction is catalyzed by azobisisobutyronitrile (*58*). The reaction of triphenyltin hydride with group IV triphenylmetal azides has been studied (*232a*).

Tri-*n*-butyltin hydride reacts with alkyl isocyanides in the presence of a radical initiator (*193*):

$$(n\text{-}C_4H_9)_3SnH + RN{=}C: \longrightarrow RH + (n\text{-}C_4H_9)_3SnCN \quad (107d)$$

The following reaction scheme was suggested:

$$R_3SnH \xrightarrow{\text{Initiator}} R_3Sn\cdot$$
$$R_3Sn\cdot + R'N{=}C: \longrightarrow R'N{=}\dot{C}SnR_3 \longrightarrow R'\cdot + R_3SnCN$$
$$R'\cdot + R_3SnH \longrightarrow R'H + R_3Sn\cdot$$

The reactions of triphenyltin hydride with sulfur-containing functional groups have been studied by Pang and Becker (*181*). The results are summarized in Table 15. The reaction between allyl mercaptan and di-*n*-butyltin dihydride affords bis(allythio)di-*n*-butyltin (*111*). Organotin hydrides react with sulfur and selenium to produce organotin sulfides and selenides (*246b*).

Insertion products result from the reaction of trialkyltin hydrides with aliphatic diazo compounds in the presence of ultraviolet light (*88*) or copper powder (*88*, *112*):

$$R_3SnH + RCHN_2 \longrightarrow R_3SnCH_2R + N_2 \quad (108)$$

Insertion of the CCl_2 group can be accomplished by allowing the organotin hydride to react with sodium trichloroacetate (*233*).

Tri-*n*-butyl-*t*-butylperoxytin results from the reaction of tri-*n*-butyltin hydride with two moles of *t*-butyl hydroperoxide in decalin (*4*):

$$(n\text{-}C_4H_9)_3SnH + t\text{-}C_4H_9OOH \longrightarrow (n\text{-}C_4H_9)_3SnOC_4H_9\text{-}t + H_2O \quad (109)$$

$$(n\text{-}C_4H_9)_3SnOC_4H_9\text{-}t + t\text{-}C_4H_9OOH \longrightarrow (n\text{-}C_4H_9)_3SnOOC_4H_9\text{-}t + t\text{-}C_4H_9OH \quad (110)$$

Triethyltin hydride reacts with di-*t*-butyl peroxide (*158*) and *t*-butyl hydroperoxide (*248*) according to Eqs. (111) and (112), respectively:

$$2\,(C_2H_5)_3SnH + t\text{-}C_4H_9OOC_4H_9\text{-}t \longrightarrow 2\,t\text{-}C_4H_9OH + (C_2H_5)_6Sn_2 \quad (111)$$

$$2\,(C_2H_5)_3SnH + 2\,t\text{-}C_4H_9OOH \longrightarrow 2\,t\text{-}C_4H_9OH + [(C_2H_5)_3Sn]_2O + H_2O \quad (112)$$

Organotin hydrides strongly induce the decomposition of diacetyl and dibenzoyl peroxides (*158*, *158a*, *158b*). Oxygen-18 labeling shows that the tin atom becomes attached to the peroxide oxygen:

$$R_3Sn\cdot + R'\overset{^{18}O}{\overset{\|}{C}}O\overset{^{18}O}{\overset{\|}{C}}R' \longrightarrow R'\overset{^{18}O}{\overset{\|}{C}}OSnR_3 + R'\overset{^{18/2}O^{18/2}}{\overset{\|}{C}}O\cdot \quad (113)$$

TABLE 15

REDUCTION OF SULFUR-CONTAINING FUNCTIONAL GROUPS WITH TRIPHENYLTIN HYDRIDE (*181*)

Reactants	Conditions		Products	% Yield		
	Temp., °C	Time, h		No catalyst	AIBN catalyst	$(C_6H_5)_3B$ catalyst
CH_3SH	160	3	$(C_6H_5)_3SnSCH_3$, H_2	0	47	33
$C_6H_5CH_2SH$	85	5	$(C_6H_5)_3SnSSn(C_6H_5)_3$, H_2S, $C_6H_5CH_3$	0	55	60
C_6H_5SH	85	2	$(C_6H_5)_3SnSC_6H_5$, H_2	68	79	85
$2\text{-}C_{10}H_7SH$	120	3	$2\text{-}C_{10}H_7SSn(C_6H_5)_3$, H_2	68	72	79
$C_6H_5CH_2SCH_2C_6H_5$	130	3	$(C_6H_5)_3SnSSn(C_6H_5)_3$, $C_6H_5CH_3$	27	38	42
$C_6H_5CH_2SSCH_2C_6H_5$	130	2	$(C_6H_5)_3SnSSn(C_6H_5)_3$, H_2S, $C_6H_5CH_3$	12	17	79
$C_6H_5SSC_6H_5$	90	1	$(C_6H_5)_3SnSC_6H_5$, H_2	88	84	91
$C_6H_5CSC_6H_5$	145	4	$C_6H_5CH_2C_6H_5$,	68		
			$(C_6H_5)_3SnSSn(C_6H_5)_3$	59		
$C_6H_5SO_3H$	Exothermic		$C_6H_5SO_3Sn(C_6H_5)_3$, H_2	89		
$C_6H_5SO_2H$	Exothermic		$C_6H_5SO_2Sn(C_6H_5)_3$, H_2	100		
$C_6H_5SO_2Cl$	Exothermic		$(C_6H_5)_3SnCl$, H_2	100		
C_6H_5SCl	Exothermic		$(C_6H_5)_3SnCl$, H_2	78		
			$(C_6H_5)_3SnSC_6H_5$	60		
CS_2	155	6	$(C_6H_5)_3SnSSn(C_6H_5)_3$	13	46	54

The reactions of organotin hydrides with azo compounds and with the radicals produced by them have been studied (*166*). Triorganotin hydrides were found to be better radical scavengers than thiols. Triethyltin hydride did not affect the rate of decomposition of azobisisobutyronitrile or of compound (**22**).

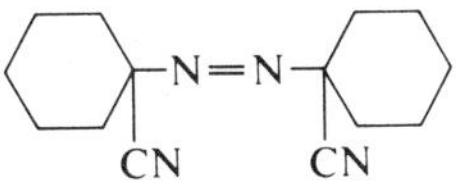

N=N
CN CN

(**22**)

On the other hand, triethyltin hydride accelerates the radical decomposition of phenyl phenylazo sulfone and dibenzyloxydiimide (*149*, *149a*).

A study of the photoreduction of acetone by tri-*n*-butyltin hydride has led to the conclusion that excited singlet acetone is much less reactive than triplet acetone (*250b*). Photochemical reduction of 1-naphthaldehyde and 2-acetonaphthone occurs in the presence of tri-*n*-butyltin hydride but not in the presence of secondary alcohols (*66*). It was postulated that the triplet energies for these compounds were not high enough to enable hydrogen abstraction from alcohols. 1-Nitronaphthalene behaves similarly, but the nitrobenzene triplet undergoes hydrogen abstraction with either tri-*n*-butyltin hydride or isopropyl alcohol (*231*).

Organotin hydrides react with phosphonic and phosphinic acids to give compounds containing Sn—O—P bonds (*55*, *190d*). A colorimetric determination of organotin hydrides depends on their reaction with ninhydrin or isatin (*57a*).

VI. Summary

Numerous methods are available for the preparation of organotin hydrides. Some particularly convenient methods are the reaction of an organotin halide with lithium aluminum hydride or sodium borohydride and the reaction of an organotin oxide with an organosilicon hydride.

Organotin hydrides undergo reactions with a large variety of organic, organometallic, and inorganic compounds. The mechanism of many of these reactions have recently been studied. Both homolytic and heterolytic cleavage of the tin-hydrogen bond can occur. The reduction of organic halides and acid halides by organotin hydrides proceeds by a free radical mechanism. The addition of organotin hydrides to 1-alkenes proceeds by a free radical mechanism leading to a β-adduct in which the tin is attached to the terminal carbon; however, an ionic mechanism leading to the α-adduct also can occur with electrophilic alkenes like acrylonitrile. The addition of organotin hydrides to alkynes can proceed either by an ionic *trans*-mechanism or by a free radical *trans*-mechanism. In the ionic mechanism the hydrogen of the organotin hydride is transferred to the triple bond as a hydride ion. The addition of organotin hydrides to isocyanates and isothiocyanates proceeds exclusively by an ionic mechanism in which the hydrogen of the organotin hydride is transferred as a hydride ion. The cleavage of tin-oxygen and tin-nitrogen bonds by organotin hydrides proceeds by an ionic mechanism in which the hydrogen of the organotin hydride is transferred to oxygen or nitrogen as a proton. The addition of organotin hydrides to the carbonyl group of an aldehyde or ketone can occur by a free radical mechanism or by an ionic mechanism. Exchange reactions between the tin-hydrogen bond and tin-element bonds probably occur by a heterolytic mechanism. The reduction of chloroboranes by tri-*n*-butyltin hydride also appears to occur by a heterolytic mechanism.

Organotin hydrides show great promise as reducing agents in organic chemistry. However, relatively few reactions have been carried out between organotin hydrides and organic compounds containing two or more reactive functional groups. Further studies along these lines are needed to provide detailed information on the reactive selectivity of organotin hydrides. The importance of such studies to organic synthesis is obvious.

Organotin hydrides are important synthetic intermediates in organotin chemistry. Their ability to add across multiple bonds, for example, enables the preparation of functionally substituted organotin compounds which could not otherwise be easily obtained. Also, they are useful for the synthesis of compounds containing tin-metal bonds.

No attempt has been made in this chapter to compare the chemistry of organotin hydrides to that of the organosilicon, -germanium, and -lead hydrides. For developments in these related areas, the reader is urged to consult the annual surveys of Seyferth and King (*209*).

APPENDIX 1

PMR AND IR DATA FOR ORGANOTIN HYDRIDES OF THE TYPE R_nSnH_{4-n}

Compound	$\tau(SnH)$ (ppm)	Solvent	$J(^{119}SnH)$ (cps)	$J(^{117}SnH)$ (cps)	$J(HSnCH)$ (cps)	Ref.	$\nu(SnH)$ (cm^{-1})	Solvent	Ref.
SnH_4	6.15	Cyclopentane	1931	1846		*(56)*	1898	$n\text{-}C_6H_{14}$	*(134)*
	6.11	CS_2	1933	1842		*(187)*	1906	$n\text{-}Bu_2O$	*(187)*
CH_3SnH_3[a]	5.86	Neopentane	1852	1770	2.7	*(56)*	1870	Cyclohexane	*(134)*
$(CH_3)_2SnH_2$[b]	5.24	Neopentane	1758	1682	2.55	*(56)*	1850	Cyclohexane	*(134)*
	5.61[c]	Neat	1797.1	1717.4	2.65	*(27)*			
	5.68[d]	Neat	1773	1696	2.67	*(235)*			
	5.55	Neopentane	1797	1717		*(134)*			
	5.59	Neat				*(71)*			
	5.73	Cyclohexane				*(198)*	1850	Cyclohexane	*(198)*
$(CH_3)_3SnH$[e]	5.27	Neopentane	1744	1664	2.37	*(56)*	1833	Cyclohexane	*(134)*
	5.39[f]	Neat	1755	1677	2.4	*(28)*			
	5.38[g]	Neat	1738	1660	2.38	*(235)*			
	5.36	CS_2				*(85)*	1830	Neat	*(85)*
$(CH_3)_2(C_2H_5)SnH$	5.27	Neat	1706.6	1630	2.6	*(28)*			
$(CH_3)_2(CF_2CF_2H)SnH$[h]	4.47	Neat	1976.6	1889.4	2.25	*(27)*	1877	Gas	*(25)*
$C_2H_5SnH_3$	5.66	Cyclopentane	1790.1	1710.8		*(134)*	1853	Cyclohexane	*(134)*
$(C_2H_5)_2SnH_2$	5.25	Cyclopentane	1691.1	1616.2		*(134)*	1822	Cyclohexane	*(134)*
	5.52	Neat			1.6	*(49)*			
	5.41	Neat				*(198)*	1835	Neat	*(198)*

[a] $\tau(CH_3)$, 9.73; $J(^{13}CH)$, 130.
[b] $\tau(CH_3)$, 9.83; $J(^{119}SnCH)$, 58.0; $J(^{117}SnCH)$, 55.5; $J(^{13}CH)$, 126.5.
[c] $\tau(CH_3)$, 9.80; $J(^{119}SnCH)$, 60.2; $J(^{117}SnCH)$, 57.6; $J(^{13}CH)$, 129.9.
[d] $\tau(CH_3)$, 9.82; $J(^{119}SnCH)$, 59; $J(^{117}SnCH)$, 56; $J(^{13}CH)$, 127.
[e] $\tau(CH_3)$, 9.82; $J(^{119}SnCH)$, 56.5; $J(^{117}SnCH)$, 54.5; $J(^{13}CH)$, 128.5.
[f] $\tau(CH_3)$, 9.92; $J(^{119}SnCH)$, 56.9; $J(^{117}SnCH)$, 54.5; $J(^{13}CH)$, 128.8.
[g] $\tau(CH_3)$, 9.81; $J(^{119}SnCH)$, 57; $J(^{117}SnCH)$, 54.5; $J(^{13}CH)$, 128.5.
[h] $\tau(CH_3)$, 9.69; $J(^{119}SnCH)$, 62.8; $J(^{117}SnCH)$, 60.1; $J(^{13}CH)$, 132.1.

APPENDIX 1 (*continued*)

Compound	τ(SnH) (ppm)	Solvent	$J(^{119}SnH)$ (cps)	$J(^{117}SnH)$ (cps)	J(HSnCH) (cps)	Ref.	ν(SnH) (cm^{-1})	Solvent	Ref.
$(C_2H_5)_3SnH$	5.00	Cyclopentane	1611.3	1539.9		(*134*)	1797	Cyclohexane	(*134*)
	5.11	Neat	1612.4	1539.6		(*104*)			
	5.17	Neat	1574	1504	2	(*49*)			
n-$C_3H_7SnH_3$	5.83	Et_2O	1790.2	1710.5		(*134*)	1860	Cyclohexane	(*134*)
$(n$-$C_3H_7)_2SnH_2$	5.48	Neat	1689.4	1614.7		(*134*)	1833	Cyclohexane	(*134*)
	5.42	CS_2				(*85*)	1835	Neat	(*85*)
	5.46	Neat				(*198*)	1830	Neat	(*198*)
$(n$-$C_3H_7)_3SnH$	5.21	Neat	1605.0	1533.5		(*134*)	1811	Cyclohexane	(*134*)
			1600	1530		(*187*)	1795	n-Bu_2O	(*187*)
	5.23	CS_2				(*85*)	1809	Neat	(*85*)
	5.17	Neat	1604.8	1533.7	1.77	(*104*)	1807	Neat	(*104*)
i-$C_3H_7SnH_3$	5.46	Et_2O	1750.0	1672.5		(*134*)	1853	Cyclohexane	(*134*)
$(i$-$C_3H_7)_2SnH_2$	5.07	Cyclopentane	1612.1	1540.3		(*134*)	1820	Cyclohexane	(*134*)
	5.50	Neat				(*198*)	1831	Neat	(*198*)
$(i$-$C_3H_7)_3SnH$	4.82	Cyclopentane	1505.8	1439.4		(*134*)	1794	Cyclohexane	(*134*)
	3.03	CS_2				(*85*)	1838	Neat	(*85*)
	4.86	Neat	1505.0	1437.8		(*104*)	1807	Neat	(*104*)
n-$C_4H_9SnH_3$	5.98	CS_2	1800	1720		(*187*)	1865	n-Bu_2O	(*187*)
			1796.1	1716.5		(*134*)	1862	Cyclohexane	(*134*)
	5.71	CS_2				(*85*)	1870	Neat	(*85*)
	5.67	Neat				(*196*)	1861	Neat	(*196*)
	5.76	Neat				(*49*)	1880	Gas	(*71*)
$(n$-$C_4H_9)_2SnH_2$[i]	6.46	Neat	1680	1602	2.0	(*235*)			
	5.23	CS_2	2219	2119		(*187*)	1842	n-Bu_2O	(*187*)
			1639.9	1618.6		(*134*)	1835	Cyclohexane	(*134*)
	5.43	CS_2				(*85*)	1832	Neat	(*85*)
	5.42	Neat				(*197*)	1835	Neat	(*197*)
	5.47	Neat	1682	1612	2.75	(*49*)			
	5.36	Neat				(*71*)			

[i] $\tau(n$-$C_4H_9)$, 9.17.

$(n\text{-}C_4H_9)_3SnH$	5.22	Neat	1609	1532	1.8	*(28)*			
	7.93	CS_2	1722	1650		*(187)*	1808	$n\text{-}Bu_2O$	*(187)*
			1610.6	1539		*(134)*	1813	Cyclohexane	*(134)*
	5.22	CS_2				*(85)*	1807	Neat	*(85)*
	5.16	Neat	1607.8	1536.4	1.79	*(104)*	1813	Neat	*(104)*
	5.16	Neat				*(196)*	1814	Neat	*(196)*
	5.2	Neat	1609	1524	1.71	*(49)*			
	5.21	Neat				*(71)*			
$(i\text{-}C_4H_9)_2SnH_2$	5.53	Neat	1692	1618	2.2	*(49)*	1845	Neat	*(154)*
$(i\text{-}C_4H_9)_3SnH$	5.12	Neat	1604.8	1532.6	1.80	*(104)*	1817	Neat	*(104)*
	5.2	Neat	1608	1535	1.6	*(49)*			
$(t\text{-}C_4H_9)_2SnH_2$	4.70	Neat	1554	1484		*(49)*	1812	Cyclohexane	*(157)*
$n\text{-}C_8H_{17}SnH_3$	5.73	Neat	1798	1720	2.1	*(49)*	1863	Cyclohexane	*(140)*
$(n\text{-}C_8H_{17})_2SnH_2$	5.42	Neat	1694	1618	1.85	*(49)*	1836	Cyclohexane	*(140)*
	5.46	Neat				*(198)*	1830	Neat	*(198)*
$(n\text{-}C_8H_{17})_3SnH$	5.16	Benzene	1604.7	1534.4	1.80	*(104)*	1805	Neat	*(104)*
	5.2	Neat	1600	1534		*(49)*			
$(cyclo\text{-}C_6H_{11})_2SnH_2$	5.15	Neat				*(198)*	1816	Neat	*(198)*
$(C_6H_5CH_2)_3SnH$[j]	4.488	Neat	1770	1692		*(244)*			
$C_6H_5SnH_3$	5.07	Neat	1921.5	1836.7		*(5)*	1880	Cyclohexane	*(150)*
	4.98	Et_2O	1916.5	1833.1		*(5)*			
$(C_6H_5)_2SnH_2$	3.91	Neat	1927.8	1842.0		*(5)*	1855	Cyclohexane	*(150)*
	3.98	Et_2O	1916.7	1831.4		*(5)*			
	4.31[k]	Neat	1928	1844		*(235)*			
	4.27	Neat				*(198)*	1849	Neat	*(198)*
$(C_6H_5)_3SnH$	3.17	Neat	1935.8	1850.8		*(5)*	1843	Cyclohexane	*(134)*
	3.16	Et_2O	1926.0	1843.2		*(5)*			
	3.03	CS_2				*(85)*	1838	Neat	*(85)*
$(CH_3C_6H_4)_3SnH$	2.92	Neat	1911.2	1825.9		*(104)*	1835	Nujol	*(104)*

[j] $\tau(CH_2)$, 8.158; $J(^{119}SnCH)$, 62; $J(^{117}SnCH)$, 59.8; $J(^{13}CH)$, 131.
[k] $\tau(C_6H_5)$, 3.69.

REFERENCES

1. N. A. Adrova, M. M. Koton, and V. A. Klages, *Vysokomolekul. Soedin.*, **5**, 1817 (1963); *CA*, **60**, 6936 (1964).
2. M. Akhtar and H. C. Clark, *Can. J. Chem.*, **46**, 633 (1968).
3. M. Akhtar and H. C. Clark, *Can. J. Chem.*, **46**, 2165 (1968).
4. D. L. Alleston and A. G. Davies, *J. Chem. Soc.*, 2465 (1962).
4a. L. J. Altman and B. W. Nelson, *J. Am. Chem. Soc.*, **91**, 5163 (1969).
5. E. Amberger, H. P. Fritz, C. G. Kreiter, and M.-R. Kula, *Chem. Ber.*, **96**, 3270 (1963).
6. E. Amberger and M.-R. Kula, *Chem. Ber.* **96**, 2560 (1963).
7. E. Amberger, R. Römer, and A. Layer, *J. Organometal. Chem.*, **12**, 417 (1968).
8. T. Ando, F. Namigata, H. Yamanaka, and W. Funasaka, *J. Am. Chem. Soc.*, **89**, 5719 (1967).
9. C. Barnetson, H. C. Clark, and J. T. Kwon, *Chem. Ind. London*, 458 (1964).
9a. W. E. Barnett and R. F. Koebel, *Chem. Commun.*, 875 (1969).
9b. Yu. I. Baukov, G. S. Burlachenko, I. Yu. Belavin, and I. F. Lutsenko, *Zh. Obshch. Khim.*, **38**, 1899 (1968); CA, **70**, 4252b (1969).
10. G. A. Baum and W. J. Considine, *J. Org. Chem.*, **29**, 1267 (1964).
10a. B. Bellegarde, M. Pereyre, and J. Valade, *Bull. Soc. Chim. Fr.*, 3082 (1967).
11. H. A. Bent, *Can. J. Chem.*, **38**, 1235 (1960).
12. H. A. Bent, *Chem. Rev.*, **60**, 275 (1960).
13. A. D. Beveridge, H. C. Clark, and J. T. Kwon, *Can. J. Chem.*, **44**, 179 (1966).
14. E. R. Birnbaum and P. H. Javora, *Inorg. Syntheses*, **12**, 45 (1970).
15. E. R. Birnbaum and P. H. Javora, *J. Organometal. Chem.*, **9**, 379 (1967).
15a. M. N. Bochkarev, L. P. Sanina, and N. S. Vyazankin, *Zh. Obshch. Khim.*, **39**, 135 (1969); CA, **70**, 96876 (1969).
16. A. E. Borisov and A. N. Abramova, *Izv. Akad. Nauk SSSR, Ser. Khim.*, 844 (1964); *CA*, **61**, 5680 (1964).
17. A. E. Borisov, A. N. Abramova, and Z. N. Parnes, *Izv. Akad. Nauk SSSR, Ser. Khim.*, 941 (1964); *CA*, **61**, 5470 (1964).
17a. J. Braun, *Compt. Rend.*, **260**, 218 (1965).
18. R. Breslow, J. T. Groves, and G. Ryan, *J. Am. Chem. Soc.*, **89**, 5048 (1967).
19. R. Breslow and G. Ryan, *J. Am. Chem. Soc.*, **89**, 3073 (1967).
20. R. H. Bullard and R. A. Vingee, *J. Am. Chem. Soc.*, **51**, 892 (1929).
21. R. Calas, J. Valade, and J. C. Pommier, *Compt. Rend.*, **255**, 1450 (1962).
21a. F. A. Carey and H. S. Tremper, *Tetrahedron Letters*, 1645 (1969).
22. D. J. Carlsson and K. U. Ingold, *J. Am. Chem. Soc.*, **90**, 1055 (1968).
23. D. J. Carlsson and K. U. Ingold, *J. Am. Chem. Soc.*, **90**, 7047 (1968).
24. R. F. Chambers and P. C. Scherer, *J. Am. Chem. Soc.*, **48**, 1054 (1926).
25. H. C. Clark, S. G. Furnival, and J. T Kwon, *Can. J. Chem.*, **41**, 2889 (1963).
26. H. C. Clark and J. T. Kwon, *Can. J. Chem.*, **42**, 1288 (1964).
27. H. C. Clark, J. T. Kwon, L. W. Reeves, and E. J. Wells, *Can. J. Chem.*, **41**, 3005 (1963).
28. H. C. Clark, J. T. Kwon, L. W. Reeves, and E. J. Wells, *Inorg. Chem.*, **3**, 907 (1964).
29. W. J. Considine and J. J. Ventura, *Chem. Ind. London*, 1683 (1962).
30. H. M. J. C. Creemers, *Hydrostannolysis*, Schotanus and Jens, Utrecht, 1967.
31. H. M. J. C. Creemers, A. J. Leusink, J. G. Noltes, and G. J. M. van der Kerk, *Tetrahedron Letters*, 3167 (1966).

32. H. M. J. C. Creemers and J. G. Noltes, *J. Organometal. Chem.*, **7**, 237 (1967).
33. H. M. J. C. Creemers and J. G. Noltes, *Rec. Trav. Chim.*, **84**, 382 (1965).
34. H. M. J. C. Creemers and J. G. Noltes, *Rec. Trav. Chim.*, **84**, 590 (1965).
35. H. M. J. C. Creemers and J. G. Noltes, *Rec. Trav. Chim.*, **84**, 1589 (1965).
36. H. M. J. C. Creemers, J. G. Noltes, and G. J. M. van der Kerk, *Rec. Trav. Chim.*, **83**, 1284 (1964).
37. H. M. J. C. Creemers, J. G. Noltes, and G. J. M. van der Kerk, *J. Organometal. Chem.*, **14**, 217 (1968).
38. H. M. J. C. Creemers, F. Verbeek, and J. G. Noltes, *J. Organometal. Chem.*, **8**, 469 (1967).
39. H. M. J. C. Creemers, F. Verbeek, and J. G. Noltes, *J. Organometal. Chem.*, **15**, 125 (1968).
40. S. J. Cristol and R. V. Barbour, *J. Am. Chem. Soc.*, **90**, 2832 (1968).
40a. S. J. Cristol and A. L. Noreen, *J. Am. Chem. Soc.*, **91**, 3969 (1969).
41. S. J. Cristol, R. M. Sequeira, and C. H. DePuy, *J. Am. Chem. Soc.*, **87**, 4007 (1965).
42. W. R. Cullen and G. E. Styan, *Inorg. Chem.*, **4**, 1437 (1965).
43. W. R. Cullen and G. E. Styan, *J. Organometal. Chem.*, **6**, 117 (1966).
44. W. R. Cullen and G. E. Styan, *J. Organometal. Chem.*, **6**, 633 (1966).
45. G. J. Del Franco, P. Resnick, and C. R. Dillard, *J. Organometal. Chem.*, **4**, 57 (1965).
45a. M. Delmas, J. C. Maire, and R. Pinzelli, *J. Organometal. Chem.*, **16**, 83 (1969).
46. D. B. Denney, R. M. Hoyte, and P. T. MacGregor, *Chem. Commun.*, 1241 (1967).
47. R. E. Dessey, T. Hieber, and F. Paulik, *J. Am. Chem. Soc.*, **86**, 28 (1964).
48. R. S. Drago, *Physical Methods in Inorganic Chemistry*, Reinhold, New York, 1965, p. 48.
48a. H. Dreeskamp and Chr. Schumann, *Chem. Phys. Lett.*, **1**, 1399 (1968).
49. J. Dufermont and J. C. Maire, *J. Organometal. Chem.*, **7**, 415 (1967).
50. C. Eaborn, A. R. Thompson, and D. R. M. Walton, *Chem. Commun.*, 1051 (1968).
51. F. W. Evans, R. J. Fox, and M. Szwarc, *J. Am. Chem. Soc.*, **82**, 6414 (1960).
52. R. M. Fantazier and M. L. Poutsma, *J. Am. Chem. Soc.*, **90**, 5490 (1968).
53. A. F. Finholt, A. C. Bond, Jr., K. E. Wilzbach, and H. J. Schlesinger, *J. Am. Chem. Soc.*, **69**, 2692 (1947).
54. R. H. Fish, H. G. Kuivila, and I. J. Tyminski, *J. Am. Chem. Soc.*, **89**, 5861 (1967).
55. E. E. Flagg, Division of Inorganic Chemistry, 155th Meeting, American Chemical Society, San Francisco, Calif., April 1968.
56. N. Flitcroft and H. D. Kaesz, *J. Am. Chem. Soc.*, **85**, 1377 (1963).
57. R. J. Fox, R. W. Evans, and M. Szwarc, *Trans. Faraday Soc.*, **57**, 1915 (1961).
57a. M. Frankel, D. Wagner, D. Gertner, and A. Zilkha, *Israel J. Chem.*, **4**, 183 (1966); CA, **66**, 52043 (1967).
58. M. Frankel, D. Wagner, D. Gertner, and A. Zilkha, *J. Organometal. Chem.*, **7**, 518 (1967).
59. G. Fritz and H. Scheer, *Z. Anorg. Allgem. Chem.*, **338**, 1 (1965).
60. G. Fritz and H. Scheer, *Z. Naturforsch.*, **19b**, 537 (1964).
61. R. Fuchs and H. Gilman, *J. Org. Chem.*, **22**, 1009 (1957).
62. R. F. Fulton, Ph.D. Dissertation, Purdue Univ., Lafayette, Indiana, 1960; *Dissertation Abstr.*, **22**, 3397 (1962).
63. T. C. Gibb and N. N. Greenwood, *J. Chem. Soc.*, 43 (1966).
64. H. Gilman and J. Eisch, *J. Org. Chem.*, **20**, 763 (1955).
64a. G. L. Grady and H. G. Kuivila, *J. Org. Chem.*, **34**, 2014 (1969).
65. F. D. Greene and N. C. Lowry, *J. Org. Chem.*, **32**, 882 (1967).
66. G. S. Hammond and P. Leermakers, *J. Am. Chem. Soc.*, **84**, 207 (1962).

67. K. Hayashi, J. Iyoda, and I. Shihara, *J. Organometal. Chem.*, **10**, 81 (1967).
67a. T. Hayashi, S. Kikkawa, and S. Matsuda, *Kogyo Kagaku Zasshi*, **70**, 1389 (1967); CA, **68**, 59672 (1968).
68. M. C. Henry and J. G. Noltes, *J. Am. Chem. Soc.*, **82**, 56 (1960).
69. M. C. Henry and J. G. Noltes, *J. Am. Chem. Soc.*, **82**, 558 (1960).
70. R. H. Herber, *Ann. Rev. Phys. Chem.*, **17**, 261 (1966).
71. R. H. Herber and G. I. Parisi, *Inorg. Chem.*, **5**, 769 (1966).
72. R. H. Herber, H. A. Stöckler, and W. T. Reichle, *J. Chem. Phys.*, **42**, 2447 (1965).
73. J. R. Holmes and H. D. Kaesz, *J. Am. Chem. Soc.*, **83**, 3903 (1961).
74. R. K. Ingham, S. D. Rosenberg, and H. Gilman, *Chem. Rev.*, **60**, 459 (1960).
75. K. Issleib and B. Walther, *J. Organometal. Chem.*, **10**, 177 (1967).
75a. K. Itoh, S. Sakai, and Y. Shi, *Yuki Gosei Kagaku Kyokai Shi*, **24**, 729 (1966); CA, **65**, 16998 (1966).
76. K. Itoi and S. Kumano, *Kogyo Kagaku Zasshi*, **70**, 82 (1967); *CA*, **67**, 11556v (1967).
76a. G. S. Jackel and W. Gordy, *Phys. Rev.*, **176**, 443 (1968).
76b. B. B. Jarvis and J. B. Yount, tert., *Chem. Commun.*, 1405 (1969).
76c. B. B. Jarvis and J. B. Yount, tert., *J. Chem. Soc. D*, 1405 (1969).
77. F. R. Jensen and D. B. Patterson, *Tetrahedron Letters*, 3837 (1966).
77a. F. Johnson, N. A. Starkovsky, and A. A. Carlson, *J. Am. Chem. Soc.*, **87**, 4612 (1965).
78. W. L. Jolly, *Angew. Chem.*, **27**, 268 (1960).
79. W. L. Jolly, *J. Am. Chem. Soc.*, **83**, 335 (1961).
80. L. Kaplan, *J. Am. Chem. Soc.*, **88**, 1833 (1966).
81. L. Kaplan, *J. Am. Chem. Soc.*, **88**, 4531 (1966).
82. L. Kaplan, *J. Am. Chem. Soc.*, **88**, 4970 (1966).
82a. L. Kaplan, *Chem. Commun.*, 106 (1969).
83. M. Karplus and D. H. Anderson, *J. Chem. Phys.*, **30**, 6 (1959).
84. K. Kawakami, T. Saito, and R. Okawara, *J. Organometal. Chem.*, **8**, 377 (1967).
85. Y. Kawasaki, K. Kawakami, and T. Tanaka, *Bull. Chem. Soc. Japan*, **38**, 1102 (1965).
85a. M. A. Kazankova, N. P. Protzenko, and J. F. Lutsenko, *Zh. Obshch. Khim.*, **38**, 106 (1968); *CA*, **69**, 67501 (1968).
86. L. E. Khoo and H. H. Lee, *Tetrahedron Letters*, 4351 (1968).
87. H. Kimmel and C. R. Dillard, *Spectrochim. Acta*, **24A**, 909 (1968).
88. K. A. W. Kramer and A. N. Wright, *J. Chem. Soc.*, 3604 (1963).
89. C. A. Kraus and W. N. Greer, *J. Am. Chem. Soc.*, **44**, 2629 (1922).
90. C. G. Krespan and V. A. Engelhardt, *J. Org. Chem.*, **23**, 1565 (1958).
90a. H. Kriegsmann and K. Ulbricht, *Z. Anorg. Allgem. Chem.*, **328**, 90 (1964).
90b. K. Kühlein, W. P. Neumann, and H. Mohring, *Angew. Chem.*, **80**, 438 (1968).
91. H. G. Kuivila, *J. Org. Chem.*, **25**, 284 (1960).
92. H. G. Kuivila, in *Advances in Organometallic Chemistry* (F. G. A. Stone and R. West, eds.), Vol. 1, Academic, New York, 1964, p. 47.
93. H. G. Kuivila, *Accounts Chem. Res.*, **1**, 299 (1968).
94. H. G. Kuivila and O. F. Beumel, Jr., *J. Am. Chem. Soc.*, **80**, 3250 (1958).
95. H. G. Kuivila and O. F. Beumel, Jr., *J. Am. Chem. Soc.*, **80**, 3798 (1958).
96. H. G. Kuivila and O. F. Beumel, Jr., *J. Am. Chem. Soc.*, **83**, 1246 (1961).
97. H. G. Kuivila and P. L. Levins, *J. Am. Chem. Soc.*, **86**, 23 (1964).
98. H. G. Kuivila and L. W. Menapace, *J. Org. Chem.*, **28**, 2165 (1963).
99. H. G. Kuivila, L. W. Menapace, and C. R. Warner, *J. Am. Chem. Soc.*, **84**, 3584 (1962).
100. H. G. Kuivila, W. Rahman, and R. H. Fish, *J. Am. Chem. Soc.*, **87**, 2835 (1965).
101. H. G. Kuivila, A. K. Sawyer, and A. G. Armour, *J. Org. Chem.*, **26**, 1426 (1961).

102. H. G. Kuivila and R. Sommer, *J. Am. Chem. Soc.*, **89**, 5616 (1967).
103. H. G. Kuivila and E. J. Walsh, Jr., *J. Am. Chem. Soc.*, **88**, 571 (1966).
104. M.-R. Kula, E. Amberger, and H. Rupprecht, *Chem. Ber.*, **98**, 629 (1965).
105. M.-R. Kula, J. Lorberth, and E. Amberger, *Chem. Ber.*, **97**, 2087 (1964).
106. E. J. Kupchik and R. J. Kiesel, *Chem. Ind. London*, 1654 (1962).
107. E. J. Kupchik and R. J. Kiesel, *J. Org. Chem.*, **29**, 764 (1964).
108. E. J. Kupchik and R. J. Kiesel, *J. Org. Chem.*, **29**, 3690 (1964).
109. E. J. Kupchik and R. J. Kiesel, *J. Org. Chem.*, **31**, 456 (1966).
110. E. J. Kupchik and T. Lanigan, *J. Org. Chem.*, **27**, 3661 (1962).
110a. J. C. Lahournére and J. Valade, *J. Organometal. Chem.*, **22**, C3 (1970).
111. B. R. Laliberte, W. Davidsohn, and M. C. Henry, *J. Organometal. Chem.*, **5**, 526 (1966).
111a. M. F. Lappert and N. F. Travers, *Chem. Commun.*, 1569 (1968).
111b. F. C. Leavitt and L. U. Matternas, *J. Polymer Sci.*, **62**, 568 (1962).
112. M. Lesbre and R. Buisson, *Bull. Soc. Chim. France*, 1204 (1957).
113. A. J. Leusink, *Hydrostannation*, Schotanus and Jens, Utrecht, 1966.
114. A. J. Leusink and H. A. Budding, *J. Organometal. Chem.*, **11**, 533 (1968).
115. A. J. Leusink, H. A. Budding, and W. Drenth, *J. Organometal. Chem.*, **9**, 295 (1967).
116. A. J. Leusink, H. A. Budding, and W. Drenth, *J. Organometal. Chem.*, **11**, 541 (1968).
117. A. J. Leusink, H. A. Budding, and W. Drenth, *J. Organometal. Chem.*, **13**, 155 (1968).
118. A. J. Leusink, H. A. Budding, and W. Drenth, *J. Organometal. Chem.*, **13**, 163 (1968).
119. A. J. Leusink, H. A. Budding, and J. W. Marsman, *J. Organometal. Chem.*, **9**, 285 (1967).
120. A. J. Leusink, H. A. Budding, and J. G. Noltes, *Rec. Trav. Chim.*, **85**, 151 (1966).
121. A. J. Leusink, W. Drenth, J. G. Noltes, and G. J. M. van der Kerk, *Tetrahedron Letters*, 1263 (1967).
122. A. J. Leusink and J. W. Marsman, *Rec. Trav. Chim.*, **84**, 1123 (1965).
123. A. J. Leusink, J. W. Marsman, and H. A. Budding, *Rec. Trav. Chim.*, **84**, 689 (1965).
124. A. J. Leusink, J. W. Marsman, H. A. Budding, J. G. Noltes, and G. J. M. van der Kerk, *Rec. Trav. Chim.*, **84**, 567 (1965).
125. A. J. Leusink and J. G. Noltes, *Rec. Trav. Chim.*, **84**, 585 (1964).
126. A. J. Leusink and J. G. Noltes, *Tetrahedron Letters*, 335 (1966).
127. A. J. Leusink and J. G. Noltes, *Tetrahedron Letters*, 2221 (1966).
127a. A. J. Leusink and J. G. Noltes, *J. Organometal. Chem.*, **16**, 91 (1969).
128. A. J. Leusink, J. G. Noltes, H. A. Budding, and G. J. M. van der Kerk, *Rec. Trav. Chim.*, **83**, 609 (1964).
129. A. J. Leusink, J. G. Noltes, H. A. Budding, and G. J. M. van der Kerk, *Rec. Trav. Chim.*, **83**, 1036 (1964).
130. D. H. Lorenz and E. I. Becker, *J. Org. Chem.*, **27**, 3370 (1962).
131. D. H. Lorenz and E. I. Becker, *J. Org. Chem.*, **28**, 1707 (1963).
132. D. H. Lorenz, P. Shapiro, A. Stern, and E. I. Becker, *J. Org. Chem.*, **28**, 2332 (1963).
133. E. Y. Lukevits and M. G. Voronkov, *Organic Insertion Reactions of Group IV Elements*, Plenum, New York, 1966.
134. M. L. Maddox, N. Flitcroft, and H. D. Kaesz, *J. Organometal. Chem.*, **4**, 50 (1965).
135. E. N. Mal'tseva, V. S. Zavgorodnii, I. A. Maretina, and A. A. Petrov, *Zh. Obshch. Khim.*, **38**, 203 (1968); *CA*, **69**, 52240x (1968).
135a. E. N. Mal'tseva, V. S. Zavgorodnii, and A. A. Petrov, *Zh. Obshch. Khim.*, **39**, 152 (1969); *CA*, **70**, 106625t (1969).
135b. L. May and J. J. Spijkerman, *J. Chem. Phys.*, **46**, 3272 (1967).
136. J. Meinwald, J. W. Wheeler, A. A. Nimetz, and J. S. Liu, *J. Org. Chem.*, **30**, 1038 (1965).

137. L. W. Menapace and H. G. Kuivila, *J. Am. Chem. Soc.*, **86**, 3047 (1964).
137a. E. Müller, R. Sommer, and W. P. Neumann, *Ann.*, **1**, 718 (1968).
137b. A. N. Nesmeyanov and A. E. Borisov, *Dokl. Akad. Nauk SSSR.* **174**, 96 (1967); *CA*, **67**, 90903 (1967).
137c. A. N. Nesmeyanov and A. E. Borisov, *Izv. Akad. Nauk SSSR, Ser. Khim.*, 226 (1967); *CA*, **66**, 95147 (1967).
137d. A. N. Nesmeyanov, A. E. Borisov, and N. V. Novikova, *Dokl. Akad. Nauk SSSR* **172**, 1329 (1967); *CA*, **67**, 3127 (1967).
137e. A. N. Nesmeyanov, A. E. Borisov, and Shi-Hua Wang, *Izv. Akad. Nauk SSSR, Ser. Khim.*, 1141 (1967); *CA*, **68**, 29807 (1968).
138. W. P. Neumann, *Angew. Chem. Intern. Ed.*, **2**, 165 (1963).
139. W. P. Neumann, *Angew. Chem.*, **76**, 849 (1964).
140. W. P. Neumann, *Die Organische Chemie des Zinns*, Ferdinand Enke Verlag, Stuttgart, Germany, 1967.
141. W. P. Neumann, H. J. Albert, and W. Kaiser, *Tetrahedron Letters*, 2041 (1967).
142. W. P. Neumann and E. Heymann, *Angew. Chem. Intern. Ed.*, **2**, 100 (1963).
143. W. P. Neumann and E. Heymann, *Ann.*, **683**, 11 (1965).
144. W. P. Neumann and E. Heymann, *Ann.*, **683**, 24 (1965).
145. W. P. Neumann and K. König, *Angew. Chem. Intern. Ed.*, **1**, 212 (1962).
146. W. P. Neumann and K. König, *Angew. Chem. Intern. Ed.*, **3**, 751 (1964).
147. W. P. Neumann and K. König, *Ann.*, **677**, 1 (1964).
148. W. P. Neumann and K. König, *Ann.*, **677**, 12 (1964).
149. W. P. Neumann and H. Lind, *Angew. Chem. Intern. Ed.*, **6**, 76 (1967).
149a. W. P. Neumann, H. Lind, and G. Alester, *Chem. Ber.*, **101**, 2845 (1968).
150. W. P. Neumann and H. Niermann, *Ann.*, **653**, 164 (1964).
151. W. P. Neumann, H. Niermann, and B. Schneider, *Angew. Chem. Intern. Ed.*, **2**, 547 (1963).
152. W. P. Neumann, H. Niermann, and B. Schneider, *Ann.*, **707**, 15 (1967).
153. W. P. Neumann, H. Niermann, and R. Sommer, *Angew. Chem.*, **73**, 768 (1961).
154. W. P. Neumann, H. Niermann, and R. Sommer, *Ann.*, **659**, 27 (1962).
155. W. P. Neumann and J. Pedain, *Ann.*, **672**, 34 (1964).
156. W. P. Neumann and J. Pedain, *Tetrahedron Letters*, 2461 (1964).
157. W. P. Neumann, J. Pedain, and R. Sommer, *Ann.*, **694**, 9 (1966).
158. W. P. Neumann, K. Rübsamen, and R. Sommer, *Angew. Chem.*, **77**, 733 (1965).
158a. W. P. Neumann, K. Rübsamen, and R. Sommer, *Chem. Ber.*, **100**, 1063 (1967).
158b. W. P. Neumann, K. Rübsamen, R. Sommer, and U. Frommer, *Angew. Chem.*, **79**, 1006 (1967).
159. W. P. Neumann and B. Schneider, *Angew. Chem. Intern. Ed.*, **3**, 751 (1964).
160. W. P. Neumann and B. Schneider, *Ann.*, **707**, 20 (1967).
161. W. P. Neumann, B. Schneider, and R. Sommer, *Ann.*, **692**, 1 (1966).
162. W. P. Neumann and R. Sommer, *Angew. Chem. Intern. Ed.*, **2**, 547 (1963).
163. W. P. Neumann and R. Sommer, *Angew. Chem. Intern. Ed.*, **3**, 133 (1964).
164. W. P. Neumann and R. Sommer, *Ann.*, **675**, 10 (1964).
165. W. P. Neumann and R. Sommer, *Ann.*, **701**, 28 (1967).
166. W. P. Neumann, R. Sommer, and H. Lind, *Ann.*, **688**, 14 (1965).
167. W. P. Neumann, R. Sommer, and E. Müller, *Angew. Chem. Intern. Ed.*, **5**, 514 (1966).
168. H. C. Newsom and W. G. Woods, *Inorg. Chem.*, **7**, 177 (1968).
169. J. G. Noltes, *Rec. Trav. Chim.*, **83**, 515 (1964).
170. J. G. Noltes and M. J. Janssen, *Rec. Trav. Chim.*, **82**, 1055 (1963).
171. J. G. Noltes and M. J. Janssen, *J. Organometal. Chem.*, **1**, 346 (1964).

172. J. G. Noltes and G. J. M. van der Kerk, *Chem. Ind. London*, 294 (1959).
173. J. G. Noltes and G. J. M. van der Kerk, *Chimia*, **16**, 122 (1962).
174. J. G. Noltes and G. J. M. van der Kerk, *Rec. Trav. Chim.*, **80**, 623 (1961).
175. J. G. Noltes and G. J. M. van der Kerk, *Rec. Trav. Chim.*, **81**, 41 (1962).
176. H. Nöth and K.-H. Hermannsdörfer, *Angew. Chem.*, **76**, 377 (1964).
177. M. Ohara and R. Okawara, *J. Organometal. Chem.*, **3**, 484 (1965).
178. J. P. Oliver and U. V. Rao, *J. Org. Chem.*, **31**, 2696 (1966).
178a. J. P. Oliver, U. V. Rao, and M. T. Emerson, *Tetrahedron Letters*, 3419 (1964).
179. D. H. Olson and R. E. Rundle, *Inorg. Chem.*, **2**, 1310 (1963).
179a. I. Omae, S. Matsuda, S. Kikkawa, and R. Sato, *Kogyo Kagaku Zasshi*, **70**, 705 (1967); *CA*, **68**, 13107 (1968).
179b. I. Omae, S. Ohnishi, and S. Matsuda, *Kogyo Kagaku Zasshi*, **70**, 1755 (1967); *CA*, **68**, 87371 (1968).
180. F. Paneth and K. Fürth, *Ber. Deut. Botan. Ges.*, **52**, 2020 (1919).
181. M. Pang and E. I. Becker, *J. Org. Chem.*, **29**, 1948 (1964).
182. R. V. Parish and R. H. Platt, *Chem. Commun.*, 1118 (1968).
182a. M. Pereyre, G. Colin, and J. Valade, *Tetrahedron Letters*, 4805 (1967).
182b. M. Pereyre, G. Colin, and J. Valade, *Bull. Soc. Chim. Fr.*, 3358 (1968).
183. M. Pereyre and J. Valade, *Bull. Soc. Chim. France*, 1928 (1967).
184. M. Pereyre and J. Valade, *Compt. Rend.*, **258**, 4785 (1964).
185. M. Pereyre and J. Valade, *Compt. Rend.*, **260**, 581 (1965).
186. J. C. Pommier and J. Valade, *Bull. Soc. Chim. France*, 975 (1965).
187. P. E. Potter, L. Pratt, and G. Wilkinson, *J. Chem. Soc.*, 524 (1964).
188. W. A. Pryor, *Free Radicals*, McGraw-Hill, New York, 1966, p. 40.
189. W. Rahman and H. G. Kuivila, *J. Org. Chem.*, **31**, 772 (1966).
190. N. F. Ramsey, *Phys. Rev.*, **91**, 303 (1953).
190a. L. W. Reeves, *J. Chem. Phys.*, **40**, 2128 (1964).
190b. G. H. Reifenberg and W. J. Considine, *J. Organometal. Chem.*, **9**, 505 (1967).
190c. G. H. Reifenberg and W. J. Considine, *J. Am. Chem. Soc.*, **92**, 2401 (1969).
190d. R. E. Ridenour and E. E. Flagg, *J. Organometal. Chem.*, **16**, 393 (1969).
191. L. A. Rothman and E. I. Becker, *J. Org. Chem.*, **24**, 294 (1959).
192. L. A. Rothman and E. I. Becker, *J. Org. Chem.*, **25**, 2203 (1960).
192a. G. A. Russell and G. W. Holland, *J. Am. Chem. Soc.*, **91**, 3968 (1969).
193. T. Saegusa, S. Kobayashi, Y. Ito, and N. Yasuda, *J. Am. Chem. Soc.*, **90**, 4182 (1968).
194. A. K. Sawyer, *J. Am. Chem. Soc.*, **87**, 537 (1965).
195. A. K. Sawyer, *3rd Intern. Organometal. Symp.*, Munich, Germany (1967).
196. A. K. Sawyer and J. E. Brown, *J. Organometal. Chem.*, **5**, 438 (1966).
197. A. K. Sawyer, J. E. Brown, and E. L. Hanson, *J. Organometal. Chem.*, **3**, 464 (1965).
198. A. K. Sawyer, J. E. Brown, and G. S. May, *J. Organometal. Chem.*, **11**, 192 (1968).
199. A. K. Sawyer and H. G. Kuivila, *J. Am. Chem. Soc.*, **82**, 5958 (1960).
200. A. K. Sawyer and H. G. Kuivila, *J. Am. Chem. Soc.*, **85**, 1010 (1963).
201. A. K. Sawyer and H. G. Kuivila, *Chem. Ind. London*, 260 (1961).
202. A. K. Sawyer and H. G. Kuivila, *J. Org. Chem.*, **27**, 610 (1962).
203. A. K. Sawyer and H. G. Kuivila, *J. Org. Chem.*, **27**, 837 (1962).
204. A. K. Sawyer, G. S. May, and R. E. Scofield, *J. Organometal. Chem.*, **14**, 213 (1968).
205. U. Schmidt, K. Kabitzke, K. Markau, and W. P. Neumann, *Chem. Ber.*, **98**, 3827 (1965).
206. B. Schneider and W. P. Neumann, *Ann.*, **707**, 7 (1967).
207. D. Seyferth, J. M. Burlitch, H. Dertouzos, and H. D. Simmons, Jr., *J. Organometal. Chem.*, **7**, 405 (1967).

208. D. Seyferth, T. F. Jula, H. Dertouzos, and M. Pereyre, *J. Organometal. Chem.*, **11**, 63 (1968).
209. D. Seyferth and R. Bruce King, eds., *Annual Surveys of Organometallic Chemistry*, Academic, New York; Vols. 1–3 (1964–1966) have appeared in book form; the series is now continued as *Organometallic Chemistry Reviews*, Section B.
210. D. Seyferth, G. Raab, and K. A. Brändle, *J. Org. Chem.*, **26**, 2934 (1961).
211. D. Seyferth, Y. Sata, and M. Takamizawa, *J. Organometal. Chem.*, **2**, 367 (1964).
212. D. Seyferth, H. D. Simmons, Jr., and L. J. Todd, *J. Organometal. Chem.*, **2**, 282 (1964).
213. D. Seyferth and M. Takamizawa, *Inorg. Chem.*, **2**, 731 (1963).
214. D. Seyferth and L. G. Vaughan, *J. Organometal. Chem.*, **1**, 138 (1963).
215. D. Seyferth, H. Yamazaki, and D. L. Alleston, *J. Org. Chem.*, **28**, 703 (1963).
216. P. Shapiro and E. I. Becker, *J. Org. Chem.*, **27**, 4668 (1962).
217. W. B. Simpson, *Chem. Ind., London*, 854 (1966).
218. K. Sisido, S. Kozima, and K. Takizawa, *Tetrahedron Letters* 33 (1967).
219. R. Sommer and H. G. Kuivila, *J. Org. Chem.*, **33**, 802 (1968).
220. R. Sommer and W. P. Neumann, *Angew. Chem. Intern. Ed.*, **5**, 515 (1966).
221. R. Sommer, W. P. Neumann, and B. Schneider, *Tetrahedron Letters*, 3875 (1964).
222. R. Sommer, B. Schneider, and W. P. Neumann, *Ann.*, **692**, 12 (1966).
223. W. F. Stack, G. A. Nash, and H. A. Skinner, *Trans. Faraday Soc.*, **61**, 2122 (1965).
224. A. Stern and E. I. Becker, *J. Org. Chem.*, **27**, 4052 (1962).
225. A. Stern and E. I. Becker, *J. Org. Chem.*, **29**, 3221 (1964).
226. R. W. Taft, Jr., in *Steric Effects in Organic Chemistry* (M. S. Newman, ed.), Wiley, New York, 1956, p. 619.
227. C. Tamborski and E. J. Soloski, *J. Am. Chem. Soc.*, **83**, 3734 (1961).
228. C. Tamborski, F. E. Ford, and E. J. Soloski, *J. Org. Chem.*, **28**, 181 (1963).
229. C. Tamborski, F. E. Ford, and E. J. Soloski, *J. Org. Chem.*, **28**, 237 (1963).
230. F. J. A. Des Tombe, G. J. M. van der Kerk, H. M. J. C. Creemers, and J. G. Noltes, *Chem. Commun.*, 914 (1966).
231. W. Trotter and A. C. Testa, *J. Am. Chem. Soc.*, **90**, 7044 (1968).
232. T. T. Tsai, A. Cutler, and W. L. Lehn, *J. Org. Chem.*, **30**, 3049 (1965).
232a. T. T. Tsai, W. L. Lehn, and C. J. Marshall, Jr., *J. Organometal. Chem.*, **22**, 387 (1970).
233. C.-L. Ts'eng, J.-H. Cho, and S.-C. Ma, *K'o Hsueh T'ung Pao*, **17**, 77 (1966); *CA*, **66**, 28862u (1967).
234. J. Valade and J. C. Pommier, *Bull. Soc. Chim. France*, 199 (1963).
235. G. P. van der Kelen, L. Verdonck, and D. van de Vondel, *Bull. Soc. Chim. Belges*, **73**, 733 (1964).
236. G. J. M. van der Kerk, J. G. A. Luijten, and J. G. Noltes, *Angew. Chem.*, **70**, 298 (1958).
237. G. J. M. van der Kerk, J. G. A. Luijten, and J. G. Noltes, *Chem. Ind. London*, 352 (1956).
238. G. J. M. van der Kerk and J. G. Noltes, *Ann. N.Y. Acad. Sci.*, **125**, 25 (1965).
239. G. J. M. van der Kerk and J. G. Noltes, *J. Appl. Chem.*, **9**, 106 (1959).
240. G. J. M. van der Kerk, J. G. Noltes, and J. G. A. Luijten, *J. Appl. Chem.*, **7**, 356 (1957).
241. G. J. M. van der Kerk, J. G. Noltes, and J. G. A. Luijten, *J. Appl. Chem.*, **7**, 366 (1957).
242. G. J. M. van der Kerk, J. G. Noltes, and J. G. A. Luijten, *Chem. Ind. London*, 1290 (1958).

243. G. J. M. van der Kerk, J. G. Noltes, and J. G. A. Luijten, *Rec. Trav. Chim.*, **81**, 853 (1962).

244. L. Verdonck and G. P. van der Kelen, *J. Organometal. Chem.*, **5**, 532 (1966).

245. T. Vladimiroff and E. R. Malinowski, *J. Chem. Phys.*, **42**, 1 (1965).

245a. V. M. Vlasov, R. G. Mirskov, and V. N. Petrova, *Zh. Obshch. Khim.*, **37**, 954 (1967); *CA*, **68**, 13114 (1968).

246. E. Vogel, W. Grimme, and S. Korte, *Tetrahedron Letters*, 3625 (1965).

246a. N. S. Vyazankin, M. N. Bochkarev, and L. P. Sanina, *Zh. Obshch. Khim.*, **36**, 1154 (1966); *CA*, **65**, 10617 (1966).

246b. N. S. Vyazankin, M. N. Bochkarev, and L. P. Sanina, *Zh. Obshch. Khim.*, **36**, 1961 (1966); *CA*, **66**, 76114 (1967).

246c. N. S. Vyazankin, M. N. Bochkarev, and L. P. Sanina, *Zh. Obshch. Khim.*, **37**, 1037 (1967); *CA*, **68**, 13099t (1968).

247. N. S. Vyazankin, M. N. Bochkarev, and L. P. Sanina, *Zh. Obshch. Khim.*, **38**, 414 (1968); *CA*, **69**, 96844b (1968).

248. N. S. Vyazankin and V. T. Bychkov, *Zh. Obshch. Khim.*, **35**, 684 (1965); *CA*, **63**, 4320 (1965).

248a. N. S. Vyazankin, G. S. Kalinina, O. A. Kruglaya, and G. A. Razuvaev, *Zh. Obshch. Khim.*, **38**, 906 (1968); *CA*, **69**, 77387 (1968).

248b. N. S. Vyazankin, G. S. Kalinina, O. A. Kruglaya, and G. A. Razuvaev, *Zh. Obshch. Khim.*, **39**, 2005 (1969); *CA*, **72**, 31946d (1970).

249. N. S. Vyazankin, G. A. Razuvaev, and S. P. Korneva, *Zh. Obschch. Khim.*, **34**, 2787 (1964); *CA*, **61**, 14700 (1964).

250. N. S. Vyazankin, G. A. Razuvaev, S. P. Korneva, O. A. Kruglaya, and R. F. Galiulina, *Dokl. Akad. Nauk SSSR*, **158**, 884 (1964); *CA*, **62**, 2788 (1965).

250a. N. S. Vyazankin, G. A. Razuvaev, O. A. Kruglaya, and G. S. Semchikova, *J. Organometal. Chem.*, **6**, 474 (1966).

250b. P. J. Wagner, *J. Am. Chem. Soc.*, **89**, 2503 (1967).

251. M. Wahren, P. Hädge, H. Hübner and M. Mühlstädt, *Isotopenpraxis*, **1**, 65 (1965).

252. C. Walling, J. H. Cooley, A. A. Ponaras, and E. J. Racah, *J. Am. Chem. Soc.*, **88**, 5361 (1966).

253. E. J. Walsh, Jr. and H. G. Kuivila, *J. Am. Chem. Soc.*, **88**, 576 (1966).

253a. E. J. Walsh, Jr., R. L. Stoneberg, M. Yorke, and H. G. Kuivila, *J. Org. Chem.*, **34**, 1156 (1969).

254. J. Warkentin and E. Sanford, *J. Am. Chem. Soc.*, **90**, 1667 (1968).

255. C. R. Warner, R. J. Strunk, and H. G. Kuivila, *J. Org. Chem.*, **31**, 3381 (1966).

256. S. Weber and E. I. Becker, *J. Org. Chem.*, **27**, 1258 (1962).

257. E. Wiberg, E. Amberger, and H. Cambensi, *Z. Anorg. Allgem. Chem.*, **351**, 164 (1967).

258. S. Winstein, H. M. Walborsky, and K. C. Schreiber, *J. Am. Chem. Soc.*, **72**, 5795 (1950).

259. G. Wittig, F. J. Meyer, and G. Lange, *Ann.*, **571**, 167 (1951).

260. V. S. Zavgorodnii and A. A. Petrov, *Zh. Obshch. Khim.*, **35**, 1313 (1965); *CA*, **63**. 11601 (1965).

3. ORGANOTIN HALIDES

G. P. VAN DER KELEN, E. V. VAN DEN BERGHE, AND L. VERDONCK

Rijksuniversiteit Gent
Laboratorium voor Algemene
en Anorganische Chemie—B
Gent, Belgium

I. Introduction

The organotin halides occupy a key position in the chemistry of the organic compounds of tin. Indeed, they are used as starting materials for the synthesis of many series of other organotin compounds. Apart from their practical importance, the vast amount of information available on these compounds as regards physicochemical properties and reaction kinetics has provided new insight into the chemical bonding of organotins as a whole. Also complex formation between organotin halides and various bases is a subject of particular interest in the theory of chemical bonding between tin and the ligands. This topic is, therefore, also treated in this review, although some aspects of it will also be covered in other chapters on organotin compounds containing Sn—O, Sn—N, and Sn—S bonds.

II. Synthesis

The synthesis of $(C_2H_5)_3SnBr$ in 1852 by Löwig (*326*) was soon followed by the development of a series of methods for the synthesis of such compounds by various reactions and by the use of different starting materials. In several books on organometallic compounds (*105*, *133*, *294*) and in several reviews (*194*, *221*, *328*) the synthetic methods have been described and a classification proposed. In the following sections these methods will be discussed, the main emphasis being on the most recent developments in this field. A review of the many organotin halides known today is presented in Appendices 1 to 7.

A. The Direct Synthesis

By direct synthesis is meant the reaction between tin and an alkyl or aryl halide to yield an organotin halide. The simplest reaction would be

$$Sn + 2\,RX \longrightarrow R_2SnX_2$$

This reaction would obviously offer the most economic and therefore the most interesting method for synthesizing organotin halides on an industrial scale. In the early experiments, however, the yields of this reaction proved to be very poor, and recent studies therefore try to outline conditions for obtaining optimal yields.

In the earliest experiments (*76*, *81*, *154*, *193*, *237*) tin and an alkyl halide were heated in a sealed tube at temperatures ranging from 130° to 220°C for 20–40 h. R_2SnX_2 was found to be predominantly formed, along with small amounts of R_3SnX. The reaction was found to proceed more easily for chlorides than for bromides or iodides, in this order; higher alkyls are more difficult to incorporate.

With methylene bromide or chloride, Kocheshkov (*265*) obtained CH_3SnBr_3 and CH_3SnCl_3, respectively. The reaction of $BrCH_2COOC_2H_5$ with Sn yielded $(C_2H_5OOCCH_2)_2SnBr_2$ (*172*).

In 1953, Smith and Rochow (*483*) developed a new technique. They allowed the gaseous alkyl halide to bubble through molten tin at 350°–450°C. With methyl chloride, for instance, a 10% yield of a methyltin halides mixture was obtained consisting of $(CH_3)_2SnCl_2$ (75%) and CH_3SnCl_3 (20%); with CH_3Br, $(CH_3)_2SnBr_2$ was obtained in good yield, whereas CH_3I yielded mainly CH_3SnI_3. It was found that copper and zinc added to the melt in catalytic amounts, considerably enhanced the yields (*484*, *529*, *530*).

Another procedure was tried by Irmscher et al.: tin powder and methyl bromide, together with some methanol, were heated in an autoclave at 100°C and 19 atm; a high yield of $(CH_3)_2SnBr_2$ was obtained (*225*). For this kind of reaction between Sn and RI, with R ranging from methyl to hexyl, different catalysts were tried (*350–354*). It was found that some metals like magnesium and zinc in the presence of such alcohols as butanol, cyclohexanol or with THF, possessed strong catalytic capacities. Oakes and Hutton (*391*, *392*), however, obtained Bu_2SnI_2 in 90% yield by refluxing BuI with Sn in an open system, using lithium or lithium bromide as a catalyst, together with either butanol, 2-5 hexanedione, or 2-ethoxyethanol. Considerably lower yields were obtained for Bu_2SnBr_2 and $(Oct)_2SnBr_2$. Sisido et al. (*476*, *480*) used organic ammonium halides or a mixture of an organic base and iodine as catalysts for the syntheses of Bu_2SnBr_2, Bu_2SnCl_2, $(n\text{-Pr})_2SnBr_2$, and di-allyltin dibromide. Di- or tribenzyltin chlorides could be prepared in high yield without a catalyst by treating tin powder in suspension in either toluene or water with benzyl chloride at 108–114°C (*479*). The catalytic direct synthesis also proved to be successful for $X_2Sn(p\text{-iso-Pr } C_6H_4CH_2)_2$, $XSn(p\text{-iso-Pr } C_6H_4CH_2)_3$ (*114*), and $X_2Sn(CHRCHR'CONHCH_2CO_2Et)_2$, $X_2Sn(CH_2CHMeCONRR')_2$. $X_2Sn(CHMeCH_2CONRR')_2$ for X = I, Br and R, R′ = H, Me (*202*, *203*).

A different method for the direct synthesis of higher alkyltin halides was developed by Kocheshkov and co-workers (*2–4*, *280*). A mixture of tin and the appropriate halide, sealed in a glass capsule, was heated and in the meantime irradiated with γ radiation from a ^{60}Co source. Thus Bu_2SnBr_2 was obtained from 1-bromobutane and tin powder. This radiation-induced synthesis was extensively studied by Wyant et al. (*147*, *148*, *538*) for the preparation of Bu_2SnBr_2. Attention was focused on the influence of surface conditions of the Sn particles, particle size, catalysts, and temperature. It was shown that apart from Bu_2SnBr_2, which was the main product (80% yield), Bu_3SnBr, $BuSnBr_3$, and $SnBr_4$ were also formed.

It is also worth mentioning that some alkyltin halides could be prepared by the electrolysis (*17*, *18*) of alkyl halides in a solvent medium, using a magnesium cathode and a soluble tin anode.

As a conclusion of this survey it can be restated that the direct synthesis yields R_2SnX_2 as the main product, that the reaction proceeds more smoothly for iodides than for bromides and chlorides, and that the introduction of higher alkyl groups is only feasible under appropriate reaction conditions.

Some other methods of synthesis are also closely related to the direct synthesis that it seems logical to discuss them in this section.

By treating $SnCl_2$ with CH_3Cl at high temperature, CH_3SnCl_3 is formed in good yield. When, however, SnO is used with copper as a catalyst, mainly $(CH_3)_3SnX$ (X = Cl, Br, I) is obtained (*484*).

The reaction between an organic halide RX and a Sn—Na or a Sn—Mg alloy is also considered to be a means of direct synthesis. It is, however, found that the results are quite different from those discussed above in the reactions with pure tin. Löwig (*326*) was the first to report the reaction between Sn—Na and C_2H_5I; he obtained a mixture of $(C_2H_5)_2SnI_2$, $(C_2H_5)_3SnI$, $(C_2H_5)_4Sn$, and $(C_2H_5)_2Sn$. Cahours (*76*) and Ladenburg (*311*) found the same reaction products; analogous results were obtained with CH_3I. According to Harada (*195*, *196*, *199*) activation of the Sn-Na alloy with Zn results in the formation of R_3SnI and R_4Sn only, for R = CH_3, C_2H_5, and C_3H_7.

In 1954 Van der Kerk and Luijten (*251*) prepared $(C_2H_5)_3SnBr$ and $(C_2H_5)_3SnCl$ together with some $(C_2H_5)_4Sn$ and $(C_2H_5)_2SnX_2$ by treating a Sn—Mg alloy, with stoichiometry corresponding to Mg_2Sn, with the appropriate ethyl halide in cyclohexane. Faulker (*146*), however, prepared R_3SnCl by treating a Sn—Mg alloy with mercury or mercury salts as catalysts in an autoclave at 150°C. Under the same reaction conditions Laine et al. (*312*) synthesized R_3SnBr and R_2SnBr_2 (R = *n*-propyl, *n*-butyl). With a Sn—Na alloy activated with zinc, mainly R_3SnCl, together with R_4Sn, was obtained (*550*) (R = *n*-propyl, *n*-butyl, *n*-amyl).

Since 1960, this procedure has been completely abandoned, because the yields are low and the reaction is difficult to control due to the various possible side reactions.

B. Halogenation of Tetraorganostannanes

Among the tetraorganostannanes two classes can be distinguished: first, the symmetrical R_4Sn and second, the unsymmetrical or mixed R_3SnR', $R_2SnR'_2$, $R_2SnR'R''$.

By partial halogenation of R_4Sn compounds the three possible types of organotin halides, R_3SnX, R_2SnX_2, and $RSnX_3$, have been obtained. With mixed tetraorganostannanes the preparations of $R_2R'SnX$, $RR'SnX_2$, and $RR'R''SnX$ type compounds have been reported. In the latter studies the relative reactivity toward I_2 and Br_2 of a series of organic groups bonded to tin could be determined.

The most widely used halogenating agent is, of course, the pure halogen, but other reagents such as halogen acids, and metal-, alkyl-, and aryl halides are also effective. A particular case of this type of agent is the SnX_4 molecule itself. The reactions with this halide will be discussed in Sec. II.B.4.

A brief critical survey of the various methods for halogenation will be outlined in the following sections.

1. *Halogenation with Free Halogens*

Although the four halogens have been found to be active halogenating agents, I_2 and Br_2 are generally preferred to Cl_2 and F_2 because the latter are corrosive gases and cause some difficulties in their use. Moreover the reactions with I_2 and Br_2 are easier to control as they are slower.

The first two organic groups in a tetraorganostannane can be replaced gradually under appropriate reaction conditions. On further halogenation, however, both remaining organic groups are replaced simultaneously (*336*). The reaction sequence can be written as follows:

$$R_4Sn + X_2 \longrightarrow R_3SnX + RX$$
$$R_3SnX + X_2 \longrightarrow R_2SnX_2 + RX$$
$$R_2SnX_2 + 2\,X_2 \longrightarrow SnX_4 + 2\,RX$$

Alkyl- as well as aryltin halides can be synthesized in this way. For the tetraaryltin compounds, however, in an inert solvent, two aryl groups are at first simultaneously replaced. The halogens are added either as the pure substance or dissolved in an appropriate solvent such as carbon tetrachloride, chloroform, or diethyl ether, etc.

The halogenation of tetraphenyltin with Cl_2 by Aronheim (*20*) in 1878 was the first reaction of this type from which diphenyltin dichloride was the sole product isolated. Later, several investigators studied this reaction (*48*, *291*, *417*). In an inert solvent, even at low temperature, two aryl groups are removed. In pyridine solvent, however, the triaryltin halides could be obtained also. Krause et al. (*286*, *290*) prepared $(CH_3)_3SnBr$ and $(CH_3)_2SnBr_2$ by the bromination reaction of $(CH_3)_4Sn$ with Br_2 at 10°C and 50–60°C, respectively, and Seyferth (*465*) obtained an 88% yield of $(CH_3)_3SnI$ by iodination of $Sn(CH_3)_4$ in benzene. The halogenation of $(C_2H_5)_4Sn$ to yield $(C_2H_5)_3SnX$ and $(C_2H_5)_2SnX_2$ was elaborated by Harada (*196*).

In a comparative study of the reaction

$$R_4Sn + I_2 \longrightarrow R_3SnI + RI$$

wherein R groups ranging from CH_3 to groups with 6 C atoms were considered, Manulkin (*336*, *339*) found that the reactivity in ether solution decreased with an increasing number of carbon atoms; $(n\text{-}C_7H_{15})_4Sn$ and $(n\text{-}C_8H_{17})_4Sn$ did not react with I_2 in ether, but reacted in boiling toluene and xylene. In the latter solvents their reactivity was found to be about equal. Good yields of R_3SnX (X = Cl, Br, I) could be obtained by this method for R = *p*-tolyl (*49*, *293*), *o*-tolyl, *p*-xylyl (*293*); R_3SnX and R_2SnX_2 (X = Br, I) were obtained from $(C_6H_{11})_4Sn$ (*296*), R_3SnX with X = F, Cl, Br, and I, from $(CH_2CH_2CN)_4Sn$ (*236*). $(p\text{-}FC_6H_4)_4Sn$ yielded $(p\text{-}FC_6H_4)_3SnBr$ (*155*). $(p\text{-}ClC_6H_4)_3SnBr$ was synthesized by the bromination of $(p\text{-}ClC_6H_4)_4Sn$ in pyridine at $-15°C$; in CCl_4 however, at room temperature, $(p\text{-}ClC_6H_4)_2SnBr_2$ was obtained (*299*).

The bromination of tetraallyltin is reported (*523*) to result in the formation of a mixture of the mono- and the dibromide, which could not be separated. Decreasing the reaction temperature to $-50°C$ for the bromination of R_3SnR' (R = allyl) allowed stepwise replacement of allyl groups (*445*). By bromination of tetraneophyltin, $[C_6H_5C(CH_3)_2CH_2]_4Sn$, Reichle (*439*) obtained impure monobromide. Tetranaphthyltin and tetramesityltin are apparently indifferent toward I_2 and Br_2 (*24*). Tetrakis(trimethylsilylmethyl)tin, $(Me_3SiCH_2)_4Sn$, does not react with I_2 in boiling xylene (*464*), but when the solvent is removed and the temperature is raised above 175°C one trimethylsilylmethyl group is split off, yielding $[(CH_3)_3SiCH_2]_3SnI$.

Many investigators have studied the halogenation of unsymmetrical tetraorganostannanes in solvents such as $(C_2H_5)_2O$, CCl_4, $CHCl_3$, and C_6H_6 (*13*, *68*, *70*, *71*, *192*, *193*, *254*, *261*, *283*, *295*, *310*, *337*, *426*, *445*, *446*, *463*, *465*, *466*, *486*). Their results show that the ease of removal of one organic group decreases in the following sequence: *o*-tolyl, *p*-tolyl, phenyl, benzyl, vinyl, methyl, propyl, isobutyl, butyl, isoamyl, hexyl, heptyl, octyl. In fact when $R_2SnR'_2$ or R_3SnR' reacts with I_2, sometimes only one of two different groups present in the molecule is split off. Of the two possible reactions (a) and (b), only

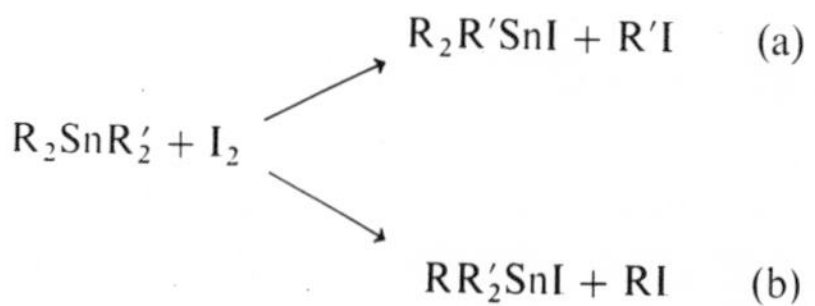

one actually seems to proceed. In this way the mutual reactivity of the groups R and R′ toward the halogen can be estimated. This selectivity as regards the reaction paths (a) or (b) is, however, dependent on the reaction conditions. Changing the solvent and the temperature can cause both reactions to occur simultaneously (*328*, *448*, *465*). Moreover thermal redistribution of the halogenated compounds formed may be a second complicating factor (*254*, *466*):

$$2\,R_2R'SnI \longrightarrow R_3SnI + RR'_2SnI$$

The isolation of the reaction products by fractional distillation must therefore be carried out at as low a temperature as is feasible. These remarks apply to a lesser extent to R_3SnX, R_2SnX_2, and $RSnX_3$ compounds because thermal redistribution into new compounds, R_nSnX_{4-n}, in this case is only important at higher temperatures.

For functionally substituted organic groups the reactivity toward replacement by halogen was found by Van der Kerk and Noltes (*257*) to depend on the position of the substituent, i.e., in the α or β position with respect to the tin atom. In α-substituted organotin nitriles and esters the functionally substituted group is easily removed; with a substituent in the β position, or in a position more distant from the tin atom, however, other groups are preferentially split off, as is demonstrated in the following reactions:

$$(C_3H_7)_3SnCH_2COOC_2H_5 + Br_2 \longrightarrow (C_3H_7)_3SnBr + BrCH_2COOC_2H_5$$

$$(C_4H_9)_3SnCH_2CH_2COOCH_3 + Br_2 \longrightarrow (C_4H_9)_2SnBrCH_2CH_2COOCH_3 + C_4H_9Br$$

For several groups the reactivity toward I_2 or Br_2 has been determined only approximately; these groups could not therefore be fitted into the reactivity sequence mentioned above. The α thienyl group (*298*) and the allyl group (*445*) were found to be more reactive than phenyl. The cyanomethyl group is situated between phenyl and *n*-butyl (*255*). The cyanoethyl group (*440*) is also less reactive than the phenyl group as is the 1-cyclopentadienyl group (*169*). The neopentyl group is less reactive than butyl (*551*).

The reactivity of the trimethylsilylmethyl group in $(Me_3SiCH_2)_2Sn(CH_3)_2$ and $(Me_3SiCH_2)_2Sn(C_4H_9)_2$ (*464*) was studied; in reactions with I_2 this group is less reactive than either methyl or butyl, but in reactions with Br_2 both possible groups are replaced by Br. This group is, however, always less reactive than phenyl, both in reactions with I_2 and Br_2. The bromination of diethyl cyclopentamethylene tin $(C_2H_5)_2(CH_2)_5Sn$ causes cleavage of the ring with the formation of $(C_2H_5)_2SnBr(CH_2)_5Br$ (*254*).

The reaction kinetics of iodination and bromination of R_4Sn and $R_3R'Sn$ compounds have been extensively studied (*50, 52, 64, 65, 160–166, 235, 366, 446*). For both aliphatic and aromatic substituents, the most probable reaction scheme could be established. Although a detailed discussion of these results is beyond the scope of this monograph, it is interesting to point out that reactivity sequences such as those mentioned above, depend heavily upon such properties of the solvent as the dielectric constant, polarizability, and nucleophilic character.

2. *Halogenation with Halogen Acids*

In the reaction between a halogen acid HX and a symmetrical tetraorganostannane, only one organic group is generally removed from tin. In a few cases, however, the dihalide is also obtained:

$$R_4Sn + HX \longrightarrow R_3SnX + RX$$
$$R_3SnX + HX \longrightarrow R_2SnX_2 + RX$$

This reaction can be carried out by treating the R_4Sn compound directly with pure gaseous halogen acid, or by bubbling gaseous HX through an ether solution of the tetraorganotin compound. Chloroform and benzene are also good solvent media for this reaction. In some cases the reaction even proceeds by refluxing the tetraorganostannane with an aqueous solution of the halogen acid. This method is, however, of little importance for preparative purposes.

By reaction of hydrochloric acid with tetramethylstannane in boiling chloroform, Manulkin (*340*) obtained $(CH_3)_3SnCl$. In the same way, but at room temperature, a whole series of R_3SnCl compounds were formed, with $R = C_6H_5$ (*23*), *n*-dodecyl, *n*-tetradecyl, *n*-hexadecyl, and *n*-octadecyl (*357*).

Changing the relative amounts of hydrochloric acid, so that two moles of acid are made to react with one mole of tetraorganostannane resulted in the formation of, predominantly, diphenanthryl- and dibiphenyl-2-tin dichloride. With a 4 : 1 ratio of acid to stannane even tin (IV) chloride was obtained (*24*).

In the unsymmetrical tetraorganostannanes, the reactivity of the Sn—C bond for cleavage by HX has not been studied as thoroughly as was the reaction with chlorine, bromine, or iodine. Nevertheless, the few studies devoted to this subject (*26*, *42*, *66*, *69*, *92*, *261*, *340*, *463*, *464*, *553*) have shown that, for some groups, the reactivity is the same as that with pure halogens and that they fit into the same sequence as mentioned in the preceding paragraph. In many cases, however, this adherence to the previously established sequence is not obeyed. The reaction conditions are apparently of great importance. For instance, the vinyl group has the same relative reactivity toward HCl and HBr as it has toward I_2 (*463*). In $(C_2H_5)_2Sn(CH_3)_2$ and in $(C_2H_5)_2Sn(n\text{-}C_3H_7)_2$, however, two different R groups are simultaneously removed by the action of gaseous HCl at 100–140°C (*69*). From $(CH_3)_3SnC_2H_5$ only one CH_3 group is split off by HCl in boiling chloroform (*340*). The trimethylsilyl group, which is known to be split off with difficulty by I_2, is easily replaced by Br due to the reaction with HBr (*464*). In tribenzylethyltin, a benzyl group is removed by I_2, whereas with HCl the ethyl group is replaced by chlorine (*26*). In the compounds $(C_6H_5)_3SnR$ and $(C_6H_5)_2SnR_2$, wherein R = indenyl and fluorenyl (*553*) it is the indenyl group which is replaced in the indenyl compound, but in the fluorenyl compounds a phenyl group is removed. The corresponding tetraorganotin compounds are, however, both left unattacked by HCl. Buchmann et al. (*66*) made an extensive investigation on the reaction kinetics of the Sn—C cleavage in aryltin compounds by HCl in methanol solution.

3. *Halogenation with Metal Halides, Alkyl Halides, and Acyl Halides*

Generally the reactions between tetraorganostannanes and these halides are of little importance for preparative purposes. These reactions were mainly studied with the aim of investigating the reactivity of the Sn—C bond. From the preparative viewpoint they nevertheless are sometimes important for the synthesis of alkyl or phenyl compounds of other metals. The reactions are generally carried out in solution in chloroform, ethanol, or diethyl ether at reflux temperature.

Among the metallic halides, $HgCl_2$ and $HgBr_2$ have been studied with respect to their reactions with various tetraorganostannanes. In an alkaline medium $HgCl_2$ is reported to successively replace the phenyl groups in $(C_6H_5)_4Sn$ by chlorine (*277, 379*). This compound is more reactive toward alkyltin compounds than is HCl (*13, 340*). No reaction occurs, however, with $(C_6H_5CH_2)_4Sn$ (*340*). With tetracyclopropyltin and $HgCl_2$, tricyclopropyltin chloride in 66% yield was obtained and $HgBr_2$ yielded the corresponding bromide in 77% yield (*467*).

In the unsymmetrical alkylvinyltin compounds, the vinyl group is preferentially split off (*466*). For a series of unsymmetrical tetraorganostannanes the attack of the Sn—C bonds by metal halides is found to be analogous to that by HCl (*279, 340, 467*). In $[(CH_3)_3SiCH_2]_2Sn(CH_3)_2$ one methyl group bonded to tin is replaced by Br in the reaction with $HgBr_2$ (*464*).

Manulkin (*341*) has investigated the reaction between $AlCl_3$ and Ph_4Sn, Et_4Sn and Bu_4Sn and found that the reaction products containing tin were $SnCl_4$, Et_2SnCl_2 and a mixture of Bu_3SnCl and Bu_2SnCl_2, respectively.

With $FeCl_3$, tetraethyltin yielded only Et_3SnCl (*341*).

$BiCl_3$ reacts with Et_4Sn to form Et_3SnCl and Et_2SnCl_2 in a 5 : 1 mole ratio. With Bu_4Sn, Bu_2SnCl_2 is formed and with Pr_3SnBu only $PrBuSnCl_2$ is obtained (*342*). In vinyl-alkyltin compounds it is the vinyl group which is split off (*466*). Analogous results are obtained with $SbCl_3$ and $AsCl_3$. With ICl_3 and Ph_4Sn, Ph_2SnCl_2 is reported to be obtained (*343*).

Other metal halides whose reactivity toward tetraethyltin (*17*) has been investigated are $SnCl_2$, $SnBr_2$, SnI_2, $CuBr_2$, AgCl, AgBr, $KAuCl_4$, HgI_2, $TiCl_4$, $VoCl_3$, $TaCl_5$, $PdCl_2$, and PdI_2. The reaction product was always the triethyltin halide. With $TlCl_3$ (*47, 177*) however, $(C_2H_5)_2SnCl_2$ is mainly obtained. With Ph_4Sn (*177*) the reaction products were Ph_3SnCl and Ph_2SnCl_2, and with Bu_4Sn (*47*) a mixture of the three butyltin halides was formed.

With BF_3, PF_5, AsF_5, and SiF_4 (*472a, 512a*), Ph_4Sn can be easily converted into Ph_3SnF. Two methyl groups in $(CH_3)_4Sn$ are replaced by chlorine (*158a*) using BCl_3.

The controlled reaction between Bu_4Sn and $GeCl_4$ or between $SnCl_4$ and

Bu_4Ge proceeds to the formation of the redistribution products Bu_3SnCl and $BuSnCl_3$ (*329*).

The cleavage of a Sn—C bond in tetraorganostannanes can also be realized with such halogen transfer agents as acyl chlorides (*48, 319, 482*), chlorosulfonic acid (*486*), benzenesulfonyl chloride, sulfuryl chlorides (*48, 482*), and even with alkyl halides (*48, 319, 437, 438*) using appropriate reaction conditions. The reaction products are found to be either R_3SnX or R_2SnX_2 or a mixture of both.

4. *Halogenation by Redistribution Reactions*

One of the most successful methods for the preparation of organotin halides is the redistribution reaction between SnR_4 molecules and the appropriate tin tetrahalide. This method was first introduced by Kocheshkov (*266, 268, 269, 270, 273, 274*). Depending on the molar ratio of the reactants one out of the three following reactions seems to proceed:

$$3\,R_4Sn + SnX_4 \longrightarrow 4\,R_3SnX \qquad (a)$$

$$R_4Sn + SnX_4 \longrightarrow 2\,R_2SnX_2 \qquad (b)$$

$$R_4Sn + 3\,SnX_4 \longrightarrow 4\,RSnX_3 \qquad (c)$$

For R = C_6H_5, $C_6H_4CH_3$ (*256, 273, 274*), $CH_2=CH—$ (*446, 471*), and X = Cl and Br very good yields of the various reaction products could be obtained. When R is CH_3, C_2H_5, n-C_3H_7 (*268–270*) and $CH_3—CH=CH—$ (*523*) only the reactions of types (a) and (b) are effective. Initially the same behavior was also found for the butyltin compounds (*232*). In 1953, however it was shown that when the redistribution between Bu_4Sn and $SnCl_4$ is carried out at temperatures between 0° and 20°C an equimolar mixture was obtained of Bu_3SnCl and $BuSnCl_3$, which corresponds to the following reaction equation:

$$Bu_4Sn + SnCl_4 \longrightarrow Bu_3SnCl + BuSnCl_3$$

Further investigations (*36, 184, 386*) then proved that the latter reaction was not exceptional, but that, instead, it is always the first step in the consecutive reactions occurring in the Kocheshkov redistribution synthesis. The occurrence of this intermediate reaction stage could be shown by gas chromatographic analysis of the reactions proceeding in the synthesis of $(C_2H_5)_2SnCl_2$ from $(C_2H_5)_4Sn$ and $SnCl_4$ (*386*). By means of NMR spectroscopy (*36, 184*) and using various ratios of reactants and observing spectra at different temperatures, it was possible to follow the stepwise reactions in the redistribution between $(CH_3)_4Sn$ and SnX_4 (X = Cl, Br, I). The reactions (d) and (e) could be shown to occur

$$\begin{array}{lcl} (CH_3)_4Sn + SnX_4 & \longrightarrow & (CH_3)_3SnX + CH_3SnX_3 \\ (CH_3)_3SnX + CH_3SnX_3 & \longrightarrow & 2\,(CH_3)_2SnX_2 \end{array} \qquad (d)$$

$$(CH_3)_4Sn + SnX_4 \longrightarrow (CH_3)_3SnX + CH_3SnX_3$$
$$(CH_3)_3SnX + CH_3SnX_3 \longrightarrow 2\,(CH_3)_2SnX_2 \qquad (e)$$
$$2\,(CH_3)_4Sn + 2\,(CH_3)_2SnX_2 \longrightarrow 4\,(CH_3)_3SnX$$

When R is CH_3, C_2H_5 or n-C_3H_7, however, the first reaction of sequence (d) is not of interest for synthetic purposes because the reaction rate constant for this reaction is so high that the second reaction becomes the only controllable step. Moreover it can be shown (*36*) that the overall reaction rates decrease in going from chlorides, to bromides and iodides.

Equilibrium constants and formation enthalpies for $(CH_3)_3SnCl$, $(CH_3)_2SnCl_2$, and CH_3SnCl_3 were also calculated from NMR data (*184*). Calorimetric measurements by Skinner et al. (*407*) on the redistribution system $(CH_3)_4Sn$ and $SnBr_4$ allowed the evaluation of the formation enthalpy for $(CH_3)_3SnBr$.

Other compounds obtained by redistribution reactions are dicyclopropyltin dichloride and dibromide (*467*), $(C_6F_5)_3SnCl$, and $(C_6F_5)_2SnCl_2$ (*214*). The reaction between tetraneophyltin and $SnBr_4$ (*439*) yielded only impure monobromide.

When excess amounts of either SnX_4 or SnR_4 are used, or when the first reaction steps are sufficiently slow, the reaction products can start new reactions with the starting materials (*268*, *386*, *446*, *463*, *466*, *474*) such as these depicted below:

$$2\,R_3SnX + SnX_4 \longrightarrow 3\,R_2SnX_2$$
$$2\,RSnX_3 + SnR_4 \longrightarrow 3\,R_2SnX_2$$
$$R_2SnX_2 + SnR_4 \longrightarrow 2\,R_3SnX$$
$$R_2SnX_2 + SnX_4 \longrightarrow 2\,RSnX_3$$

As a matter of fact, redistribution reactions are generally uninteresting for the preparation of organotin halides with different R groups in the same molecule (*221*).

C. Halogenation of Hexaorganodistannanes $(R_3Sn)_2$ and Diorganostannanes $(R_2Sn)_n$

By the halogenation of $(R_3Sn)_2$ and $(R_2Sn)_n$ compounds with free halogens, R_3SnX and R_2SnX_2, respectively, are obtained in a very pure state (*250*, *293*, *307*, *326*, *513*, *551*), with R either alkyl or aryl and X being Cl, Br, or I:

$$(R_3Sn)_2 + X_2 \longrightarrow 2\,R_3SnX$$
$$(R_2Sn)_n + n\,X_2 \longrightarrow n\,R_2SnX_2$$

These reactions are, nevertheless, of little importance from the viewpoint of the synthetic chemist, because the organotin halides so obtained are the starting materials for the synthesis of the $(R_3Sn)_2$ and $(R_2Sn)_n$ compounds. The tin-tin bond in the latter compounds is also easily cleaved by alkyl halides (*309, 411, 478, 513*) and halogen acids (*513, 522*):

$$(R_3Sn)_2 + 2\ RX \longrightarrow 2\ R_3SnX + R_2$$
$$(R_3Sn)_2 + R'X \longrightarrow R_3SnR' + R_3SnX$$
$$(R_2Sn)_n + n\ RX \longrightarrow n\ R_3SnX$$
$$(R_3Sn)_2 + 2\ HX \longrightarrow 2\ R_3SnX + H_2$$
$$(R_2Sn)_n + n\ HX \longrightarrow n\ R_2SnX_2 + n\ H_2$$

Nasielski and co-workers (*51*) studied the reaction kinetics of the cleavage by I_2 of the tin-tin bond in hexamethyl- and hexabutylditin.

Dialkyltins and hexaalkyldistannanes reduce $HgCl_2$ or RHgCl to metallic mercury, with the formation of R_3SnX, R_2SnX_2, or R_4Sn (*279, 380*):

$$(R_3Sn)_2 + HgCl_2 \longrightarrow 2\ R_3SnCl + Hg$$
$$(R_2Sn)_n + n\ HgCl_2 \longrightarrow n\ R_2SnCl_2 + n\ Hg$$
$$(R_3Sn)_2 + RHgCl \longrightarrow R_4Sn + R_3SnCl + Hg$$

The redistribution reaction between $SnCl_4$ and $[(C_2H_5)_3Sn]_2$ (*436*) yielded Sn, $(C_2H_5)_3SnCl$, $(C_2H_5)_2SnCl_2$, and $(C_2H_5)_4Sn$, but with $[(C_6H_5)_3Sn]_2$ (*513*) the final products were found to be $SnCl_2$ and $(C_6H_5)_3SnCl$. Hexaphenylditin also did not react with $SnCl_2$ or $ZnCl_2$; with $CuCl_2$ formation of CuCl and $(C_6H_5)_3SnCl$ occurred.

D. Halogenation of Organotin Hydroxides and Oxides

By alkaline hydrolysis, organotin halides are converted into the corresponding organotin hydroxide or organotin oxide. The latter compounds can then be made to react with a suitable halogenating agent to yield the desired organotin halide. This is obviously the best way for changing a halogen into an organotin halide. This reaction is also of great value for the synthesis of organotin halides that cannot be obtained in a pure form or in good yields by other classical methods (*221*).

The halogen acids are the most obvious reagents for the halogenation of organotin hydroxides and oxides:

$$R_3SnX + MOH \longrightarrow R_3SnOH + MX$$
$$R_3SnOH + HX' \longrightarrow R_3SnX' + H_2O$$
$$R_2SnX_2 + 2\ MOH \longrightarrow R_2Sn(OH)_2 + 2\ MX$$
$$R_2Sn(OH)_2 + 2\ HX' \longrightarrow R_2SnX'_2 + 2\ H_2O$$
$$RSnX_3 + 2\ H_2O \longrightarrow RSnOOH + 3\ HX$$
$$RSnOOH + 3\ HX' \longrightarrow RSnX'_3 + 2\ H_2O$$

PCl_5, PCl_3, or $SnBr_4$ are also suitable halogenating agents for compounds containing Sn—O bonds (*14*, *76*). In many cases, the halogenation of an alkanestannonic acid, RSnOOH, is the best method for the preparation of the alkyltin trihalides. Alkanestannonic acids are, in fact, easily obtained by the reaction of potassium stannite with an alkyl or aryl halide (*413*):

$$K_2SnO_2 + RX \longrightarrow RSnOOK + KX$$

Higher alkyltin halides, however, could not be obtained by this reaction; for instance allylstannonic acid reacts with HBr to form the very stable allylpentabromostannic acid: $H_2[C_3H_5SnBr_5]$ or $C_3H_5SnBr_3 \cdot 2\ HBR$ (*232*). On the other hand, it has been reported (*328*) that the product isolated from a reaction mixture of C_2H_5SnOOH with HBr could not be $C_2H_5SnBr_3$ as mentioned by Druce (*130*).

The replacement of one halogen atom by another can, however, also be performed directly by the action of an appropriate metal halide on the organotin halide. Thus, $(CH_3)_2SnCl_2$ could be converted to $(CH_3)_2SnI_2$ with NaI (*276*, *378*, *408*, *469*) and $(n\text{-}Bu)_2SnCl_2$ to $(n\text{-}Bu)_2SnF_2$ with NaF (*9*). The reaction of sodium fluoride in alkaline-aqueous solution with organotin oxides is reported to be an excellent method for the preparation of organotin fluorides (*25*, *261*, *293*, *296*, *298a*, *299*, *300*).

E. Alkylation and Arylation of Tin Halides

Tin(II) chloride, tin(IV) chloride and organotin halides can be alkylated or arylated with such reagents as the Grignard reagent, alkyl- and aryllithiums, organomercury compounds, aluminum alkyls, etc. In most instances, however, a mixture is obtained, from which the individual organotin halides are separated with difficulty. Nearly always considerable amounts of R_4Sn are obtained as well.

The reaction between a Grignard reagent and tin(IV) chloride has been studied for the preparation of organotin halides (*221*), according to the reactions:

$$RMgX + SnX_4 \longrightarrow RSnX_3 + MgX_2$$
$$2\ RMgX + SnX_4 \longrightarrow R_2SnX_2 + 2\ MgX_2$$
$$3\ RMgX + SnX_4 \longrightarrow R_3SnX + 3\ MgX_2$$

It was found that in most instances a mixture of R_4Sn, R_3SnX, R_2SnX_2, and $RSnX_3$ is formed. When excess Grignard reagent is used, this procedure has proved to be a very useful method for the synthesis of tetraalkyltin and tetraaryltin compounds (*357*, *416*, *423*, *523*). In the field of the synthesis of organotin halides, the Grignard method has been of most use for linking large organic groups, such as 1-naphthyl (*299*), 2-diphenyl (*24*), and neophyl (*439*) to tin. Perfluorophenyltin halides were obtained from $SnCl_4$ and a

Grignard reagent (*92, 215, 408*), an organolithium reagent (*501*), or an organomercury reagent (*72, 92*).

A Würtz-type reaction, such as is used usually for the synthesis of R_4Sn compounds between an alkyl halide and Sn(IV) halide with sodium, has been also successfully applied to the synthesis of Bu_2SnCl_2 (*402*). Starting from C_4H_9Na and $SnCl_4$, $(C_4H_9)_2SnCl_2$ was obtained in a 41% yield (*388*). Aronheim (*20*) reported the synthesis of Ph_2SnCl_2 from Ph_2Hg and $SnCl_4$.

Starting in 1960, several authors (*108, 138, 227, 229, 385, 535, 547, 548*) have published reports on the investigation of the reactions between aluminum alkyls and SnX_4 (X = Cl, Br, I)

$$SnX_4 + R_3Al \longrightarrow R_3SnX + AlX_3$$
$$3\ SnX_4 + 2\ R_3Al \longrightarrow 3\ R_2SnX_2 + 2\ AlX_3$$
$$3\ SnX_4 + R_3Al \longrightarrow 3\ RSnX_3 + AlX_3$$

The first two reactions proceed smoothly and give good yields on the condition that the aluminum chloride formed during the reaction is removed from the reaction mixture either by complexation with ether or with an amine. In this way good results were also obtained for alkyltin trihalides when dialkylaluminum alkoxides were used instead of the trialkylaluminum (*549*).

By the reaction of diazomethane with tin(IV) chloride in benzene chloromethyltin trichloride was obtained (*540, 541*):

$$SnCl_4 + CH_2N_2 \longrightarrow ClCH_2SnCl_3 + N_2$$

In place of Sn(IV) halides, alkyltin halides can also be used and the reaction products are then α-halogenoalkylalkyltin halides. When, however, bezenediazonium chloride and tin(IV) chloride are mixed (*278, 381*) a complex compound is obtained with stoichiometry $[C_6H_5N_2Cl]_2 \cdot SnCl_4$. This compound, when treated with copper, zinc or tin in boiling ethylacetate, decomposes and diphenyltin dichloride is formed:

$$[C_6H_5N_2Cl]_2 \cdot SnCl_4 + 2\ M^{II} \longrightarrow (C_6H_5)_2SnCl_2 + N_2 + 2\ M^{II}Cl_2$$

Tin(II) chloride has been shown to reduce organomercury compounds with the formation of organotin compounds and precipitation of mercury. Nesmeyanov and Kocheshkov (*377*) obtained diphenyltin dichloride by treating tin(II) chloride with diphenylmercury in boiling acetone:

$$SnCl_2 + R_2Hg \longrightarrow R_2SnCl_2 + Hg$$

This reaction was successfully performed with R = phenyl, benzyl, *p*-tolyl, α-naphthyl, β-naphthyl (*377*) and *p*-chloro-, *p*-bromo-, and *p*-iodophenyl (*276*). In boiling ethanol the same reaction is found to proceed, except for R = *p*-tolyl, α-naphthyl, or β-naphthyl where the reaction seemed to be

$$SnCl_2 + (\alpha\text{-}C_{10}H_7)_2Hg + 2\ C_2H_5OH \longrightarrow 2\ C_{10}H_8 + Hg + (C_2H_5O)_2SnCl_2$$

Diphenyltin dichloride is obtained also with phenylmercury chloride in boiling acetone:

$$2\,SnCl_2 + 2\,C_6H_5HgCl \longrightarrow (C_6H_5)_2SnCl + 2\,Hg + SnCl_4$$

In boiling ethanol, however, and with R = *p*-tolyl, α-naphthyl, or β-naphthyl there is again formation of the tin alkoxychloride and the aromatic hydrocarbon

$$SnCl_2 + RHgCl + C_2H_5OH \longrightarrow RH + Hg + C_2H_5OSnCl_3$$

Generally, these reactions cannot be applied to aliphatic compounds. The only known exception until now is for the synthesis of (2-chlorovinyl)tin dichloride (*370*).

From diethyllead dichloride and tin(II) chloride, Kocheshkov and Freidlina (*271*) synthesized diethyltin dichloride and obtained a yield of 51 % in boiling dry ethanol. In a similar way $(CH_3)_2SnCl_2$ and $(C_6H_5)_2SnCl_2$ were obtained with yields of 47 % and 13.6 %, respectively. Analogous reactions were reported with R_2TlX (*374, 376*). The reaction of stannous chloride with R_2Zn or RMgX results in the formation of R_2Sn (*410*).

Small yields (21–25 %) of CH_3SnCl_3 (*484*) and of CH_3SnI_3 (*412*) were obtained when $SnCl_2$ and CH_3Cl or CH_3I were heated with $SnCl_2$.

As a modification of this method we consider the reaction between $K(SnCl_3)$ (*316, 508*) and an alkyl iodide or phenyl iodide. When methyl, ethyl, *n*-propyl, *n*-butyl, or phenyl iodide is used, alkyl- or aryl-tin triiodides are formed instead of the alkyltin or phenyltin trichlorides which one would expect; only with isopropyl iodide does the reaction result in the formation of isopropyltin trichloride.

Tin(II) chloride does not form double salts with diazonium compounds. Nesmeyanov and Makarova (*382*), however, found that organotin compounds were formed by the decomposition of aryldiazonium fluoborates with zinc powder in the presence of tin(II) chloride.

III. Chemical Properties

In the synthesis of organotin compounds containing tin-other element bonds, the organotin halides are frequently used as starting materials. Reaction of the organotin halides with either an alkali or a silver derivative of these elements is then applied. In order to avoid overlap with other chapters where these reactions will be treated extensively, we will confine this discussion to the enumeration of the characteristic types of reaction and refer for more detailed comments to the other appropriate chapters.

A. Reductions to Organotin Hydrides

Organotin halides are reduced to the corresponding organotin hydride with lithium aluminum hydride, lithium hydride, and dialkylaluminum hydrides (see the chapter on compounds containing Sn—H bonds).

Examples:

$$4\ R_nSnX_{4-n} + (4\text{-}n)\ LiAlH_4 \longrightarrow 4\ R_nSnH_{4-n} + (4\text{-}n)\ LiX + (4\text{-}n)\ AlX_3$$

and

$$R_nSnCl_{4-n} + (4\text{-}n)\ R_2'AlH \longrightarrow R_nSnH_{4-n} + (4\text{-}n)\ R_2'AlCl$$

(R = organic radical, X = Cl, Br, I, $n = 1,2,3$)

B. Reactions with Alkali and Alkali Earth Metals

Alkali metals react with R_3SnX compounds in inert solvents with the formation of R_3SnSnR_3 and R_3SnM (M = Li, Na, K):

$$2\ R_3SnX + 2M \longrightarrow R_3SnSnR_3 + 2\ MX$$

$$R_3SnSnR_3 + 2\ M \longrightarrow 2\ R_3SnM$$

R_2SnX_2 type compounds yield R_2SnM_2 with sodium and lithium in liquid ammonia (R = butyl, phenyl, methyl):

$$R_2SnX_2 + 4\ M \longrightarrow R_2SnM_2 + 2\ MX$$

Metallization of organotin halides also proceeds with alkali earth metals:

$$(C_6H_5)_3SnCl + 2\ M \longrightarrow (C_6H_5)_3Sn\text{—}M\text{—}Sn(C_6H_5)_3 + MCl$$

(M = Mg, Ca, Sr, Ba)

(see the chapters on Sn—Sn and Sn—M bonds).

C. Reactions with Formation of Sn-Element Group IV Bonds

New tin-carbon bonds can be formed, by the reaction of organotin halides with other organometallic compounds such as RMgX, R_2Zn, RLi and RNa, resulting in the formation of either symmetrical compounds R_4Sn or asymmetrical compounds $R'R_3Sn$ and $R_2'R_2Sn$ (R and R' can be either alkyl or aryl groups). If in these reactions alkali metal derivatives of organotin, -germanium or -silicon compounds are used, compounds are obtained with Sn—Sn, Sn—Ge or Sn—Si bonds:

$$R_3SnCl + R_3'SiM \longrightarrow R_3Sn\text{—}SiR_3' + MCl$$

(M = Li, K)

Organoditin compounds, however, are best synthesized using the reaction of an organotin halide with sodium:

$$2\ R_3SnX + 2\ Na \longrightarrow R_3Sn\text{—}SnR_3 + 2\ NaX$$

With R_2SnX_2, however, diorganotins are obtained (R_2Sn). These compounds are polymeric and are dealt with in the chapter on compounds with Sn—Sn bonds.

Although the five first compounds of Appendix 7 were claimed as ditins by the authors listed in (*230*) and (*231*) it was subsequently shown by several workers (*11a*, *158b*, *231a*) that these were in fact stannoxanes $\diagdown\!\!\!\diagup$—Sn—O—Sn$\diagdown\!\!\!\diagup$—, containing an oxygen between the two tin atoms.

Tetra-*n*-butylditin 1,2-dichloride has been obtained by treatment of tetra-*n*-butylditin 1,2-diacetate with hydrogen chloride in ether (*456a*).

Also Neumann et al. have prepared tetraalkylditin 1,2-dichlorides from dialkyltin chloride hydrides by catalytic decomposition with amines (*387a*, *489a*).

D. Reactions with Formation of Sn-Element Group V Bonds

These bonds result from the reaction between organotin halides with metallated nitrogen compounds such as amides, pyrrole, pyrrazole, hydrazine, and derivatives:

$$R_nSnX_{4-n} + (4\text{-}n)\ MNR'_m \longrightarrow R_nSn(NR'_m)_{4-n} + (4-n)\ MX$$
$$(M = Li,\ Na;\ n = 0,1,2,3;\ m = 1,2)$$

In a similar way compounds with Sn—P, Sn—As, Sn—Sb, and Sn—Bi bonds can be obtained, as shown in the following example for phosphines, R_nPH_{3-n}:

$$R_nSnCl_{4-n} + (4\text{-}n)\ HPR_2 \longrightarrow R_nSn(PR_2)_{4-n} + (4\text{-}n)\ HCl$$
$$(n = 0,1,2,3)$$

E. Reactions with Formation of Sn-Element Group VI Bonds

A great number of compounds containing the Sn—O bond are obtained by simple hydrolysis (1) or alcoholysis reactions of organotin halides eventually followed by condensation reactions:

$$(1)\quad R_3SnX + HOH \longrightarrow R_3SnOSnR_3 + 2\ HX$$
$$R_2SnX_2 + H_2O \longrightarrow (R_2SnO)_x + 2HX$$
$$(2)\quad R_3SnX + R'OH \longrightarrow R_3SnOR' + HX$$

Organotin alkoxyperoxides are obtained from the halide and sodium tertiary butyl peroxide:

$$R_3SnCl + NaOOC(CH_3)_3 \longrightarrow R_3SnOOC(CH_3)_3 + NaCl$$

Stannosiloxanes and stannogermanoxanes are obtained by the reaction of an organotin halide with siloxides or germoxides:

$$(CH_3)_2SnCl_2 + 2\ LiOSi(CH_3)_3 \longrightarrow [(CH_3)_3SiO]_2Sn(CH_3)_2 + 2\ LiCl$$

Several organotin compounds with Sn—S bonds can be prepared by the reaction of the organotin halide with H_2S, alkali sulfides, mercaptans, and even by direct reaction with sulfur:

$$R_2SnX_2 + H_2S \longrightarrow R_2SnS + 2\,HX$$

$$R_3SnX + MSR' \longrightarrow R_3SnSR' + MX$$

$$(C_4H_9)_3SnCl + \tfrac{1}{8}\,S_8 \longrightarrow (C_4H_9)_2SnCl(SC_4H_9)$$

$$2(C_4H_9)_2SnCl(SC_4H_9) \longrightarrow Cl{-}\underset{C_4H_9}{\overset{C_4H_9}{Sn}}{-}S{-}\underset{C_4H_9}{\overset{C_4H_9}{Sn}}{-}Cl + (C_4H_9)_2S$$

The reaction of metallated organosilicon, -germanium, and -lead sulfides $LiSMR_3$ (M = Si, Ge, Pb) with an organotin halide results in formation of organotin sulfides containing the Sn—S—Si, Sn—S—Ge, and Sn—S—Pb bonds, respectively. In a similar way compounds can be prepared where S is replaced by Se and Te:

$$(C_6H_5)_3SnCl + LiMGe(C_6H_5)_3 \longrightarrow (C_6H_5)_3Sn{-}M{-}Ge(C_6H_5)_3 + LiCl \quad (M = Se,\ Te)$$

The organostannyl selenides can also be prepared by using sodium selenides:

$$2\,(C_6H_5)_3SnCl + Na_2Se \longrightarrow (C_6H_5)_3SnSeSn(C_6H_5)_3 + 2\,NaCl$$

F. Reactions with Formation of Sn-Transition Metal Bonds

Alkali metal carbonyls of Co, Fe, Mn, and W react with organotin halides to yield compounds containing the Sn—Co, Sn—Fe, Sn—Mn, and Sn—W bonds, respectively:

$$(CH_3)_2SnCl_2 + NaMn(CO)_5 \longrightarrow (CH_3)_2SnClMn(CO)_5 + NaCl$$

$$(CH_3)_2SnClMn(CO)_5 + NaMo(CO)_3C_5H_5 \longrightarrow (CO)_5Mn\underset{CH_3}{\overset{CH_3}{Sn}}Mo(CO)_3C_5H_5 + NaCl$$

G. Scrambling of Halogens

In a Raman spectroscopic investigation of mixtures of various tin(IV) halides Delwaulle et al. (*119a*) found that a rapid exchange of halogen atoms proceeds in these mixtures, resulting in a complete randomization of halogens:

$$SnX_4 \rightleftharpoons SnX_3Y \rightleftharpoons SnX_2Y_2 \rightleftharpoons SnXY_3 \rightleftharpoons SnY_4$$

This random exchange, or scrambling, of the halogen atoms Cl, Br, and I was confirmed by Burke and Lauterbur by ^{119}Sn nmr spectroscopy of similar systems (*73*).

An analogous rapid exchange of halogen between alkyltin halides was first observed by Alleston and Davies (*9*). Starting from dibutyltin dichloride and dibromide they were able to isolate $(n\text{-Bu})_2SnClBr$. Van den Berghe et al. (*38*) made a pmr investigation of the halogen exchange reactions in $(CH_3)_3SnX$, $(CH_3)_2SnX_2$, and CH_3SnX compounds (X = Cl, Br, I) dissolved in either chloroform or carbon tetrachloride. The existence of analogous equilibria to those found for the Sn(IV) halides could be shown. A pentavalent intermediate was assumed

$$(CH_3)_3SnX \rightleftharpoons (CH_3)_3SnX,Y \rightleftharpoons (CH_3)_3SnY$$

$$(CH_3)_2SnX_2 \rightleftharpoons (CH_3)_2SnXY \rightleftharpoons (CH_3)_2SnY_2$$

$$CH_3SnX_3 \rightleftharpoons CH_3SnX_2Y \rightleftharpoons CH_3SnXY_2 \rightleftharpoons CH_3SnY_3$$

in the trimethyltin halide systems. The rate of exchange depends on the nature and the number of halogen atoms, the temperature, and the nature of the solvent.

H. Reactions with Formation of Various Organotin Salts

Several salts, particularly the silver salts, or inorganic oxo-acids, react with organotin halides to yield the corresponding organotin salts.

$$R_{4-n}SnX_n + n\ AgY \longrightarrow R_{4-n}SnY_n + n\ AgX$$

In this way organotin isocyanates, isothiocyanates, and selenocyanates could be prepared. Silver hexafluorophosphate, tetrafluoroborate, hexafluoroarsenate, and hexafluoroantimonate react with $(CH_3)_3SnBr$ in liquid sulfur dioxide to form the corresponding trimethyltin compounds

$$(CH_3)_3SnBr + AgX \longrightarrow (CH_3)_3SnX + AgBr$$

$(X = PF_6, BF_4, AsF_6, SbF_6)$.

By the same method also perchlorates and nitrates could be prepared. The reaction of some organotin halides with sodium azide or NH_3 in aqueous solution results in the formation of organotin azides.

IV. Physical Properties and Chemical Bonding

A. Spectroscopy

1. *Infrared and Raman*

Infrared spectroscopy and also to a lesser extent Raman spectroscopy have greatly contributed to a better understanding of the structure of organotin halides and of the chemical bonding.

The ir and Raman spectra of $(CH_3)_nSnX_{4-n}$ (*74, 95, 97, 101, 137, 302–304, 320, 362, 398, 502, 534, 542*) (X = F, Cl, Br, I and $n = 1, 2, 3$) have been extensively studied and the fundamental frequencies were assigned. Characteristic absorption regions for the Sn—C, Sn—Cl, Sn—Br, and Sn—I bonds of several representative alkyltin halides are collected in Table 1 (*74*). For the Sn—F bond, stretching vibration absorption bands are observed at $(CH_3)_3SnF$ (355 cm^{-1}), $(CH_3)_2SnF_2$ (373 cm^{-1}), $(n\text{-Bu})_3SnF$ (330 cm^{-1}), $(n\text{-Pr})_2SnF_2$ (330 cm^{-1}), and Et_2SnF_2 (555 cm^{-1}). In Table 2 the assignment to the fundamental vibration modes of the observed frequencies, as proposed by Edgell and Ward (*137*), is shown for the series $(CH_3)_nSnCl_{4-n}$. These authors suggest that the two Sn—Cl stretching vibrations in $(CH_3)_2SnCl_2$ and in CH_3SnCl_3 coincide through accidental degeneration. This argument is questioned by Taimsalu and Wood (*502*) who state that this apparent accidental coincidence in the series of methyltin chlorides should mainly be due to strong interactions between the molecules resulting in a considerable band broadening. Strong interaction was also postulated by Kriegsmann and Pischtschan (*304*) for $(CH_3)_3SnCl$. They found that the Sn—Cl symmetrical stretching vibration frequency is very sensitive to the aggregational state (see Table 3). Moreover the trend observed in the melting points of the trimethyltin compounds (see Table 4) also suggests partial association with formation of Sn—Cl—Sn bridges.

The force constants for the Sn—Cl and Sn—C stretching vibrations calculated by Taimsalu and Wood (*502*) from the experimental values of ν_sSn—Cl, and ν_sSn—C using a general valence force field (Table 5) show a regular decrease in the series CH_3SnCl_3, $(CH_3)_2SnCl_2$, $(CH_3)_3SnCl$. Although the significance of these data could be reduced, due to the fact that the Sn—Cl vibrations are obviously sensitive to environmental changes (see remarks above), the sequence observed is clearly that of increasing electron release

TABLE 1

VIBRATIONAL FREQUENCIES OF THE Sn—X AND Sn—C BONDS IN DIALKYLTIN HALIDES

R_2SnX_2 compounds (R = Me, Et, n-Pr, n-But and n-Octyl and X = Cl, Br or I)				
ν_{as} Sn—Cl = 355 ± 5 cm^{-1}			ν_s Sn—Cl = 347 ± 9 cm^{-1}	
ν_{as} Sn—Br = 253 ± 1 cm^{-1}			ν_s Sn—Br = 240 ± 2 cm^{-1}	
ν_{as} Sn—I = 200 ± 2 cm^{-1}			ν_s Sn—I = 182 ± 1 cm^{-1}	
	Methyl	n-Propyl	n-Butyl	n-Octyl
ν_{as} Sn—C	558 ± 3 cm^{-1}	594 ± 5 cm^{-1}	597 ± 5 cm^{-1}	606 cm^{-1}
ν_s Sn—C	523 ± 1 cm^{-1}	509 ± 8 cm^{-1}	508 ± 3 cm^{-1}	518 cm^{-1}

TABLE 2

THE FUNDAMENTAL VIBRATIONAL FREQUENCIES OF THE METHYLTIN CHLORIDES

Vibration (Freq. in cm^{-1})	Symmetry class	Me_3SnCl (C_{3v}) ir	Me_3SnCl (C_{3v}) Raman	Me_2SnCl_2 (C_{2v}) ir	Me_2SnCl_2 (C_{2v}) Raman	$MeSnCl_3$ (C_{3v}) ir	$MeSnCl_3$ (C_{3v}) Raman
$SnMe_3$ deform.	a_1						
$SnMe_3$ deform.	e	}145	150				
MeSnCl bending	e						
Sn—Cl deform.	a_1					(133)	142
Sn—Cl_3 deform.	e					(123)	112
MeSn—Cl bending	e					152	142
Cl—Sn—Cl bending	a_1			124			
Me—Sn—Cl bending	a_2			inactive			
Me—Sn—Cl bending	b_1			146	}135		
Me—Sn—Cl bending	b_2			129			
Me—Sn—Cl bending	a_1			158			
Sn—Cl stretching	a_1	325	318	307	}344	366	363
Sn—Cl stretching	b_1			332			
Sn—Cl stretching	e					384	363
Sn—Me stretching	a_1	(514)	518	(515)	531	542	550
Sn—Me stretching	e	(545)	548				
Sn—Me stretching	b_2			(567)	566		

Note: braces at numbers refer to frequencies with uncertain assignment.

by the methyl groups, causing an increased tendency toward ionization of the Sn—Cl bond. Indeed, all the $(CH_3)_3SnX$ compounds are known to dissolve in water with the formation of hydrated $(CH_3)_3Sn^+$ ions and X^- ions.

A study of the normal vibrations and the thermodynamic properties of methylstannane and methyltin trichloride was made by Galasso et al. (*156*).

Kriegsmann (*304*) reported that no band was observed for the Sn—F vibration in the ir or Raman spectra of the $(CH_3)_3SnF$ molecule and concluded that the structure for this compound is ionic. Recent ir and Raman data (see above) and an x-ray diffraction pattern (*101*, *542*) suggested a structure wherein $(CH_3)_3Sn$ groups are linked together by bridging F atoms, with the formation of a nonlinear Sn—F—Sn chain. The weak band in the ir spectrum at about 515 cm^{-1} is accepted as evidence that the $(CH_3)_3Sn$ group is not planar (*101*). A similar structure with Sn—F—Sn bridges also seems to be present in $(CH_3)_2SnF_2$ crystals (*95*, *534*) (see Sec. IV.B.1). From the ir spectra of autocomplexes $(CH_3)_3SnMX_n$ (M = B, As, Sb, Al and X = F, Cl, Br), it is also seen that these complex molecules form a chain structure with planar trimethyltin groups and bridging MX_n groups. This matter is discussed more extensively in the paragraph on autocomplexation (see Sec. V.D.3).

TABLE 3

COMPARISON OF SPECTRAL PARAMETERS AND BOND ORDER FOR ANALOGOUS TRIORGANOSILICON AND TRIORGANOTIN CHLORIDES

Compound	ν_s(Sn—Cl) cm^{-1} solid	liquid	solution in CS_2	f(mdyn/Å) single bond	f(mdyn/Å) calculated from ν_s	N (bond order)
$(CH_3)_3SnCl$	288	315	331	1.82	1.8	1
$(C_6H_5)_3SnCl$	329	—	328	1.82	1.8	1
$(CH_3)_3SiCl$		470		2.35	2.50	1
$(C_6H_5)_3SiCl$		550		2.35	3.7	1.5

Several studies were also made of the vibrational spectra of $(C_2H_5)_nSnX_{4-n}$ (*74*, *128*, *322*, *453*, *502*, *524*), longer chain alkyltin halides (*55*, *74*, *112*, *113*, *150*, *157*, *347*, *358*, *502*) and benzyltin halides (*90*, *189*). A *trans-gauche* isomerism seems to exist with respect to the tin atom in the alkyl chains with more than three carbon atoms (*112*, *113*, *157*, *502*). In the crystals, the *trans* configuration always exists in equilibrium with the *gauche*. For the $(n\text{-Bu})_nSnCl_{4-n}$ series, Geissler and Kriegsmann (*157*) assign the bands at about 500 and 600 cm^{-1} to the Sn—C vibrations of rotational isomers. The butyl groups are mostly in the extended conformations (*150*). In the case of the benzyltin derivatives, the variation of the frequency of the CH_2 deformation is shown to be dependent on the nature of the substituent in both the benzyl halides and the benzyltin compounds (*90*).

The phenyltin halides (*55*, *135*, *187*, *301*, *314*, *418*, *419*, *485*, *493*) have been the subject of special attention. The ν_sSn—Cl value (Table 3) for $(C_6H_5)_3SnCl$ is not influenced by the aggregational state (*301*). In the halide series $(C_6H_5)_3SnX$ (X = Cl, Br, I) the chloride is not found to have a higher melting

TABLE 4

MELTING POINTS OF TRIMETHYLTIN COMPOUNDS

Compound	mp, °C
$(CH_3)_3SnH$	−70
$(CH_3)_3SnOH$	118 (subl.)
$(CH_3)_3SnF$	375 (dec.)
$(CH_3)_3SnCl$	39.5
$(CH_3)_3SnBr$	27.5

TABLE 5

FORCE CONSTANTS FOR Sn—C AND Sn—Cl SYMMETRICAL STRETCHING VIBRATIONS IN THE METHYLTIN CHLORIDES

Compound	f_{Sn-Cl}, dyn/cm	f_{Sn-CH_3}, dyn/cm
CH_3SnCl_3	2.8×10^5	2.35×10^5
$(CH_3)_2SnCl_2$	1.6×10^5	2.25×10^5
$(CH_3)_3SnCl$	1.5×10^5	2.25×10^5

point than the other substances. Both observations are accepted as evidence against association of the triphenyltin chloride in contrast to the trimethyltin chloride. Because of the inductive effect of the phenyl group, an increase of the Sn—Cl stretching vibration frequency could be expected in the series $(C_6H_5)_3SnCl$ to $C_6H_5SnCl_3$. For all these three chlorides this frequency is, however, found to be practically unchanged. Apparently the inductive influence of the phenyl group on the Sn—Cl bond therefore seems negligible. In Table 3 the force constant data, calculated by Kriegsmann (*301*), using a simplified valence force field, developed by Siebert (*475*), are collected for some parallel organotin and organosilicon halides. By comparison of these values with theoretical values for single bonding, the actual bond order was derived. The high values for the silicon compounds confirm the theory advanced by Ingold (*222*) according to which a maximum orbital overlap and therefore the strongest bond is obtained when the two atoms joining in the bond have the same orbital symmetry and comparable energies, i.e., when they have the same quantum number. This is the case for Si and Cl, but not for Sn and Cl. The high bond order of 1.5 observed for the Si—Cl bond in $(C_6H_5)_3SiCl$ is thought to be due to conjugation effects in the phenyl compounds.

On the assumption (*187*) that the relationship between the frequency of some C—H modes for monosubstituted benzenes and the square of the electronegativity of the substituent atom, also holds for the compounds Ph_nSnCl_{4-n}, one can conclude that the electronegativity of tin increases in the order $Ph_4Sn < Ph_3SnCl < Ph_2SnCl_2 < PhSnCl_3$. This is explained by the simultaneous action of the inductive effect and the mesomeric effect of the phenyl group. Characteristic Sn—Cl and Sn—Ph frequencies are listed in Table 6.

For Ph_3SnX, the Sn—Br frequencies observed in ir and Raman spectra are found at 256 and 231 cm^{-1}, respectively, for Sn—I at 170 cm^{-1} and for Sn—F at 350 cm^{-1}. For $(C_6F_5)_2SnF_2$, the Sn—F absorption band is found at 328 cm^{-1}. The ionic structure proposed for Ph_3SnF (*301*), on the supposed absence of a Sn—F band, therefore, is to be discarded.

TABLE 6

SELECTED VIBRATIONAL FREQUENCY ASSIGNMENTS FOR THE PHENYLTIN CHLORIDES

Vibration (Freq. in cm^{-1})	$(C_6H_5)_3SnCl$	$(C_6H_5)_2SnCl_2$	$(C_6H_5)SnCl_3$
ν(Sn—Cl)	442		
ν_s(Sn—Cl)		356, 350	363
ν_{as}(Sn—Cl)		364	383, 439
ν(Sn—Ph)			250, 248
ν_s(Sn—Ph)	239	230, 226	
ν_{as}(Sn—Ph)	270	279, 276, 274	

The comparison of the characteristic Sn—H stretching vibration frequencies for the alkyltin halide hydrides (*241*, *454*, *455*) with those of the corresponding alkyltin hydrides furnishes information concerning the influence of halogen substituents on the spectroscopic parameters. In each series of R_2SnHX compounds (regardless of the nature of R and with X = H, I, Br, Cl, F) the ν_{Sn-H} is found to increase with increasing electronegativity of the X-substituent, as for instance $(n\text{-}C_4H_9)_2SnHX$:

$$(X = H)1835\ cm^{-1}, \quad (X = I)1838\ cm^{-1}, \quad (X = Br)1847\ cm^{-1},$$
$$(X = Cl)1852\ cm^{-1}, \quad (X = F)1875\ cm^{-1}.$$

In several studies on organometallic compounds of group IV elements, ir data are quoted on various alkyl- (*139*, *140*), alkenyl- (*140*), aryl- (*140*, *200*, *333*), and vinyltin (*317*) compounds. The inductive effect of the R_3M groups was defined and found to increase, in going from C to Sn, proportionally to the ionization potential of M (*140*). Infrared spectra of $[Cl(n\text{-}C_3H_7)_2Sn]_2O$ and of $R_2SnCl(OH)$ compounds were also studied (*543*).

A new series of compounds was recently synthesized by Patmore and Graham (*403*), the mono(tetracarbonylcobalt) tin(IV) derivatives with general formula $R_nX_{3-n}SnCo(CO)_4$ ($R = C_6H_5$; X = Cl, Br, I; $n = 0, 1, 2$) and their infrared spectra were analyzed.

Infrared spectroscopy has been of great value in studying complex formation between organotin halides and Lewis bases, and this matter is treated extensively in the paragraph on coordination compounds (see Sec. V.A through D).

The interaction of chloride and bromide ions with the dimethyltin(IV) ions was investigated by Raman and ir spectroscopy (*145*, *304*, *356*). The integrated Raman intensities of the Sn—Cl stretching vibrations as a function of chloride ion concentration were interpreted in terms of inner and outer sphere complex formation (*145*). A parallel study for CH_3SnCl_3 and CH_3SnBr_3, combined with measurements of pH, revealed that the former yield hexavalent tin complexes more easily than the latter (*303*). Solvent effects on the lattice vibrations of alkyltin chlorides were also studied (*503*). The factors influencing the Sn—Cl stretching vibration: dipole interaction with the surroundings, change of coordination number around tin, and formation of chlorine bridges, are found to cause a decrease of this frequency. This frequency shift is in the order: nujol solution > benzene solution > liquid (pure) > solid.

A very interesting correlation (*56*) was shown to exist between the NMR coupling constants J_{Sn-C-H} and the symmetric and the antisymmetric Sn—C stretching vibration frequencies, suggesting that the Sn—C stretching frequencies should be also linearly related to the percentage of *s*-character in the tin orbital used in the bond. Similarly, Brown and Puckett (*57*) have shown that for some selected tin compounds there is a linear relationship between ν_{C-H} and $J_{^{13}C-H}$. Using a modified valence-force-field potential function for the CH_3 group they found that the C—H stretching force constant K_r could be expressed in terms of J by

$$K_r = 4.74 + 8.8 \times 10^{-3}\,(J - 125) \quad \text{mdyn/Å}$$

2. *Ultraviolet*

Griffiths and Derwish (*186*) studied the uv absorption spectra of the series of phenyltin chlorides Ph_nSnCl_{4-n}. The absorption bands at ±260 mμ and 210 mμ were assigned to the *B* and *K* bands of the phenyl group respectively. The band at about 220 mμ could be ascribed either to the electron transition in the Sn—Cl bond or to the absorption caused by photodecomposition products. From a discussion of these data, there resulted the opinion that in phenyltin chlorides, both inductive and resonance effects play a role in the bonding. With an increasing number of chlorine atoms the bond electronegativity of tin increases. The interaction between the π-orbitals of the aromatic ring and the empty *d*-orbitals of tin therefore becomes easier and so resonance structures such as:

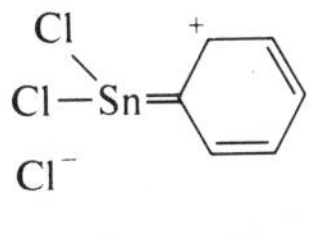

Cl
Cl—Sn= (+)
Cl⁻

may yield a contribution to the actual structural picture of these compounds. Marrot et al. (*344*) and Lapkin et al. (*314*) reached similar conclusions in a comparative study of the uv spectra of phenyltin- and phenylgermanium chlorides.

In the uv spectra of some ethyltin compounds (*442*), a continuous absorption was observed that was ascribed to dissociation. The energy at the boundary of the continuous phase minus the energy of activation of the radicals which are formed, is related to the energy necessary to split the first Sn—C bond. The uv spectrum of tribenzyltin chloride (*188*) in cyclohexane and chloroform solution yields indications that photodecomposition occurs, the limit being probably at 2100 Å in cyclohexane solution.

For CH_3SnI_3 (*12*) and several vinyltin (*317*) derivatives of the group IV elements the uv spectra were recorded with the aim of studying the relative importance of p_π-d_π contribution to the bonding.

Ultraviolet spectrophotometry proved to be a very powerful method in the investigation of organotin halide complexes. These results are treated in detail in the chapter on organotin halide complexes (see Sec. V.A through D).

3. *Nuclear Magnetic Resonance*

The technique of nmr spectroscopy has proved to be of great importance in the study of bonding in organometallic compounds, of the structure, of reaction kinetics, and of the formation of addition compounds.

In the pmr studies on organotin compounds, great interest was paid to tin-proton coupling constants ($^{117/119}$Sn—H): both the direct coupling and the coupling over two and three bonds; the ^{13}C—H, ^{1}H—^{1}H, and $^{117/119}$Sn—^{13}C coupling constants were also used to yield additional information.

The first compounds investigated in this field were the methyltin (*37, 96, 144, 213, 249, 460, 520*) halides (Table 7). The $^{117/119}$Sn—C—H coupling constants were mainly interpreted in terms of isovalent rehybridization of the tin atom on progressive halogenation. In fact the ^{13}C—H couplings are

TABLE 7

PMR Data for the Methyltin Halides

Compound	$J_{^{119}Sn—C—H}$ (in cps)				Δ (TMS) (in cps)			
		X				X		
		Cl	Br	I		Cl	Br	I
$(CH_3)_4Sn$	54.3				4.2			
$(CH_3)_3SnX$		58.5	58.5	58.5		37	45	55
$(CH_3)_2SnX_2$		70	66	62.5		70	83	97
CH_3SnX_3		98	89	73		95	110	140

rather insensitive to halogenation. A correlation was established between Sn—H coupling over two bonds and either the ir symmetric and antisymmetric Sn—C vibration modes (*56, 159*) and between the ^{13}C—H coupling and the symmetric and antisymmetric C—H vibrations (*57*). The chemical shift data were correlated to the inductive deshielding effect on progressive halogenation (*37, 59, 93, 249*) and to the neighbor anisotropy effect of the tin-halogen bond (*37*).

In the pmr spectra of the ethyltin compounds, the coupling over two bonds J_{Sn-H_α} as well as over three bonds J_{Sn-H_β} (*158, 323, 325, 477, 515, 517*) are observed. A study of these spectra (*515, 517*) showed that both J_{Sn-H_α} and J_{Sn-H_β} are essentially dependent on the Fermi-contact interaction term, but the apparent anomaly $J_{Sn-H_\alpha} < J_{Sn-H_\beta}$ could only be explained by the assumption of a supplementary contribution of the electron-orbital term. This has been confirmed by the pmr data of the ethyltin chloride pyridine complexes (*40*). The chemical shifts were again shown (*325, 515, 517*) to be determined by the simultaneous influence of the electronic inductive effect and the neighbor anisotropy effect.

In the pmr spectra of the phenyltin chlorides (*332, 334, 514*) the relative chemical shifts for *o*-, *m*- and *p*-protons furnish evidence for the assumption of considerable p_π-d_π interaction between the aromatic ring and the tin atom. The $J_{Sn-H(ortho)}$ values (*514*) are in agreement with this hypothesis. In an analogous way, the J_{Sn-CH_2} values observed on the benzyltin halides (*516, 518*) can be explained by a slight delocalization of the π-electronic system of the phenyl group toward the tin atom. The CH_2-chemical shifts in these compounds are, to a large extent, determined by the ring current effect of the phenyl group.

A comparative study of a series of isopropyl-, *n*-propyl-, *n*-butyl-, and β-functionally substituted ethyltin compounds (*519*) yielded strong confirmation to the theory that the Fermi-contact term is not the sole mechanism contributing to J_{Sn-H}, but that a supplementary contribution of the electron orbital term is very probable. On the other hand the data (*259, 518*) from the pmr spectra of the methyl-, ethyl,- and the halogenomethyltin halides allowed an empirical equation (*518*) to be drawn for the J_{Sn-H} values:

$$J_{Sn-H_\alpha} \simeq \Phi^2_{Sn}(0)\Phi^2_{H}(0)\,\frac{1}{\lambda(\text{Sn—C})}$$

showing that the Sn—H_α coupling constants are mainly determined by the Fermi-contact term.

Moreover, some data are mentioned in the nmr spectra of vinyltin halides (*60*), trineophyltin halides (*439*), dialkylhalogenotin hydrides (*241, 454–456*), and a $R^1R^2R^3$SnCl compound (*405*).

The ^{119}Sn resonance spectra (*73, 220*) of several alkyltin halides were determined. The dependence of the chemical shift on the electronegativity of X in the series (alkyl)$_3$SnX is noted. Some shielding due to $(p \rightarrow d)_\pi$ bonding in the series $(n\text{-Bu})_n\text{SnCl}_{4-n}$ is suggested. The tin chemical shift has been measured as a function of concentration and solvent for some simple methyltin bromides and halides (*220*).

Valuable information was gained by the pmr study of addition compounds of organotin halides with Lewis bases TMSO, DMA, DMF (*355*), DMSO (*263*), pyridine (*39, 40*), and of acetylacetonate (*242–245, 247, 248*), and 8-hydroxyquinoline (*262, 533*) complexes. The formation and structure of these complexes will be discussed in another paragraph (see Sec. V. A through D). Nuclear magnetic resonance studies of methyltin chlorides in H_2O (*35, 145*) and in HCl (*34, 145, 356*) solution and a study of the solvent effect on $J_{\text{Sn–C–H}}$ in $(CH_3)_nSnX_{4-n}$ (*46, 58, 160, 161, 220, 349*) were also undertaken with the aim of providing evidence for complex formation. These investigations showed that the solvent effect was essentially due to weak or strong complex formation, accompanied by a rehybridization around tin with a resultant increase in the percentage of *s*-character in the tin-carbon bonds, as revealed by a parallel increase in the tin-proton coupling constants (Table 8).

The nmr technique was also used to study rapid exchange reactions in equilibrium systems, such as those occurring in mixtures of $Sn(CH_3)_4$ and $SnCl_4$ (*38, 184*) and mixtures of methyltin halides (*36*). Reaction mechanisms and relative exchange rates could thereby be determined. Also mixtures of dimethyltin dihalide with dimethyltin disulfide were studied by nmr (*360*).

4. *Nuclear Quadrupole Resonance*

The nuclear quadrupole resonance (nqr) frequencies of the ^{35}Cl atoms in some organotin halides were measured (*185, 493, 498*). From the resonance frequencies the quadrupole coupling constants $|e^2Qq_{zz}|$ listed in Table 9 were calculated.

TABLE 8

SOLVENT EFFECT ON THE TIN-PROTON COUPLING CONSTANT IN $(CH_3)_3SnBr$

Solvent	$J_{^{119}\text{Sn–C–H}}$	$J_{\text{solv.}}$-J_{CCl_4}
CCl_4	57.8	0.0
Dioxane	62.4	4.6
Acetone	64.3	6.5
Methanol	67.3	9.5
Pyridine	68	10.3
H_2O	69.3	11.5
DMSO	69.8	12.0
DMF	69.8	12.0

TABLE 9

^{35}Cl Nuclear Quadrupole Resonance Data for Some Organotin Chlorides

Compound	$\lvert e^2Qq_{zz}\rvert$, Mc/sec
$SnCl_4$	48.200
$(n\text{-}C_4H_9)SnCl_3$	43.124
$(n\text{-}C_4H_9)_2SnCl_2$	34.546
$(C_6H_5)SnCl_3$	42.332
$(C_6H_5)_2SnCl_2$	33.894
$(C_6H_5)_3SnCl$	27

The pronounced decrease of the quadrupole coupling constants with progressive substitution of Cl by organic radicals and the fact that no constant ratio can be observed between the coupling constants of analogous carbon and tin compounds have led to the assumption that in the Sn—Cl bond, both the double bond character and the ionic character of the bond are changed on progressive substitution (*185*, *498*). Srivastava (*493*) also explains the continuous decrease of $\lvert e^2Qq_{zz}\rvert$ in going from $SnCl_4$ to $(C_6H_5)_3SnCl$, by the concomittant increase of the ionic and double bond character in the tin-chlorine bond. He believes, however, that the change in the ionic character is the dominant factor.

Several attempts (*62*, *461*) were made to correlate nqr frequencies with σ_I (induction constant) and σ_C (conjugation or coupling constant), the Taft-Hammett parameters, for the series of tetrahedral molecules of group IV. Some nqr and Mössbauer spectral data on organotin bromides and iodides were also used to derive a correlation of the nqr frequency ν with the tin quadrupole splitting Δ in the Mössbauer spectra (*62*).

In a study (*185*) of the temperature dependence of the nqr frequencies of ^{35}Cl in R_2SnCl_2 (R = CH_3, *n*-Bu) a positive temperature coefficient was found. This is explained by the existence of intermolecular bonds.

5. *Mössbauer*

Several studies of the Mössbauer resonance absorption of 23.8-keV radiation of a $^{119m}SnO_2$ or gray Sn emitter by organotin compounds appeared first in the Russian literature. Spectra (*5*, *6*, *63*) were investigated for Bu_2SnX_2 (X = F, Cl, Br, I), Ph_nSnCl_{4-n}, R_2SnCl_2 (R = Et, Pr, Bu), Bu_3SnX (X = Cl, Br), $BuSnCl_3$, and $(CNCH_2CH_2)_3SnX$ (X = F, Br, I). The isomeric shifts (I.S.) relative to SnO_2 and the quadrupole interaction constants were determined. The latter constants, measured by the extent of the quadrupole splitting (Q.S.) of the absorption lines, are due to the interaction of the nuclear quadrupole moment with the electric field gradient in the immediate vicinity of the absorbing tin nucleus. As the valence *p*-electrons yield the main contribution of these gradients, quadrupole splitting is only observed in covalent compounds with an unequal population of the three *p*-tin orbitals, in agreement with the Townes and Dailey approximation theory. Other reports (*109, 119, 151, 205, 207–209, 363, 367, 439, 495, 496, 554*) on Mössbauer spectra of organotin compounds deal with Ph_3SnCl, $(4\text{-}ClC_6H_4)_3$ SnCl, R_3SnX (R = $C_6H_5C(CH_3)_2CH_2$; X = F, Cl, Br, I), R_nSnX_{4-n} (R = C_6H_5, C_6F_5, CH_3; X = F, Cl, Br, I), $(n\text{-}Bu)_2SnHCl$ and some penta- and hexacoordinate organotin (IV) complexes.

Two reviews have been published covering the literature through August 1963 (*178*) and through 1965 (*153*), respectively.

The isomeric shift is calculated to be proportional to

$$\frac{\delta R}{R}\,[\Phi_A^2(0) - \Phi_S^2(0)]$$

where $\delta R/R$ is the fractional change of the charge radii of the ^{119}Sn nucleus in the excited and in the ground state, respectively, and $\Phi_A^2(0)$ and $\Phi_S^2(0)$ are the probabilities of the electron densities at the absorbing and the source nuclei respectively. The experimental data, however, rather indicated that in inorganic tin compounds the I.S. is almost completely due to the 5*s* electron density around the tin atom and that for tetracovalent compounds, the I.S. is proportional to the degree of ionic character of the Sn—X bonds (*178, 180, 204*). This proportionality was then extended to organotin compounds. Substitution of an organic group with electronegativity x_R by an inorganic substituent X with electronegativity x_X will remove $0.25I$ electrons from tin, wherein $I = (x_X - x_{Sn}) - (x_R - x_X)$. However, secondary changes in the shielding by the inner electrons can compensate for this, so that the I.S. becomes less sensitive to inorganic substitution (*109, 205*), (Table 10). Nevertheless it was estimated, on the basis of the observed spread of I.S. values, that the ionic character of a tin-carbon bond should be about $15\% \pm 5\%$. In aryltin compounds neither *meta*- nor *para*-substitution seems to change the percentage of *s*-character of the Sn—C bond (*209*).

Due to some recent studies, however, in which an attempt was made to correlate some Mössbauer spectral parameters in a series of Sn(II) (*554*) and Sn(IV) (*125, 554*) compounds, it is believed that no unequivocal nor general relationship exists between I.S. values and the ionic character of tin bonds (Table 11). The Q.S. constants are theoretically proportional to the imbalance of the *p*-valence electron density equal to $0.75I$. Obviously the splitting should increase with the increasing value of x_X and the decreasing value of x_R (*205*). The experimental data, however, for the $(CH_3)_3SnX$ compounds (X = H, Na, I, Br, F) do not agree completely with this simple model (*109*). Appar-

TABLE 10

MÖSSBAUER SPECTRAL PARAMETERS FOR SOME TRIORGANOTIN HALIDES

Compound	I.S., mm/sec (Ref. gray tin)	Q.S., mm/sec	x_X	
			Pauling	Allred-Rochow[a]
$(CH_3)_3SnI$	−0.63	3.19	2.65	2.21
$(CH_3)_3SnBr$	−0.65	3.40	2.95	2.74
$(CH_3)_3SnCl$	−0.70	3.55	3.15	2.83
$(CH_3)_3SnF$	−0.85	4.03	3.90	4.10
$(C_6F_5)_3SnCl$	−0.99	1.55		
$(C_6H_5)_3SnCl$	−0.75	2.55		
$(CH_3)_3SnCl$	−0.70	3.50		

[a] Data as mentioned by F. A. Cotton and G. Wilkinson, *Advanced Inorganic Chemistry* (2nd ed.), Wiley, New York, 1966.

ently, apart from the partial ionic character, other effects determine the Q.S. values, for instance, small changes in the hybridizational state of tin and back donation of electrons into empty tin orbitals. In compounds of the type R_nSnX_{4-n} the largest quadrupole splitting and isomer shift is for $n = 2$. This behavior could be interpreted as being due to polarization effects via the p_π-p_π and p_π-d_π bonding and the influence of microcrystalline structural effects (*495*).

The ratio Q.S./I.S. has been thought to be a measure of the hybridizational changes around tin; a value of 2.2 or higher is believed to indicate five coordination (*209*). In trineophyltin fluoride a value of 2.09 is found. This is accepted as evidence that the heavy, bulky neophyl group diminishes the tendency toward polymerization through Sn—F—Sn bridges (*439*). This

TABLE 11

COMPARISON OF I.S. DATA FOR ORGANOTIN COMPOUNDS AS A FUNCTION OF THE NATURE AND NUMBER OF THE SUBSTITUENTS

Compound	I.S., mm/sec	Compound	I.S., mm/sec
Neo_3SnNeo[a]	1.34	$(n\text{-}C_4H_9)_4Sn$	1.3
Neo_3SnI	1.40	$(n\text{-}C_4H_9)_3SnCl$	1.65
Neo_3SnBr	1.42	$(n\text{-}C_4H_9)_2SnCl_2$	1.5–1.6
Neo_3SnCl	1.39	$(n\text{-}C_4H_9)SnCl_3$	1.7
Neo_3SnF	1.33	$SnCl_4$	0.70–0.78

[a] Neo = (2-methyl-2-phenyl-1-propyl).

criterion is, however, open to serious criticism (*367*) since it is found that for $R_3SnX \cdot B$ (B = aza-aromatic base) complexes, the spectroscopic parameters (I.S., Q.S. or their ratio) bear no relation whatsoever to the structural or electronic properties of the aromatic ligands bound to the tin atom. Nevertheless, formation of five- and six-coordinated organotin(IV) ions shift, in every case, the I.S. to lower values (*119*). A study (*151*) of the Mössbauer spectra of hexacoordinated complexes of tin(IV) containing two tin-carbon bonds revealed that the quadrupole splitting, ΔE, for a *trans*-arrangement was twice the ΔE value for a *cis*-arrangement.

Solvents with strong solvating power cause a more or less pronounced distortion of the field symmetry of the tin nucleus in organotin compounds. The change of the electric field gradient can be estimated directly by means of the Q.S. It is therefore possible to evaluate quantitatively the solvating powers toward dibutyltin dichloride by comparing Q.S. values for the different solvents after the saturation state is reached. If the solvating power (*Ps*) of diethyl ether is taken as unity, the following *Ps* values are found: THF (=7), DME (=10), HMTAP (=15.5), DMFA (=16), and DMSO (=18) (*179*). Bu_2SnBr_2 and Ph_2SnCl_2 were also investigated in various polar and nonpolar solvents. The Q.S. of these products in nonpolar solvents remains unchanged, whereas in polar solvents it increases. The solvent effect on the isomeric shifts is small in all cases (*7*).

A study of the "recoil-free" fraction of ^{119}Sn nuclei in organotin compounds seems to indicate that this fraction increases in compounds where a polymeric lattice structure is present, in which the tin atoms are directly involved (*496*).

B. Structural Analysis

1. *X ray*

The crystal structure of several organotin halides has been investigated by x-ray diffraction. Dimethyl-, diethyl-, and dipropyltin dichlorides form orthorhombic crystals (*20, 75, 211*). The crystals of trimethyltin fluoride (*101, 102, 542*) are also orthorhombic with four molecules in the unit cell (dimensions: $a = 4.32$ Å, $b = 10.85$ Å, $c = 12.84$ Å). These molecules are linked together to form a chain structure composed of planar trimethyltin groups joined by fluorine atoms. There results a nonlinear and nonsymmetrical Sn—F—Sn bond. This distorted trigonal bipyramidal configuration is incompatible with a structure held together by pure ionic interaction forces, with free $Sn(CH_3)_3^+$ ions. Instead, this is a strong indication for the covalent character of the Sn—F bond. Therefore it is believed that $(CH_3)_3SnF$ contains five coordinated tin atoms. This hybridizational state of the type sp^3d is reached by using the *d*-orbitals of the metal.

Similar chain structures with five-coordinated tin atoms have also been postulated for $(CH_3)_3SnClO_4$ (*98, 395*), $(CH_3)_3SnBF_4$ (*201*), and $(CH_3)_3SnMF_6$ (*99*) (M = As, Sb). Trimethyltin compounds thus seldom seem to have free $(CH_3)_3Sn^+$ ions. The most relevant exception to this statement is found in $(CH_3)_3SnCN$ (*458*), whose structure is best described as an arrangement of $(CH_3)_3Sn^+$ cations and CN^- anions, with possibly a very slight covalency contribution.

Dimethyltin difluoride (*30, 459*) forms crystals with two formula units per unit cell (dimensions: $a = 4.24$ Å, $c = 14.16$ Å). The tin atoms and the bridging fluorine atoms form a system of two-dimensional layers, separated by the methyl group, reaching out above and below each layer.

An x-ray diffraction study (*32, 219*) also proved the $(CH_3)_3SnCl{\cdot}C_5H_5N$ complex to be monomeric. The substituents form a trigonal bipyramid around the pentacovalent tin with the three methyl groups in the equatorial plane and a nearly linear chain formed by C—Sn—N perpendicular to it. The $[Me_2SnCl\ terpyridyl]^+[Me_2SnCl_3]^-$ adduct (*141*) was shown to have space group P_1 with lattice parameters $a = 15.71$ Å, $b = 10.04$ Å, $c = 10.30$ Å, $\alpha = 67.7°$, $\beta = 115.9°$, and $\gamma = 119.6°$; $Z = 2$.

2. *Electron Diffraction*

An electron diffraction investigation (*54*) of a series of tetramethyl compounds of group IV with tetrahedral symmetry has shown that the metal-carbon bond distance in these molecules agreed very closely with the sum of the covalent radii, as defined by Pauling (*404*); for instance $(Sn{-}C)_{exp.} = 2.18 \pm 0.03$ vs $(Sn{-}C)_{calc.} = 0.77 + 1.40 = 2.17$. On substitution of the methyl groups by halogen atoms it is found (*481*) that the tin-carbon bond distance remains practically unchanged (Table 12). The tin-halogen distance, however, is seen to decrease in the series of compounds $(CH_3)_3SnX$ to SnX_4 (X = Cl, Br, I) (Table 12). The electronegativity difference between tin and the halogen atom causes a polarization of the Sn—X bond, consequently causing a contraction of the covalent radii of the atoms and also increasing coulombic attraction (*481*). As, moreover, the halogen atoms dispose of free electron pairs in their p orbitals, these can overlap with the empty d_{xz} and d_{yz} orbitals of tin and, thus, more or less cancel the electron withdrawal around tin. This, however, causes an increase of the bond order and in this manner a shortening of the Sn—X bond length with increasing halogen substitution. This contraction is 0.07 Å for the methyltin chlorides, 0.04 Å for the bromides, and 0.08 Å for the iodides. Two characteristics with opposite effects on the atomic radius of the X-atom are important in this respect; the electronegativity and the polarizability. As a result, the largest contractions are observed for the chlorides and the iodides, respectively, with

TABLE 12

BOND DISTANCES AND BOND ANGLES FOR Me_nSnX_{4-n} ($n = 1,2,3$ AND X = Cl, Br, I) AS DERIVED FROM ELECTRON DIFFRACTION MEASUREMENTS

Compound	$X—\widehat{M}—X$	$r_{C—M}$, Å	$r_{M—X}$, Å
Me_3SnCl	108 ± 4°	2.19 ± 0.03	2.37 ± 0.03
Me_2SnCl_2	110 ± 5°	—	2.34 ± 0.03
$MeSnCl_3$	108 ± 4°	2.19 ± 0.05	2.32 ± 0.03
Me_3SnBr	~109.5°	2.17 ± 0.05	2.49 ± 0.03
Me_2SnBr_2	109 ± 3°	~2.17	2.48 ± 0.02
$MeSnBr_3$	109.5 ± 2°	~2.17	2.45 ± 0.02
Me_3SnI	~109.5°	~2.17	2.72 ± 0.03
Me_2SnI_2	109.5 ± 3°	—	2.69 ± 0.03
$MeSnI_3$	109.5 ± 2°	—	2.68 ± 0.02

the electronegativity effect being dominant in the former case and the polarizability in the latter.

In all these compounds, the bond angles approach the true tetrahedral angles fairly well and a dependence of the bond angle on the nature and the number of the X-substituents is hardly observable.

C. DIPOLE MOMENTS

The electrical dipole moments of several organotin halides (*239*, *394*, *488*, *489*, *491*) were measured in solution in order to study the relative importance of the ionic contributions to the Sn—C and the Sn—Cl bonds. The Sn—C bond was shown to have mainly covalent character, whereas for the Sn—Cl bond, the ionic character was calculated to be greater than 27% (*488*, *489*).

Lorberth and Nöth (*324*) measured the moments of a series of organotin halides R_nSnCl_{4-n} ($R = CH_3, C_2H_5, C_4H_9, C_6H_5$) with the aim of studying contributions to the bond strength in the Sn—Cl bond by back donation of unshared electron pairs of chlorine (in p_z orbitals) to empty *d*-orbitals of tin (d_{xz} and d_{yz} orbitals). Such a p_π-d_π double bond contribution would be in agreement with the well-known decrease of the Sn—Cl bond length on progressive chlorination of $(CH_3)_4Sn$. Increased p_π-d_π bonding and the increase of the electronegativity of the group R_nSnCl_{3-n} with increasing halogen substitution would both tend to decrease the polarity of the Sn—Cl bond. The experimental data, however (Table 13), seem to suggest that the Sn—Cl bond polarity increases with increasing number of chlorine atoms. According to these authors, this is accepted as evidence that the π-bonding contribution

TABLE 13

ELECTRICAL DIPOLE MOMENTS (IN DEBYE UNITS) FOR SOME RELEVANT ORGANOTIN HALIDES

Compound	μ, debye			
	(X = Cl)	(X = Br)	(X = I)	
$(CH_3)_3SnX$	3.50	3.45	3.37	
$(CH_3)_2SnX_2$	4.10	3.86	3.76	
$(CH_3)SnX_3$	3.64	3.24	2.52	
	(R = CH_3)	(R = C_2H_5)	(R = C_4H_9)	(R = C_6H_5)
R_3SnCl	3.46	3.56	3.31	3.46
R_2SnCl_2	4.14	4.32	4.38	4.21
$RSnCl_3$	3.77	4.08	4.28	4.30

is very weak and unable to counterbalance the electron withdrawal from tin by the inductive effect of the chlorine atoms. Another argument used in favor of the theory that the Sn—Cl bond polarity increases with increasing chlorine substitution is, according to these authors, found in the increase of the Sn—C—H coupling constants in the pmr spectra of $(CH_3)_nSnCl_{4-n}$ (*37*, *213*, *249*) which is generally accepted to reflect increasing percentage of *s*-character in the tin orbital used in the Sn—C bond. Lorberth and Nöth (*324*) then argue that, as a result, the polarity of the Sn—C bond decreases and the polarity of Sn—Cl bond should increase.

The dipole moments of the phenyltin chlorides are best explained by assuming overlapping of the π-electron system of the phenyl ring with the *d*-orbitals of tin. The mesomeric effect

$$(\delta-)C_6H_5\text{—}Sn^{\delta+} \longleftrightarrow {}^{+}C_6H_5\text{=}Sn^{-} \longleftrightarrow {}^{+}C_6H_5\text{=}Sn^{-}$$

then causes a decrease of the Sn—Cl bond polarity. This suggestion is confirmed by chemical evidence and by the work of Huang et al. (*217*) on vinyl derivatives. All these studies, with the exception of the work by Huang, imply the assumption, based mainly on the reported invariance of the Sn—C bond length and the conservation of nearly pure tetrahedral bond angles, that changes in the Sn—C bond moment are of negligible importance. The tin-proton coupling constants in the pmr spectra, however, clearly show that, with progressive halogenation, considerable changes also occur in the Sn—C bond. Van den Berghe (*37*) has studied the dipole moments of the series

$(CH_3)_nSnX_{4-n}$ (X = Cl, Br, I) and observed that the molecular moments were in the order:

$$\mu_{(CH_3)_nSnCl_{4-n}} > \mu_{(CH_3)_nSnBr_{4-n}} > \mu_{(CH_3)_nSnI_{4-n}}$$

This is in agreement with general theory, as it is evident that the partial tin-halogen moments should be in the order:

$$\mu_{Sn—Cl} > \mu_{Sn—Br} > \mu_{Sn—I}$$

For the chlorine derivatives, however, the ratio of the moments of the trichloride to that of the monochloride is considerably higher than unity. For methylgermanium halides an analogous anomaly is found only for the fluorides (*224*, *238*). Therefore, these data were discussed in terms of different contributions of plausible resonance structures A, B, C, and D:

$$CH_3—Sn(X)< \qquad CH_3—Sn^+< \; X^- \qquad H^+ \; CH_2{=}Sn< \; X^- \qquad CH_3—Sn({=}X^+)—X^-$$

(A) (B) (C) (D)

The sum of the contributions of resonance structures of the type C and D will increase with increasing electronegativity of the X-substituent (F > Cl > Br > I) and with increasing number of substituents. Moreover as the electronegativity of X increases, the type C structures will become more important. In this structure, no double bond character is assumed in the Sn—X bond. The CH_3Sn group is however strongly polarized and positive charge is induced in the methyl group. Obviously this polarization firstly, will be stronger with the more electronegative substituents, secondly, will increase with the number of electronegative substituents, and thirdly, will be more important in molecules of the type $(CH_3)_nMX_{4-n}$, the more easily M is polarized, i.e., in the order C < Si < Ge < Sn. Moreover, increasing contribution of type D structures will cause the partial Sn—X moment to decrease. It is believed that this theory explains fairly well the pmr data and the dipole moments. A slightly different approach to the discussion of this apparent anomaly was published by Huang et al. (*217*). They introduced two more resonance structures:

$$H^+ \; H_2C{=}Sn^-—Cl \quad \text{and} \quad CH_3—Sn^-{=}Cl^+$$

and discarded increasing importance of the second structure with progressive halogenation because it is energetically unfavorable. Thus, they finally stated that "the conclusion that Sn—Cl bond polarity increases with progressive chlorine institution, seems inescapable"!

Virtually the same conclusion has been drawn before (*37*, *324*). Calculations made by Van Hooydonck (*216*), however, allow evaluation of the changes in the Sn—C partial moments. It is shown that these yield a considerable contribution to the total molecular moment and that in the case of methyltin chlorides they could explain the apparent anomaly discussed above for the molecular moments on progressive chlorination.

In a study (*58*) of the effect of aromatic solvents on the chemical shift in the pmr spectra of $(CH_3)_nMX_{4-n}$ compounds (M = C, Si, Ge, Sn; X = halogen) a linear relationship was found between $\Delta\tau$ (the diamagnetic shift in benzene, compared to the τ value in CCl_4) and the dipole moments in each series. The aromatic solvent shift is believed to furnish a method of approximately estimating the dipole moment.

The dipole moments of a series of complexes (*399*) between R_nSnCl_{4-n} and $TiCl_4$ and $AlCl_3$, respectively, were also measured.

Closely related with these data is some work on molar refraction and bond refraction of organotin halides (*233, 457, 521, 532*). One (*457*) of these articles presents data on 147 organotin compounds.

D. Various Physical Properties

The electron impact fragmentation of tetramethyltin (*126, 212*) and of several other tetraorganotin compounds was first obtained by mass spectrometry, mainly with the aim of establishing standards for the isotopic composition of the metal.

The first trialkyltin halide to be studied by mass spectrometry was Et_3SnF (*393*). Gielen and Mayence (*159*) recently investigated the mass spectra of a large series of R_3SnX compounds (R = Et, Pr, *n*-Bu, *i*-Bu; X = F, Cl, Br, I). A fragmentation pattern was formed and it was concluded that evidence was found that the ease of cleavage of the tin-halogen bond corresponds to the bond energy in the ground state.

Bond energies of the tin-halogen bond in trimethyltin bromide and iodide have also been obtained by classical thermochemical methods (*321, 407, 544*). Dimethyltin dichloride is found to dissociate only above 555°C according to the reaction equation

$$(CH_3)_2SnCl_2 \longrightarrow SnCl_2 + 2\,CH_3$$

Apparently the thermal stability is very high. Triorganotin halides and organotin trihalides, however, are found to decompose at considerably lower temperatures. In most instances disproportionation or formation of alkyl halide (*428*) occurs. Pedley et al. (*406, 407*) investigated the thermochemistry of tin tetraalkyls in their reaction with I_2 and Br_2, with the formation of trialkyltin iodide and bromide.

The polarographic reduction of diethyltin dichloride, first studied by Riccoboni et al. (*443*), showed that the product of the reduction was diethyltin. Recently a new study (*122, 123*) was devoted to this compound and to ethyltin trichloride (*124*), in water solution as a function of pH. Trialkyltin halides dissolved in water with 30% isopropyl alcohol are shown to produce three reduction waves, the first of which was ascribed to the formation of hexaalkylditin (*110*). For similar compounds, dissolved in a water solution with 40% ethyl alcohol, the ease of the reduction was found to be a function of the nature of the organic radical in the order ethyl > propyl > butyl (*512*). These reductions are irreversible (*550*). It is believed that for R_3SnX and R_2SnX_2 compounds the first reduction wave originates from the use of one electron to form hexaalkylditin and R_2SnCl_2, respectively. In the second reduction wave two electrons are used and from both kinds of molecules dialkyltin, R_2Sn, is formed (*8*). These studies were not only continued for Ph_3SnCl and Ph_2SnCl_2 by polarography, but also by controlled electrolysis and triangular voltametry (*120*). The alternating current polarography capacity effects of a series of organotin compounds in alcohols were also investigated (*226*).

The diamagnetic susceptibility of a series of trialkyltin chlorides and alkyltin trichlorides were measured, and a linear relationship was found between the molar susceptibility and the length of the carbon chain (*1*). In an earlier study of halogenated stannanes attention was drawn to the relation between the electrical dipole moment and the diamagnetic susceptibility (*239*).

The kinetics of the reaction between $(C_6H_5)_2SnX_2$ (X = Cl, Br) and 8-quinolinol in solution (*346*) and as pure compounds (*345*), as well as of several reactions between tetraorganotin compounds and halogens, hydrochloric acid, etc., were studied (*160, 161, 305*).

V. Coordination Compounds

Many compounds are known in which the coordination number of tin is higher than four. The nature of such compounds is not always the same, and the chemical bonds involved are still a matter of discussion. Different aspects of the physicochemical properties of the partners joined in these compounds and of the complex compounds themselves will be briefly outlined in this section. As far as possible, the participating organotin halide will be the central point of the discussion.

A. Acceptor Strength of Organotin Halides

The acceptor strength of an organometallic compound is determined mainly by the nature of the metal atom and of its ligands. The differences in the electrical dipole moment values as measured in dioxane vs those measured in hexane as a solvent, for a series of organotin halides, were taken as a measure of their ability to form addition compounds and yielded the sequence

(*182*) $SnCl_4 > PhSnCl_3 > Ph_2SnCl_2 > Ph_3SnCl > Bu_3SnCl$. The coordinative bond strength in complexes between several alkyltin chlorides and 2,2′ bipyridyl, as derived from thermodynamic data (*348*) is found to decrease in the order $SnCl_4 > n\text{-}BuSnCl_3 > (CH_3)_2SnCl_2 \simeq (C_2H_5)_2SnCl_2 > (n\text{-}Bu)_2SnCl_2$, and the stability constants for methyltin oxinates (*183*, *240*) are in the order $(CH_3)_2SnClOx > CH_3SnClOx_2 \simeq (CH_3)_2SnOx_2 > (CH_3)_3SnOx$. The acidity constant of $SnCl_4$ (*528*) on substitution of one chlorine by an organic substituent is decreased by a factor of

$$4 \times 10^3(n\text{-Bu}) > 1.6 \times 10^3(\text{Me}) > 5 \times 10^2(\text{Ph})$$

The same trend is observed in the stability of the hexacoordinate organotin anions (*468*), where the sequence

$$SnCl_6^{2-} > RSnCl_5^{2-} > R_2SnCl_4^{2-} \gg R_3SnCl_3^{2-}$$

is observed and there are no reports on the existence of $R_4SnCl_2^{2-}$ ions. Anionic chloride or bromide complexes are also obtained more easily with PhSn(IV) than with MeSn(IV) (*83*). The quantitative relative acceptor strength of three organotin trichlorides and $SnCl_4$ vs aniline bases (*528*) was found to be

$$SnCl_4\ (670) > PhSnCl_3\ (12) > MeSnCl_3 > n\text{-}BuSnCl_3\ (1)$$

For the tin tetrahalides (*29*, *121*) it was found that complexes were formed with decreasing strength as the halide changed, in the sequence F > Cl > Br > I. The quadrupole splitting in the Mössbauer spectra (*492*) of Me_3SnX compounds with a series of donors was also interpreted in terms of the acceptor strength of the Me_3SnX moiety and showed that the nature of the X-substituent influenced this strength in the order

$$F^- > Cl^- > Br^- > OH^-$$

The tendency for complex formation by organotin halide systems was also studied by paper electrophoresis and anion-exchange paper chromatography (*83*). The results indicated decreasing complex stability as a function of the nature of the halogen substituent: $F^- \gg Cl^- > Br^- > I^-$.

It is therefore clear that the acceptor strength of tin in these compounds is proportional to the electronegativity of the substituents bonded on tin. The stronger the electron attracting power of the substituents, the less the electron density around tin and the acceptor strength increases accordingly.

These simple rules are, however, complicated by the fact that the apparent acceptor strength depends also on the nature of the donor (*527*). For instance the stability of the donor-acceptor complexes with nitranilines yields the sequence:

$$Me\text{-}\varphi\text{-}SnCl_3 > \varphi\text{-}SnCl_3 > Cl\text{-}\varphi\text{-}SnCl_3 > Ph\text{-}\varphi\text{-}SnCl_3$$

whereas the sequency with nitrophenylenediamines is:

$$Cl\text{-}\varphi\text{-}SnCl_3 > \varphi\text{-}SnCl_3 > Me\text{-}\varphi\text{-}SnCl_3 > Ph\text{-}\varphi\text{-}SnCl_3$$

The ΔpK values (*528*) between $SnCl_4$ and $PhSnCl_3$ are reported to be different when obtained from systems with perinaphthenone ($\Delta pK = 1.25$) as a donor, vs systems with 6-methyl-3-nitroaniline ($\Delta pK = 1.80$).

Silicon and germanium only behave as acceptor molecules when four strongly electronegative substituents are bonded to the metal atom. The acceptor strength of the group IV elements thus obviously follows the sequence:

$$Sn \gg Ge > Si$$

$RSnCl_3$ and R_2SnCl_2 usually yield stable hexacovalent complexes with nucleophilic compounds, but R_3SnCl mostly yields pentavalent compounds. Complexes with tetraorganotin compounds are not known.

B. Relative Donor Strength towards Organotin Halides

The relative nucleophilic character of a series of organic solvents toward several organotin halides has been studied mainly by nmr and Mössbauer spectroscopy.

In the nmr data, the increase of the tin-proton coupling constants $J_{^{117/119}Sn\text{—}C\text{—}H}$ (*46*, *160*, *220*, *349*, *355*) was thought to be related to the changes in the hybridization around tin on changing from the tetrahedral sp^3 state in the pure organotin compound to a trigonal bipyramidal configuration state in the pentacoordinated tin complexes and to an octahedral configuration in the hexacoordinated tin compounds. The relative nucleophilic character for a series thus established is: DMSO ~ DMF > HOH > Py > MeOH > MeCOMe ~ MeCOOMe > dioxane > MeCN ~ MeCOOH > $MeNO_2$ > PhCl ~ CCl_4 and $(C_2H_5)_2S < (C_2H_5)_2O$

$$\text{DMTA} < \text{DMA}$$

$$\text{THMPA} < \text{HMPA}$$

$$(n\text{-Bu})_3P < Py$$

and

$$\text{TMSO} \sim \text{DMA} \sim \text{DMF}$$

The latter series shows that the interaction is weaker with sulfur or phosphorus donors than with either oxygen or nitrogen donors.

The quadrupole splitting (Q.S.) observed in ^{119}Sn Mössbauer spectra apparently seems also to be influenced by complex formation. The interaction between $(n\text{-Bu})_2SnCl_2$ and a series of solvents was studied by this method (*179*). Taking the interaction parameter as being arbitrarily equal to unity for diethylether, the following sequence results: DMSO (*18*) > DMF (*16*) >

HMTAP (*15*) > DME (*10*) > THF (*7*) > Et_2O (*1*). The significance of a relationship (*208*, *209*, *496*) between the ratio, ρ = Q.S./I.S., of the quadrupole splitting (Q.S.) to the isomeric shift (I.S.) and the strength of donor-acceptor bonds as applied to a series of complexes of R_3SnX (R = neophyl, phenyl) is however open to criticism (*367*).

C. Solvolysis, Hydrolysis, and Complex Formation in Solution

In the series of organotin halides R_nSnX_{4-n} (X = F, Cl, Br, I) the fluorides occupy a somewhat particular position as regards solubility. They are practically insoluble in all common solvents, and it is believed that this is due to their polymeric structure. The other halides yield, according to molecular weight and conductometric measurements, solutions in which the molecules are present as undissociated and monomeric species in those solvents that have a low dielectric constant and that have weak electron donor properties. In solvents, however, with strong electron donor properties such as DMSO, DMF, DMA, etc. "coordination" occurs with the solvent, resulting in an increase of the coordination number around tin. These solutions are nonconducting. In solvents with a high dielectric constant, which are also efficient electron donors, the Sn—X bond may eventually be broken, resulting in the formation of solvated organotin cations and halide anions

$$R_3SnX + S \rightleftharpoons (R_3SnS)^+ + X^-$$

Water, therefore, seems to be an ideal solvent to yield such solutions. Organometallic cations are, however, strongly polarizing, i.e., they behave like Lewis acids, and the aquo-ions then become fairly strong proton acids. Water solutions of R_nSnX_{4-n} compounds, therefore, are acidic, due to proton transfer:

$$[R_3Sn(H_2O)_n]^+ + H_2O \rightleftharpoons R_3SnOH(H_2O)_{n-1} + H_3O^+$$

A thorough discussion of the occurrence, the structure and the hydrolysis of the organotin-aquo cations, as well as of their hydrolysis products is beyond the scope of this section. An excellent review has been published on this subject (*511*). In this section we will only discuss some compounds that can be isolated by the controlled hydrolysis of organotin halides and also the interaction of the $R_nSn(IV)$ group (n = 1, 2, 3) with halide ions in solution.

The complete hydrolysis of organotin dihalides yields an insoluble polymeric product with stoichiometry $(R_2SnO)_x$. Controlled hydrolysis, however, allows two intermediates to be isolated: $(R_4Sn_2X_2O)_2$ and $[R_4Sn_2X(OH)O]_2$, with R = alkyl and X = halogen. The reaction sequence for the hydrolysis of organotin dihalides therefore can be written as

$$\underset{(1)}{R_2SnX_2} \longrightarrow \underset{(2)}{[R_4Sn_2X_2O]_2} \longrightarrow \underset{(3)}{[R_4Sn_2X(OH)O]_2} \longrightarrow \underset{(4)}{(R_2SnO)_x}$$

The structure of the intermediate (**2**) has been uncertain for a long time. The ^{119}Sn nmr spectrum reveals the presence of two chemical nonequivalent tin atoms. This evidence, together with other data, suggests a dimeric structure in which two tin atoms have coordination number four and two others have coordination number five (*11*).

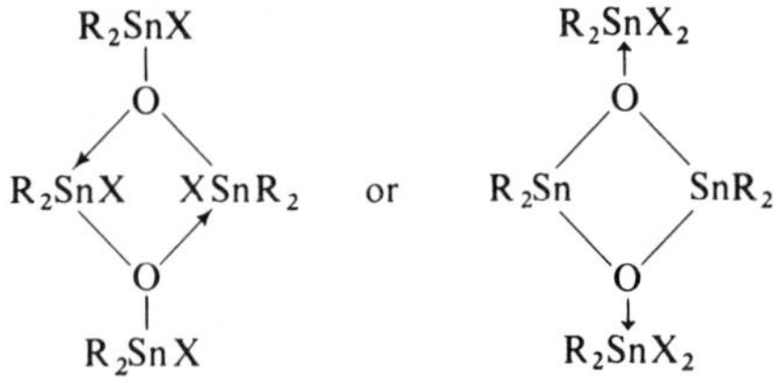

R_2SnX — O — R_2SnX $XSnR_2$ — O — R_2SnX (**5**) or R_2SnX_2 → O — R_2Sn SnR_2 — O → R_2SnX_2 (**6**)

The structure (**5**) is generally believed to be the most appropriate, as on further hydrolysis, two more X groups are replaced by OH groups. The structures of the resulting dimeric compound (**3**) (the same remarks hold for the dimeric compound obtained by controlled alcoholysis: $[XR_2SnOSnR_2OR']_2$) are probably (*10*, *398*) as represented by:

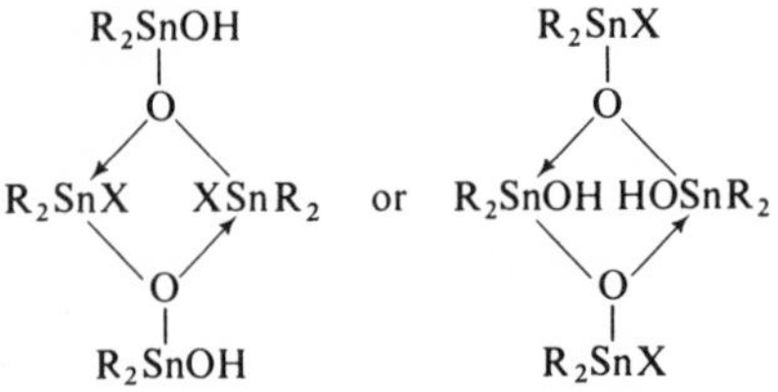

R_2SnOH — O — R_2SnX $XSnR_2$ — O — R_2SnOH (**7**) or R_2SnX — O — R_2SnOH $HOSnR_2$ — O — R_2SnX (**8**)

since in the ir absorption spectrum, only one kind of absorption band is found which corresponds to the Sn—OH (and Sn—OR′) vibration. Symmetry considerations, however, are slightly in favor of (**8**) (*398*).

Complex formation between $SnCl_4$ or Bu_2SnCl_2 and some Sn—O and Sn—S compounds were studied by means of cryoscopic, dielectrometric, and calorimetric titrations (*183*). In a pmr study of solutions of trimethyltin chloride in concentrated hydrochloric acid, no evidence could be found for the complexation of the tin compound (*34*). However, solutions of triethyltin chloride with dimethylamine hydrochloride are reported to contain the complex ion $Et_3SnCl_2^-$ (*234*). The formation of complex compounds with formulas $(CH_3)_3SnF_n^{1-n}$ ($n \leq 2$), $(CH_3)_2SnF_n^{2-n}$ ($n \leq 4$), and $CH_3SnF_n^{3-n}$ ($n \leq 5$) could also be shown in anion exchange experiments and by potentiometric titration (*84–87*). The existence of a pentacovalent ion R_3SnBrX_- (X = Cl, I) could be shown in various solvents (*161*, *167*, *190*). Distribution data of the complexes $Me_3SnCl_n^{1-n}$, $Me_2SnCl_n^{2-n}$, and $MeSnCl_n^{3-n}$ and their stability constants are also reported (*89*). The complexation of the $(CH_3)_2Sn^{2+}(aq)$ cation by Cl^- and Br^- in aqueous solutions has been extensively investigated by emf measurements, Raman and nmr spectroscopy (*34*, *145*, *304*). Stability constants for the chloro complexes are reported, $\log \beta_1 = 0.380$ and $\log \beta_2 = -0.14$. The interaction with Br^- is, however, weak. The formation of inner and outer sphere complexes is also found to be a function of the chloride ion concentration. Concentrated bromide ion solutions also yield inner-sphere complexes (*145*). The reversible formation of complexes $(CH_3)_2SnX_4^{2-n}$ ($n = 1$, 2, 3, or 4 depending on the halide ion concentration, and X = F, Cl, Br, I) and of $CH_3SnX_n^{3-n}$ and $C_6H_5SnX_n^{3-n}$ complexes ($n = 1$–5 and X = F, Cl, Br) was established by anion exchange chromatography, paper electrophoresis, and potentiometric titrations (*82*, *83*, *86–88*).

There is also strong evidence that, on potentiometric titration of $RSnCl_3$ and R_2SnCl_2 compounds ($R = C_4H_9$ and C_6H_5) with tetraethylammonium chloride in acetonitrile, pentacoordinate organotin chloride anions $RSnCl_4^-$, and $R_2SnCl_3^-$ are formed (*499*). Potentiometric titrations of $R_{4-n}SnCl_n$ ($R = CH_3, C_2H_5, C_3H_7, C_4H_9, C_6H_5$; $n = 2, 3$) with Ph_4AsCl, Et_4NCl or Et_4NBr solutions in CH_3CN also yield evidence for the formation of such complexes (*500*). The behavior in solution of the adducts $[Ph_4As][R_{4-n}SnCl_{n+1}]$ from the same series of organotin chlorides has been studied by means of conductometric measurements in CH_3CN and by molecular weight measurements in acetone. It has been shown that the adducts are strong 1:1 electrolytes and the formation constants of the anionic species therein were determined by spectrophotometric methods (*546*).

The study of the complex formation of CH_3SnCl_3 and CH_3SnBr_3 in different solvents by ir and Raman spectroscopy (*34*) and by pmr spectroscopy (*303*) also furnished indications about the existence of $CH_3SnCl_5^{2-}$ ions in concentrated hydrochloric acid solution.

D. Structure of Coordination Compounds

1. *Coordination Compounds with* R_3SnX

Complexes with stoichiometry $R_3SnX \cdot B$ are formed between trialkyltin chlorides R_3SnCl ($R = CH_3$, C_2H_5) and Lewis bases B such as pyridine (*31*, *39*, *40*) and TMSO, DMA, and DMF (*355*). The nmr and uv spectra of $(CH_3)_3SnX \cdot Py$ (*33*, *39*) in solution show evidence for dissociation, thus suggesting that the donor-acceptor bond strength is rather low. Solutions of the compounds R_3SnCl in pyridine (*510*), TMSO, DMA and DMF (*355*), respectively, are nonconducting, which proves that the molecular complexes present do not have an ionic structure. These data can only be explained by assuming a coordination number of five around tin. The relationship between the nmr coupling constants $J_{^{119}Sn-C-H}$ of a complex $(CH_3)_3SnCl \cdot B$ and the formation enthalpy (*44, 45*) of the complex allowed Bolles and Drago (*46*) to derive the following scheme:

$$B-Sn(CH_3)_3-Cl \xleftarrow{+\text{ weak donor B}} (CH_3)_3Sn-Cl \xrightarrow{+\text{ strong donor B}} B-Sn(CH_3)_3-Cl$$

A strong donor causes a profound rehybridization of the tin atom and the structure of the complex approaches a trigonal bipyramid with a $CH_3-\widehat{Sn}-Cl$ angle of about 90°. A weak donor however forms an addition compound wherein the $CH_3-\widehat{Sn}-Cl$ angle is about equal to the tetrahedral angle of 109° 28′. As a result, it is observed that generally, with weak donors in the system, steric hindrance occurs between the methyl groups bonded to tin and the donor molecule. This theory is supported by a series of experimental observations. In the ir spectra of the system $(CH_3)_3SnCl \cdot B$, remarkable differences are to be noted when compared with the spectra of the free $(CH_3)_3SnCl$ molecule; the absorption band corresponding to the symmetrical Sn—C stretching vibration disappears and the frequency of the Sn—Cl vibration is drastically decreased. A trigonal bipyramidal structure with a planar $(CH_3)_3Sn$ group with the base molecule B and the chlorine atom both in axial positions, perpendicular to this plane, explains these data fairly well. Moreover a detailed x-ray analysis (*219*) has confirmed this geometry. The hybrid orbitals in the trigonal plane should then be nearly pure sp^2 type orbitals; the resulting increase of *s*-character in these orbitals should then explain the increased J_{Sn-C-H} values (*40, 355*). On the other hand the axial bonds are supposed to result either by using hybrid tin orbitals of the $(p_z + d_{z^2})$ type or by using a three-center, four-electron molecular orbital having essentially p_z character.

For the 1 : 1 complexes of trimethyltin- and triphenyltin chloride with DMSO, it was found that the S=O stretching vibration frequency in their spectrum is lowered with respect to free DMSO; an observation which has been interpreted as indicating that coordination proceeds with the oxygen atom instead of the sulfur atom (*313*). Trimethyltin chloride and bromide give unstable diammoniates that lose one ammonia molecule spontaneously. In these compounds the ir absorption corresponding to the Sn—C symmetrical stretching vibration is not observed, so that the diammoniates are to be formulated as ionic compounds $[(CH_3)_3Sn(NH_3)_2]^+ X^-$ and the monoammoniates as molecular compounds $(CH_3)_3SnX \cdot NH_3$ (*100*).

For several complexes of R_3SnX with aza-aromatic bases Mössbauer spectra were studied; neither from the I.S. or Q.S. values nor from their ratio, $\rho =$ Q.S./I.S., was it possible to derive the coordination number of tin (*367*).

Measurements of vapor pressure, melting point, optical density, and specific conductivity were made on systems of Et_3SnCl with $NHMe_2$, $NHEt_2$, and NMe_3; it was found that in all cases there resulted 2 : 1 complexes $(Et_3SnCl)_2$·amine. For Me_2NH and Et_2NH, 1 : 1 complexes could also be detected, but not for Me_3N (*234*).

Triphenylphosphinemethylene and tetraethylammonium bromide with Me_3SnBr yield pentacoordinate tin anionic complexes (*468*) $[(CH_3)_3SnBr_2]^-$ with unknown sterical structure: the trigonal bipyramidal structure or a binuclear compound:

Br CH₃
H₃C Br CH₃
Sn Sn
H₃C Br CH₃
CH₃ Br

The $[Ph_3SnCl_2]^-$ anion in tetramethylammoniumdichlorotriphenylstannate has a trigonal bipyramidal structure with the three phenyl groups occupying equatorial positions (*494*).

2. *Coordination Compounds with* R_2SnX_2 *and* $RSnX_3$

R_2SnCl_2 and $RSnCl_3$ ($R = CH_3$, C_2H_5) form 1 : 1 addition compounds with 1,10-phenanthroline (I) and 2,2′-bipyridyl (*9, 41*) (II) but yield 1 : 2 adducts with pyridine (*39, 40*) and pyridinium hydrochloride (*531*).

The Sn—Cl stretching absorption bands in the ir for the Me_2SnCl_2 complexes are shifted to below 250 cm^{-1}; in the $MeSnCl_3$ complexes, the corresponding frequencies are found at about 320 and 280 cm^{-1} (*31*). A study of the ir absorption spectra of a series of adducts between alkyltin dichlorides with either (I) or (II) showed that the Sn—Cl stretching vibration frequency was lowered by about 80–100 cm^{-1} with respect to the free dichloride (*104*). A careful examination of these results yielded the conclusion that in all these addition compounds the coordination number of tin is increased from 4 to 6.

Absorption bands due to the symmetrical stretching Sn—C vibration are not found in the ir spectra of the Me_2SnCl_2-complexes, suggesting that the methyl groups should be present in an octahedral *trans* position (*31*). The ir spectra of a series of complexes $Me_2SnCl_2 \cdot 2B$ [B = pyridine (*40*) or triphenylphosphine oxide (*103*)] can also be interpreted in terms of octahedral structures around tin where the methyl groups are in *trans* positions and the other ligands in *cis* positions. In the compounds $(C_4H_9)_2SnX_2 \cdot (I)$ (X = Cl, Br, I) large electrical dipole moments were found and this is another argument for assuming the *trans* position of the butyl groups (*363*). ir and uv spectral data and molecular weight measurements (*507*) for a series of addition compounds between dialkyl- and diphenyltin chlorides with (I) and (II) confirmed this previous conclusion:

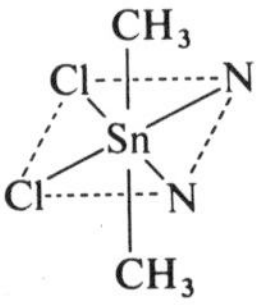

$$\begin{array}{c} CH_3 \\ Cl\text{--}|\text{------}N \\ Sn \\ Cl\text{------}|\text{--}N \\ CH_3 \end{array}$$

A subsequent confirmation for the *trans* arrangement of the alkyl groups is found in the Mössbauer spectra where the quadrupole splitting is correlated with stereochemistry (*151*). Mössbauer data for the latter series of complexes suggested the same bond polarity trend as was observed for the free butyltin dihalides, i.e.: Sn—Cl > Sn—Br > Sn—I (*363*).

The uv spectra of the 1 : 1 complexes between R_2SnCl_2 (R = Me, *n*-Bu) and (II) show an absorption maximum at about 300 mμ, indicating a chelate structure; the stability constants are shown to increase with increasing polarity of the solvent (*281*). In acetonitrile the (*n*-Bu)$SnCl_3 \cdot$(II) complex yields a 210 mμ uv band which is split into two components, whereas spectra of dialkyltin dichloride complexes with (II) do not have this fine structure. This observation is believed to be due to the fact that, as a result of the weaker coordination of the ligand molecule to tin, both pyridine rings of II are slightly twisted with respect to each other (*348*).

Coordination compounds of organotin halides with 4,4′-bipyridyl, 4-phenylpyridine and pyrazine were also studied and the Sn—Cl and Sn—Ph bands assigned (*420*).

DMSO and TMSO yield 1 : 2 hexacoordinated complexes with R_2SnCl_2 and $RSnCl_3$ (R = Me, C_6H_5). Under special conditions $PhSnCl_3$ also yields a 1 : 1 complex. The values of the SO stretching frequencies observed in their ir spectra suggest that the oxygen atom is the donor in DMSO (*313*). In $(CH_3)_2SnCl_2 \cdot 2$ DMSO the single Sn—Cl stretching vibration shows evidence for a *trans* arrangement (*95*). Nevertheless the data of a far ir spectroscopic study of the DMSO (*505*) and DMSeO (*506*) complexes of R_2SnX_2 (R = Me, Et, Ph and X = Cl, Br) are consistent with a *cis* arrangement for all except for $(CH_3)_2SnBr_2 \cdot 2$ DMSO.

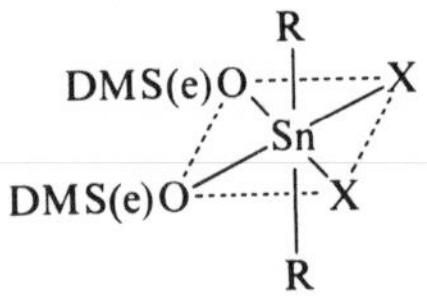

$$\begin{array}{c} R \\ DMS(e)O\text{--}|\text{------}X \\ Sn \\ DMS(e)O\text{------}|\text{--}X \\ R \end{array}$$

The pmr spectra of $(CH_3)_2SnCl_2 \cdot 2$ DMSO indicate that there is a considerable degree of rehybridization around tin, as the $J_{^{119}Sn—C—H}$ coupling constants are increased from 70 cps in $(CH_3)_2SnCl_2$ to 86 cps in the complex. For $(CH_3)_2SnCl_2$ and $(C_6H_5CH_2)_2SnCl_2$ dissolved in DMSO a value of

113 cps is found for the tin proton coupling constant (*263*). By analogy with the data for $(CH_3)_2SnCl_2$ in water (*34*), the latter values could be interpreted as evidence for a reaction of the type

$$(CH_3)_2SnCl_2 + x\,DMSO \rightleftharpoons (CH_3)_2Sn^{2+}\cdot x\,DMSO + 2\,Cl^-$$

In mixtures of $R_2SnCl_2\cdot 2$ DMSO with R_2SnCl_2 and R'_2SnCl_2 the PMR spectra show that there is an exchange of DMSO molecules with rapid ligand reorganization around tin (*263*).

Complexes of $RSnCl_3$ (R = Me, *n*-Bu, Ph) with aromatic bases such as 1,3-diamino-4-nitrobenzene and 1,4-diamino-3-nitrobenzene were investigated by uv absorption spectroscopy (*527*). The compounds have absorption bands at 398 mμ and 396 mμ, respectively, which suggests that only the amino group in *para* and *meta* positions are participating in the coordinative bond to tin (structures A, B):

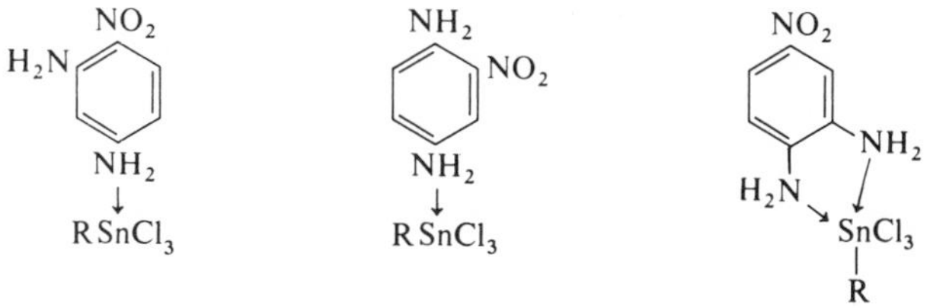

NO_2 H_2N NH_2 ↓ $RSnCl_3$ (A)

NH_2 NO_2 NH_2 ↓ $RSnCl_3$ (B)

NO_2 NH_2 H_2N $SnCl_3$ R (C)

These 1 : 1 complexes thus form examples of five coordination around tin. In 1,2-diamino-4-nitrobenzene, however, both *ortho*-amino groups coordinate with tin (structure C), forming a chelate compound with hexacoordinate tin.

1 : 1 complexes are also reported between some nitroanilines and nitrophenylene diamines and 4-chlorophenyltin, 4-tolyltin- and 4-biphenyltin trichlorides (*528*). $PhSnCl_3$ also forms addition compounds with diphenylethylene (*181*) and perinaphthenone (*361*).

The double diazonium salts (*441*) of CH_3SnCl_3 and of Et_2SnCl_2 are reported to have a constant composition:

$$[ArN_2]_2[CH_3SnCl_5] \quad \text{and} \quad [ArN_2]_2[Et_2SnCl_4]$$

where Ar is C_6H_5 or a substituted aryl nucleus.

Triphenylphosphinemethylene forms complex compounds with dimethyltin dichloride and dibromide with stoichiometry (*468*):

$$\{[(C_6H_5)_3PCH_2]_2Sn(CH_3)_2\}\{(CH_3)_2SnX_4\}$$

All these organotin anions obviously have hexacoordinated tin atoms. This is also found in the crystalline K, Cs, NH_4, NMe_4, and NEt_4 salts of Me_2SnX_2 [X = F (*534*), Cl, Br (*103*)], containing the anionic species $(Me_2SnX_4)^{2-}$. The Infrared spectra of the hexacoordinated anionic species show only one

Sn—CH_3 and Sn—X stretching frequency and thus strongly suggest a *trans* configuration. The $Me_2SnX_3^-$ spectra are more complex, indicating that the methyl groups hold equatorial positions in a trigonal bipyramidal structure, similar to the configuration established by x-ray analysis (*141*) for $(Me_2SnCl$ terpyridyl$)^+$ $(Me_2SnCl_3)^-$, an addition compound obtained on reaction of 2,2′,2″-terpyridyl with dimethyltin dichloride (*149*). In the cationic moiety of this compound the hexacoordination around tin is seriously distorted and the methyl groups keep *trans* positions.

An attempt was also made to establish the geometry of various five- and six-coordinated organotin(IV) ions from the Mössbauer quadrupole splitting values (*119*).

Tetraphenylarsonium salts with pentacoordinate tin anions are also reported to be stable (*499*):

$$[(C_6H_5)_4As]^+[RSnCl_4]^- \quad \text{and} \quad [(C_6H_5)_4As]^+[R_2SnCl_3]^-$$

3. *Autocomplexation*

$(CH_3)_3SnF$ is known to form a polymeric linear chain structure wherein $(CH_3)_3Sn$ groups are linked by bridging F-atoms. The geometrical structure is neither tetrahedral nor trigonal bipyramidal but a situation seems to exist in between these two extremes. A series of physical data (*101*, *102*, *208*, *209*, *304*, *452*, *502*, *542*) such as ir and Raman spectra and the ρ ratio gained from Mössbauer spectral parameters all point out that the tin atom is in the pentacoordinate state, that free $(CH_3)_3Sn^+$ cations do not exist in these structures, and that bridging Sn—Hal—Sn occurs also for trimethyltin chloride and bromide. However the argument that a ρ value higher than 2.1 should be characteristic for a coordination number around tin higher than 4, is liable to be seriously questioned (*367*).

The infrared spectra of trimethyltin tetrafluoroborate (*98*, *99*), hexafluoroarsenate, and hexafluoroantimonate (*97*, *99*) can be interpreted in terms of a structure where BF_4 (and MF_6 groups, respectively) and planar $(CH_3)_3Sn$ groups, join through bridging fluorine atoms, to form a more stable polymer. The hybridizational state of the five coordinate tin atom is believed to be sp^3d rather than sp^2 in the planar $(CH_3)_3Sn$ groups. For the 1 : 1 complex of $AlCl_3$ with R_nSnCl_{4-n}, $AlBr_3$ with R_nSnBr_{4-n} (R = Me, Et, *i*-Bu) the ir spectral data have also been interpreted as in favor of an analogous structure with five coordinate tin (*384*, *387*):

```
      R                  R
      |        Cl        |          Cl
····Sn····Cl   Al   Cl····Sn····Cl   Al   Cl
    / \        Cl       / \         Cl
   R   R               R   R
```

However, measurements of the dielectric permeability, magnetic susceptibility, and cryoscopic data seem to point to a hexacoordinate structure (*399*).

However, it was found that $TiCl_4$ forms 1 : 2 complexes with Et_nSnCl_{4-n}. The catalytic effects of such complexes on the polymerization of ethylene (*399*) and benzofurane (*504*) were also studied.

All the structures discussed in the above section are mainly based on the interpretation of ir spectral data. Although there is no obvious reason to doubt the assignments, x-ray analysis of some of these complexes would be welcome.

In compounds such as $(CH_3)_2SnF_2$, where the fluorine atom can act as a bridging element, the possibility of autocomplexation with the formation of hexacoordinated tin also exists (*30, 95, 534*). A recent detailed x-ray analysis (*459*) proved, moreover, that the octahedral coordination places around tin are held by four bridging fluorine atoms in the equatorial plane and by the methyl groups keeping the *trans* position in a long chain polymeric structure. Total cross-section measurements (*449*) with subthermal neutrons showed that free rotation is established around the tin-carbon axis and that the methyl groups are not in interaction of any kind with other groups in the crystal lattice.

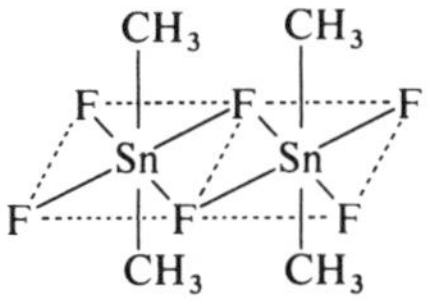

CH3 CH3
F F F
Sn Sn
F F F
CH3 CH3

A new type of autocomplexing compounds is found in the organotin halide oxinates R_nSnXOx_{3-n} (*240, 262, 526, 533*) ($n = 1, 2$; $R = CH_3, C_6H_5$; X = Cl, I; Ox = 8-hydroxyquinolinate) and kojato complexes of tin(IV) (*400*). In the monooxinates ($n = 2$) a trigonal bipyramidal configuration is found, assuming that pentacoordination around tin and the methyl groups occupy the *trans* positions.

In these complexes stabilization of the compounds is obtained most probably through d_π-p_π interaction in the Sn—X bonds; this interaction is thought to be more pronounced in the Sn—Cl than in the Sn—I bonds. The dioxinates have octahedral structures with hexacoordination around tin.

Many nmr and ir investigations (*242, 244–248*) have been devoted to the study of aryl- and alkyl-bis(acetyl-acetonato) tin halides: $R(X)Sn(acac)_2$

(R = alkyl, aryl; X = Cl, Br, I; acac = $CH_3—C(—O)=CH—C(=O)—CH_3$)

These data are in favor of a structure with a linear C—Sn—X group perpendicular to the plane of the acetylacetone chelate rings. The octahedral configuration around tin thus obtained, most probably uses almost pure sp_z hybrid orbitals for the C—Sn—X bonds whereas the Sn—O bonds are supposed to be a linear combination of tin $5p_x$ and $5p_y$ atomic orbitals. The interaction of these compounds with benzene has been investigated (*243*). To this class of compounds $PhSn(dpt)_2Cl$ (dpt = di-phenyltriazine) also belongs, and a series of compounds with the general formula $XR_2SnOOCR'$ (X = Cl; R = CH_3, Et, *n*-Pr, *n*-Bu; R′ = H, CH_3 and X = Br; R = *n*-Bu; R′ = CH_3), that are monomeric, the carboxyl group acting as a chelating agent yielding five coordination around tin (*9*, *396*, *525*):

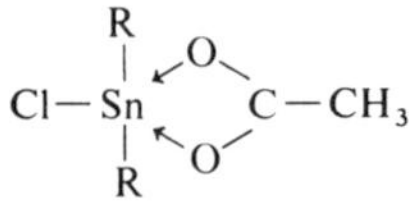

```
      R
      |  ↙O↘
Cl—Sn        C—CH₃
      |  ↖O↗
      R
```

APPENDIX 1

TRIORGANOTIN HALIDES R_3SnX

Compound	m.p., °C	b.p., °C(mm)	References
$(CH_3)_3SnI$	3.4	170(760)	(*76, 192, 195, 307, 311, 320, 336, 368, 412, 451, 481, 484, 521, 532*)
$(CH_3)_3SnBr$	27	165(760)	(*239, 270, 288, 290, 407, 481*)
$(CH_3)_3SnCl$	37–38	152–4(760)	(*16, 197, 228, 266, 284, 285, 328, 340, 401, 429, 481, 484, 509, 510*)
$(CH_3)_3SnF$	375d		(*27, 239, 292*)
$(C_2H_5)_3SnI$	−34.5	234(760)	(*14, 76, 110, 128, 192, 196, 306, 307, 311, 326, 328, 336, 338, 450, 512, 521, 531, 532*)
$(C_2H_5)_3SnBr$	−13.5	224(760)	(*14, 15, 76, 110, 152, 168, 192, 223, 251, 268, 269, 275, 306, 326, 328, 377, 450, 521, 532*)
$(C_2H_5)_3SnCl$	15.5	210(760)	(*14–16, 21, 28, 67, 76, 110, 111, 127, 146, 152, 192, 228, 232, 238, 251, 255, 269, 306, 308, 310, 326, 328, 340, 342, 451, 452, 488, 521, 532*)
$(C_2H_5)_3SnF$	302		(*14, 27, 110, 152, 292*)
$(C_3H_7)_3SnI$		140–1(15)	(*77, 79, 80, 192, 198, 199, 331, 521, 532*)
$(C_3H_7)_3SnBr$		133(13)	(*192, 198, 199, 256, 371, 390, 521, 532*)
$(C_3H_7)_3SnCl$		123(13)	*94a, 192, 328, 347, 389, 390, 521, 532, 550*)
$(C_3H_7)_3SnF$	275		(*27, 292, 389, 390*)
$(i\text{-}C_3H_7)_3SnI$		256–8(760)	(*80*)

APPENDIX 1—*continued*

Compound	m.p., °C	b.p., °C(mm)	References
$(i\text{-}C_3H_7)_3SnCl$			(*228*, *429*)
$(c\text{-}C_3H_5)_3SnI$		98(0.4)	(*467*)
$(c\text{-}C_3H_5)_3SnBr$		80.5(0.2)	(*467*)
$(c\text{-}C_3H_5)_3SnCl$		84–5(0.4)	(*467*)
$(c\text{-}C_3H_5)_3SnF$	155/760 subl.		(*467*)
$(C_4H_9)_3SnI$		168(8)	(*233*, *336*, *339*)
$(C_4H_9)_3SnBr$		163(12)	(*233*, *463*, *521*, *532*)
$(C_4H_9)_3SnCl$		145–7(5)	(*110*, *228*, *232*, *250*, *255*, *328*, *331*, *341*, *347*, *359*, *463*, *472*, *521*, *532*, *550*)
$(i\text{-}C_4H_9)_3SnI$		151(13)	(*78*, *80*, *192*, *521*, *523*)
$(i\text{-}C_4H_9)_3SnBr$		148(13)	(*192*, *521*, *523*)
$(i\text{-}C_4H_9)_3SnCl$		142(13)	(*192*, *521*, *523*)
$(i\text{-}C_4H_9)_3SnF$	244		(*27*, *292*)
$(t\text{-}C_4H_9)_3SnI$		147.5(12)	(*300*)
$(t\text{-}C_4H_9)_3SnCl$		132(12)	(*300*, *429*)
$(t\text{-}C_4H_9)_3SnF$	257d		(*300*)
$(c\text{-}C_6H_{11})_3SnI$	65		(*296*)
$(c\text{-}C_6H_{11})_3SnBr$	77		(*296*)
$(c\text{-}C_6H_{11})_3SnCl$	129–30		(*191*, *296*)
$(c\text{-}C_6H_{11})_3SnF$	305d		(*296*)
$(C_6H_5)_3SnI$	212		(*43*, *70*, *94*, *107*, *168*, *174b*, *256*, *282*, *291*, *337*, *364*, *389*, *390*)
$(C_6H_5)_3SnBr$	121–2		(*94*, *115*, *291*, *375*, *389*, *390*)
$(C_6H_5)_3SnCl$	105.5–7	249(13.5)	(*19*, *20*, *23*, *27*, *28*, *106*, *107*, *115*, *134*, *168*, *169*, *171*, *173*, *255*, *261*, *274*, *291*, *328*, *364*, *365*, *375*, *394*, *401*, *429*, *430*, *434*, *435*, *451*, *488*, *509*, *510*, *536*, *537*, *552*, *553*)
$(C_6H_5)_3SnF$	357d		(*27*, *134*, *293*, *509*)
$(C_{10}H_7)_3SnCl$	204–5		(*299*)
$(C_6F_5)_3SnCl$	106		(*214*)
$(C_5H_5)_3SnCl$			(*432*)
$(CH_2{=}CH)_3SnI$		60(1.8–9)	(*463*)
$(CH_2{=}CH)_3SnBr$		53–40.3–4	(*463*)
$(CH_2{=}CH)_3SnCl$		59–60(6)	(*446*)
$(CH_2{=}CH)_3SnF$	300		(*463*, *471*)
$(CH_2Br)_3SnBr$		165(5)	(*539*, *541*)
$(CH_2Cl)_3SnCl$		138–40(5)	(*539*)
$[(CH_3)_3SiCH_2]_3SnI$		105(0.3)	(*464*)
$[(CH_3)_3CCH_2]_3SnI$	88–91		(*551*)
$[(CH_3)_3CCH_2]_3SnBr$	112.5–113.5		(*551*)
$[(CH_3)_3CCH_2]_3SnCl$	57–58		(*551*)

APPENDIX 2

DIORGANOTIN DIHALIDES R_2SnX_2

Compound	m.p., °C	b.p., °C(mm)	References
$(CH_3)_2SnI_2$	44	228(760)	(*76, 237, 320, 336, 413, 415, 481*)
$(CH_3)_2SnBr_2$	74	209(760)	(*76, 268, 270, 287, 413, 414, 415, 481, 484*)
$(CH_3)_2SnCl_2$	107.5–8	185–90(760)	(*16, 76, 175, 210, 271, 272, 287, 340, 401, 413, 415, 427, 444, 481, 483, 484, 509, 510, 512*)
$(CH_3)_2SnF_2$	360d		(*292*)
$(CH_2Br)_2SnBr_2$	87	155–60(5)	(*539, 541*)
$(CH_2Cl)_2SnCl_2$	89.5–90		(*539, 541*)
$(C_2H_5)_2SnI_2$	44–5	245–6(760)	(*76, 128, 154, 196, 237, 260, 307, 309, 315, 326, 335, 336, 410, 415, 451, 497, 531*)
$(C_2H_5)_2SnBr_2$	63	233(760)	(*76, 154, 196, 206, 223, 238, 251, 268, 270, 275, 326, 328, 369, 377, 410, 451, 531*)
$(C_2H_5)_2SnCl_2$	89	277(760)	(*16, 28, 69, 76, 118, 127, 152, 231, 232, 253, 268–271, 275, 279, 310, 326, 328, 330, 341, 342, 375, 409, 410, 411, 443, 451, 453, 482, 488, 512, 529, 531*)
$(C_2H_5)_2SnF_2$	287–90d		(*76, 292, 540*)
$(C_3H_7)_2SnI_2$		166–7(10)	(*80, 199, 221, 237, 415*)
$(C_3H_7)_2SnBr_2$	49	263–5(760)	(*2, 198, 199, 233, 268, 270, 415*)
$(C_3H_7)_2SnCl_2$	80–1		(*80, 221, 268, 270, 347, 389, 390, 415, 451, 512*)
$(C_3H_7)_2SnF_2$	204–5		(*292*)
$(i\text{-}C_3H_7)_2SnI_2$		256–8(760)	(*80*)
$(i\text{-}C_3H_7)_2SnBr_2$	54		(*131*)
$(i\text{-}C_3H_7)_2SnCl_2$	80–4		(*80, 131, 430*)
$(1\text{-}C_3H_5)_2SnBr_2$		121–2(10)	(*372–374*)
$(1\text{-}C_3H_5)_2SnCl_2$			(*372*)
$(i\text{-}C_3H_5)_2SnBr_2$		102.5(9)	(*373, 374, 375*)
$(c\text{-}C_3H_5)_2SnI_2$	37.5–8		(*467*)
$(c\text{-}C_3H_5)_2SnBr_2$	53–4	113(0.2)	(*467*)
$(c\text{-}C_3H_5)_2SnCl_2$	59–60	105–6(1)	(*467*)
$(C_4H_9)_2SnBr_2$	20		(*2, 415*)
$(C_4H_9)_2SnCl_2$	43	135(10)	(*136, 230, 232, 250, 328, 331, 341, 342, 347, 402, 529, 550*)
$(C_4H_9)_2SnBrCl$	34–34.5	104–6(0.55)	(*218, 231*)

APPENDIX 2—*continued*

Compound	m.p., °C	b.p., °C(mm)	References
$(i\text{-}C_4H_9)_2SnI_2$		290–5(760)	(*76, 237*)
$(i\text{-}C_4H_9)_2SnBr_2$		144(15)	(*294*)
$(i\text{-}C_4H_9)_2SnCl_2$	9	135.5(15)	(*76, 294*)
$(t\text{-}C_4H_9)_2SnBr_2$		128(14)	(*300*)
$(t\text{-}C_4H_9)_2SnCl_2$	42	117(14)	(*300, 429*)
$(t\text{-}C_4H_9)_2SnF_2$	254d		(*300*)
$(C_6H_5)_2SnI_2$	71–2	176–82(2)	(*70, 107*)
$(C_6H_5)_2SnBr_2$	38		(*20, 22, 115, 275, 293, 328, 377, 417, 487*)
$(C_6H_5)_2SnCl_2$	42–4	333–7(760)	(*20, 28, 42, 91, 142, 168, 170, 176, 261, 266, 267, 271, 272, 275, 374, 377, 381, 383, 401, 433, 482, 537*)
$(C_6H_5)_2SnF_2$	>360		(*289*)
$(C_6H_5)_2SnClI$	69		(*20*)
$(C_6H_5)_2SnClBr$	39		(*20*)
$(C_6F_5)_2SnCl_2$		130(2)	(*215*)
$(c\text{-}C_6H_{11})_2SnI_2$	42		(*232, 296*)
$(c\text{-}C_6H_{11})_2SnBr_2$	58		(*232, 296*)
$(c\text{-}C_6H_{11})_2SnCl_2$	88–9		(*42, 296*)
$(c\text{-}C_6H_{11})_2SnF_2$	278		(*296*)
$(1\text{-}C_{10}H_7)_2SnI_2$	160		(*498*)
$(1\text{-}C_{10}H_7)_2SnBr_2$	142		(*275, 374, 377, 408*)
$(1\text{-}C_{10}H_7)_2SnCl_2$	137–7.5		(*42, 275, 377, 375, 408*)
$(2\text{-}C_{10}H_7)_2SnBr_2$	114–5		(*275, 377*)
$(2\text{-}C_{10}H_7)_2SnCl_2$	110–1		(*275, 377*)
$(CH_2{=}CH)_2SnI_2$			(*432*)
$(CH_2{=}CH)_2SnBr_2$		47(2)	(*376*)
$(CH_2{=}CH)_2SnCl_2$	74.5–5.5	54–6(3)	(*376, 446, 471*)
$(CH_2{=}CH)_2SnF_2$			(*432*)
$[(CH_3)_3SiCH_2]_2SnI_2$	34.6–35.4		(*464*)
$[(CH_3)_3SiCH_2]_2SnBr_2$	38.6–39.8		(*464*)
$[(CH_3)_3CCH_2]_2SnBr_2$		87(0.1)	(*551*)
$[(CH_3)_3CCH_2]_2SnCl_2$	43	65(0.1)	(*551*)

APPENDIX 3

MONOORGANOTIN HALIDES $RSnX_3$

Compound	m.p., °C	b.p., °C(mm)	References
CH_3SnI_3	85		(*129, 264, 265, 412, 413, 424, 425, 451, 481, 484, 508*)
CH_3SnBr_3	53	211(760)	(*129, 143, 264, 265, 413, 414, 424, 425, 473, 481, 484*)
CH_3SnCl_3	45–6		(*129, 210, 264, 265, 413, 414, 424, 425, 441, 481, 484*)
$BrCH_2SnBr_3$		109(5)	(*539, 541*)
$ClCH_2SnCl_3$		72.5–3(5)	(*539, 541*)
$C_2H_5SnI_3$		181–4.5(19)	(*508*)
$C_2H_5SnBr_3$		103–6(14)	(*130, 268, 328*)
$C_2H_5SnCl_3$		196–8(760)	(*127, 232, 252, 328*)
$C_3H_7SnI_3$		200(16)d	(*508*)
$C_3H_7SnBr_3$		98–9(12)	(*132*)
$C_3H_7SnCl_3$			(*132, 389, 390*)
$i\text{-}C_3H_7SnBr_3$	112		(*131*)
$i\text{-}C_3H_7SnCl_3$		75(16)	(*508*)
$c\text{-}C_3H_5SnBr_3$		56–7(0.1)	(*467*)
$c\text{-}C_3H_5SnCl_3$		59–61(1)	(*467*)
$C_5H_5SnCl_3$			(*432*)
$C_6H_5SnI_3$		220(760)d	(*266, 267, 508*)
$C_6H_5SnBr_3$		245(760)	(*266, 267, 521, 532*)
$C_6H_5SnCl_3$		182–3(29)	(*53, 115, 170, 266, 267, 328, 521, 532, 537*)
$1\text{-}C_{10}H_7SnCl_3$	77–8		(*408*)
$CH_2{=}CHSnBr_3$	119		(*490*)
$CH_2{=}CHSnCl_3$		64–5(15)	(*432, 446, 471, 490*)

APPENDIX 4

TRIORGANOTIN HALIDES $R_2R'SnX$

Compound	m.p., °C	b.p., °C(mm⁻¹)	References
$(CH_3)_2CH_2ISnI$		115(5)	(*469*)
$(CH_3)_2CH_2BrSnBr$		75–9(4)	(*469*)
$(CH_3)_2CH_2ClSnCl$		76–7(11)	(*469*)
$(CH_3)_2CH_2ISnF$	250d		(*469*)
$(CH_3)_2C_2H_5SnI$		77–8(11)	(*338*, *368*, *422*)
$(CH_3)_2C_2H_5SnBr$		175–80(760)	(*71*)
$(CH_3)_2C_2H_5SnCl$		166–8(760)	(*340*)
$(CH_3)_2CH_2{=}CHSnI$		57–9(5.2)	(*463*, *466*)
$(CH_3)_2CH_2{=}CHSnBr$		59–61(9.5)	(*463*, *466*)
$(CH_3)_2CH_2{=}CHSnCl$		73–5(27)	(*463*, *466*)
$(CH_3)_2i\text{-}C_3H_7SnI$		77–8(9.2)	(*465*)
$(CH_3)_2C_4H_9SnI$		118–20(25)	(*254*, *328*, *337*, *465*, *521*, *532*)
$(CH_3)_2i\text{-}C_4H_9SnI$		95(15)	(*337*, *521*)
		8	(*523*, *532*)
		84(5.5)	(*465*)
$(CH_3)_2(CH_2)_5BrSnBr$		168(15)	(*193*)
$(CH_3)_2(CH_3)_3SiCH_2SnBr$		50(0.35)	(*253*)
$(CH_3)_2c\text{-}C_6H_{11}SnI$		86(0.65)	(*255*)
$(CH_2Cl)_2C_4H_9SnCl$		82–7(0.3)	(*469*)
$(C_2H_5)_2CH_3SnBr$		95–8(11)	(*471*)
$(C_2H_5)_2C_3H_7SnI$		132–4(16)	(*338*, *521*, *532*)
$(C_2H_5)_2C_3H_7SnBr$		112.2(16)	(*192*, *532*)
$(C_2H_5)_2C_3H_7SnCl$		108(17)	(*192*, *521*, *532*)
$(C_2H_5)_2C_3H_7SnF$	271		(*292*)
$(C_2H_5)_2C_4H_9SnI$		134–5(13)	(*254*, *328*, *337*, *521*) *532*)
$(C_2H_5)_2i\text{-}C_4H_9SnBr$		122(17)	(*192*, *233*, *521*, *532*)
$(C_2H_5)_2C_6H_5SnBr$			(*253*)
$(C_2H_5)_2n\text{-}C_5H_{11}SnBr$		135(15)	(*193*)
$(C_2H_5)_2(CH_2)BrSnBr$		190.5(16)	(*193*)
$(C_2H_5)_2(CH_3)_3CCH_2SnBr$		64(0.1)	(*551*)
$(CH_2{=}CH)_2C_4H_9SnBr$		70–2(0.6)	(*462*)
$(CH_2{=}CH)_2C_4H_9SnCl$		82–4(3)	(*447*)
$(CH_2{=}CH)_2C_6H_5SnCl$			(*432*)
$(CH_2{=}CH)_2C_6H_5SnF$	300d		(*463*)
$(C_3H_7)_2C_4H_9SnI$		159–60(24)	(*338*, *521*, *532*)
$(C_4H_9)_2CH_3SnI$		82(0.35)	(*255*)
$(C_4H_9)_2CH_2ClSnCl$		106–10(3)	(*469*)
$(i\text{-}C_3H_7)_2CH_3SnI$		96(7.6)	(*465*)
$(C_4H_9)_2CH_2{=}CHSnI$		109(1.75)	(*463*, *466*)
$(C_4H_9)_2CH_2{=}CHSnBr$		72–3(0.03)	(*446*, *463*, *466*)
$(C_4H_9)_2CH_2{=}CHSnCl$		112–4(4)	(*447*, *463*, *466*)
$(C_4H_9)_2c\text{-}C_5H_9SnI$		125(0.4)	(*255*)

APPENDIX 4—*continued*

Compound	m.p., °C	b.p., °C(mm⁻¹)	References
$(C_4H_9)_2c\text{-}C_6H_{11}SnI$		136(0.6)	*(255)*
$(C_4H_9)_2C_6H_5SnBr$		12–4(0.001)	*(445)*
$(C_4H_9)(CH_3)_3SiCH_2SnBr$		91(0.2)	*(464)*
$(i\text{-}C_4H_9)_2CH_3SnI$		71(0.25)	*(255)*
$(i\text{-}C_4H_9)_2C_2H_5SnBr$		130.6(13)	*(192, 521, 532)*
$(C_6H_5)_2CH_2BrSnF$	290		*(469)*
$(C_6H_5)_2C_4H_9SnBr$		155.9(0.2)	*(445)*
$(C_6H_5)_2CH_2CNSnI$	99–100		*(255, 328)*
$(C_6H_5)_2CH_2{=}CHSnBr$			*(445)*
$(C_6H_5)_2CH_2{=}CHSnCl$			*(432)*
$(C_6H_5)_2CH_2{=}CHSnF$	300		*(463)*
$(C_6H_{11})_2CH_3SnI$		134(0.4)	*(255)*
$[(CH_3)_3SiCH_2]_2CH_3SnI$		85–6(0.35)	*(464)*
$[(CH_3)_3SiCH_2]_2CH_3SnBr$		73(0.18)	*(464)*
$[(CH_3)_3SiCH_2]_2C_4H_9SnCl$		82–7(0.3)	*(469)*
$[(CH_3)_3CCH_2]_2C_2H_5SnBr$		75(0.1)	*(551)*
$[(CH_3)_3CCH_2]_2C_4H_9SnBr$		94(0.1)	*(551)*

APPENDIX 5

TRIORGANOTIN HALIDES RR′R″SnX

Compound	m.p., °C	b.p., °C(mm)	References
$(CH_3)(C_2H_5)(C_3H_7)SnI$		108–11(11)	*(368, 368a, 421, 422)*
$(C_4H_9)(C_6H_5)(C_6H_5CH_2)SnF$	218		*(261)*

APPENDIX 6

DIORGANOTIN DIHALIDES RR′SnX$_2$

Compound	m.p., °C	b.p., °C(mm)	References
$(CH_3)(C_2H_5)SnCl_2$	52		*(69, 318)*
$(C_2H_5)(C_3H_7)SnCl_2$	57–8		*(69, 486)*
$(C_2H_5)(C_6H_5)SnCl_2$	45		*(310)*
$(C_2H_5)[(CH_3)_3CCH_2]SnBr_2$		65(0.1)	*(551)*
$(C_2H_5)[(CH_3)_3CCH_2]SnCl_2$		63(0.1)	*(551)*
$(C_3H_7)(C_4H_9)SnCl_2$	67–8		*(342)*
$(C_4H_9)(CH_2{=}CH)SnBr_2$			*(342)*
$(C_4H_9)(CH_2{=}CH)SnCl_2$	27–8	99–101(3)	*(432, 447)*
$(C_4H_9)(C_6H_5)SnCl_2$	50		*(261)*
$(C_5H_5)(CH_2{=}CH)SnCl_2$			*(432)*
$(C_6H_5)(CH_2{=}CH)SnCl_2$			*(432)*

APPENDIX 7

OTHER ORGANOTIN HALIDES

Compound	m.p., °C	References
$Br(C_4H_9)_2SnSn(C_4H_9)_2Br$[a]	103–4	(*231*)
$Cl(C_4H_9)_2SnSn(C_4H_9)_2Cl$[a]	110–2	(*158b, 230, 231, 387a, 455, 456a, 489a*)
$Cl(i\text{-}C_4H_9)_2SnSn(i\text{-}C_4H_9)_2Cl$	—	(*489a*)
$Cl(C_2H_5)_2SnSn(C_2H_5)_2Cl$[a]	170–6	(*231*)
$Cl(C_6H_5)_2SnSn(C_6H_5)_2Cl$[a]	185–7	(*231*)
$Cl(C_3H_7)_2SnSn(C_3H_7)_2Cl$[a]	120.5–1.5	(*231*)
$Tropylium_3$-SnCl	252	(*61*)
$(CH_3)_2$-8-Hydroxyquinoline-SnCl	147–8	(*536*)
$(C_6H_5)_2$-8-Hydroxyquinoline-SnCl	155–7	(*536*)
$(C_6H_5)_2$-8-Hydroxyquinoline-SnCl	216–7	(*536*)
$F(C_4H_9)_2SnOSn(C_4H_9)_2F$	115–7	(*10, 11*)
$Cl(CH_3)_2SnOSn(CH_3)_3Cl$	360	(*11, 397*)
$Cl(C_2H_5)_2SnOSn(C_2H_5)_2Cl$	175	(*11, 397*)
$Br(C_4H_9)_2SnOSn(C_4H_9)_2Br$	104	(*11*)
$Cl(C_3H_7)_2SnOSn(C_3H_7)_2Cl$	121–2	(*397*)
$Br(C_2H_5)_2SnOSn(C_2H_5)_2Br$	172–3	(*397*)
$Cl(C_4H_9)_2SnOSn(C_4H_9)_2Cl$	109	(*11, 397*)
$Cl(C_4H_9)_2SnOSn(C_2H_5)_3$	—	(*117*)
$Cl(C_4H_9)_2SnOSn(C_4H_9)_3$	—	(*117*)
$Cl(CH_3)_2SnOSnC_2H_5Cl_2$	89–91	(*117*)
$Cl(C_4H_9)_2SnOSnC_4H_9Cl_2$	34–35	(*117*)
$Cl(C_8H_{16})_2SnOSnC_4H_9Cl_2$	42–43	(*117*)
$Cl(C_4H_9)_2SnOSn(OH)Cl_2$	46–47	(*117*)
$Cl(C_8H_{16})_2SnOSn(OH)Cl_2$	42–44	(*117*)

[a] See Sec. III.C.

REFERENCES

1. E. W. Abel, R. P. Bush, C. R. Jenkins, and T. Zobel, *Trans. Faraday Soc.*, **60**, 1214 (1964).

2. L. V. Abramova, N. J. Sheverdina, and K. A. Kocheshkov, *Dokl. Akad. Nauk S.S.S.R.*, **123**, 681 (1958).

3. L. V. Abramova, N. J. Sheverdina, and K. A. Kocheshkov, *Chem. Abst.*, **58**, 11391b (1963).

4. L. V. Abramova, N. J. Sheverdina, and K. A. Kocheshkov, *Industrial Uses of Large Radiation Sources*, Vol. I, IAEA, Vienna, 1961, p. 83.

5. A. Yu. Aleksandrov, N. N. Delyagin, K. P. Mitrofanov, L. S. Polak, and V. S. Shpinel, *Zh. Eksp. Teor. Fiz.*, **43**, 1242 (1962).

6. A. Yu. Aleksandrov, N. N. Delyagin, K. P. Mitrofanov, L. S. Polak, and V. S. Shpinel, *Dokl. Akad. Nauk S.S.S.R.*, **148**, 126 (1963).
7. A. Yu. Aleksandrov, Ya. G. Dorfman, O. L. Lependina, K. P. Mitrofanov, M. V. Plotnikov, L. S. Polak, A. Ya. Temkin, and V. S. Shpinel, *Zh. Fiz. Khim.*, **38**, 2190 (1964).
8. R. B. Allan, Dissertation, Univ. of New Hampshire (1959).
9. D. L. Alleston and A. G. Davies, *J. Chem. Soc.*, 2050 (1692).
10. D. L. Alleston, A. G. Davies, and B. N. Figgis, *Proc. Chem. Soc.*, 457 (1961).
11. D. L. Alleston, A. G. Davies, M. Hancock, and R. F. M. White, *J. Chem. Soc.*, 5469 (1963).
11a. D. L. Alleston and A. G. Davies, *Chem. Ind.*, 949 (1961).
12. C. B. Allsopp, *Proc. Roy. Soc. A*, **158**, 167 (1937).
13. H. H. Anderson, *Inorg. Chem.*, **1**, 647 (1962).
14. H. H. Anderson, *J. Org. Chem.*, **19**, 1766 (1954).
15. H. H. Anderson, *J. Am. Chem. Soc.*, **79**, 4913 (1957).
16. B. A. Arbuzow and A. N. Pudovic, *Zh. Obshch. Khim.*, **17**, 2158 (1947).
17. L. V. Armenskaya, K. N. Korotaevskii, E. N. Lysenko, and L. M. Monastyrskii, *U.S.S.R.*, patent **184**, 853 (1966); through *CA*, **66**, 71949q (1967).
18. L. V. Armenskaya, E. N. Korotaevskii, E. N. Lysenko, L. M. Monastyrskii, and Z. S. Smolyan, *U.S.S.R.*, patent **172**, 785 (1965); through *CA*, **64**, 1662d (1966).
19. B. Aronheim, *Chem. Ber.*, **12**, 509 (1879).
20. B. Aronheim, *Ann. Chem.*, **194**, 145 (1878).
21. G. B. Bachman, G. L. Carlson, and M. Robinson, *J. Am. Chem. Soc.*, **73**, 1964 (1951).
22. H. S. Backer and J. Kramer, *Rec. Trav. Chim. Pays-Bas Belges*, **53**, 1101 (1934).
23. G. Bähr, *Z. Anorg. Allgem. Chem.*, **256**, 107 (1948).
24. G. Bähr and R. Gelius, *Chem. Ber.*, **91**, 812 (1958).
25. G. Bähr and G. Zoche, *Chem. Ber.*, **88**, 1450 (1955).
26. J. C. Bailie, *Iowa State Coll. J. Sci.*, **14**, 8 (1939); through *CA*, **34**, 6241 (1940).
27. H. Ballczo and H. Schiffner, *Z. Anal. Chem.*, **152**, 3 (1956).
28. R. Barbieri, U. Belluco, and G. Tagliavini, *Ann. Chim. Rome*, **48**, 940 (1958).
29. I. R. Beattie, *Quart. Rev. London*, **17**, 382 (1963).
30. I. R. Beattie and T. Gilson, *J. Chem. Soc.*, 2585 (1961).
31. I. R. Beattie and G. P. McQuillan, *J. Chem. Soc.*, 1519 (1963).
32. I. R. Beattie, G. P. McQuillan, and R. Hulme, *Chem. Ind.*, 1429 (1962).
33. E. V. Van den Berghe, unpublished results.
34. E. V. Van den Berghe and G. P. Van der Kelen, *Ber. Bunsenges. Phys. Chem.*, **68**, 652 (1964).
35. E. V. Van den Berghe and G. P. Van der Kelen, *Bull. Soc. Chim. Belges*, **74**, 479 (1965).
36. E. V. Van den Berghe and G. P. Van der Kelen, *J. Organometal. Chem.*, **6**, 522 (1966).
37. E. V. Van den Berghe and G. P. Van der Kelen, *J. Organometal. Chem.*, **6**, 515 (1966).
38. E. V. Van den Berghe, G. P. Van der Kelen, and Z. Eeckhaut, *Bull. Soc. Chim. Belges*, **76**, 79 (1967).
39. E. V. Van den Berghe and G. P. Van der Kelen, *J. Organometal. Chem.*, **11**, 479 (1968).
40. E. V. Van den Berghe, L. Verdonck, and G. P. Van der Kelen, *J. Organometal. Chem.*, **16**, 497 (1969).
41. D. Blake, G. E. Coates, and J. M. Tate, *J. Chem. Soc.*, 756 (1961).
42. T. S. Bobrashinskaya and K. A. Kocheshkov, *J. Gen. Chem. U.S.S.R.*, **8**, 1850 (1938).
43. J. Boescken and J. J. Rutgers, *Rec. Trav. Chim. Pays-Bas Belges*, **42**, 1017 (1923).
44. T. F. Bolles and R. S. Drago, *J. Am. Chem. Soc.*, **87**, 5015 (1965).

45. T. F. Bolles and R. S. Drago, *J. Am. Chem. Soc.*, **88**, 392 (1966).
46. T. F. Bolles and R. S. Drago, *J. Am. Chem. Soc.*, **88**, 5730 (1966).
47. A. E. Borisov and N. V. Novikova, *Izv. Akad. Nauk S.S.S.R.*, 1670 (1959).
48. R. W. Bost and P. Borgstrom, *J. Am. Chem. Soc.*, **51**, 1922 (1929).
49. R. W. Bost and H. R. Baker, *J. Am. Chem. Soc.*, **55**, 1112 (1933).
50. R. A. Bott, C. Eaborn, and J. A. Waters, *J. Chem. Soc.*, 681 (1963).
51. S. Boué, M. Gielen, and J. Nasielski, *Bull. Soc. Chim. Belges*, **73**, 864 (1964).
52. S. Boué, M. Gielen, and J. Nasielski, *J. Organometal. Chem.*, **9**, 443 (1967).
53. E. M. Brainina and R. K. Freidlina, *Bull. Acad. Sci., URSS Classe Sci. Chim.*, 623 (1947).
54. L. O. Brockway and H. O. Jenkins, *J. Am. Chem. Soc.*, **58**, 2036 (1936).
55. D. H. Brown, A. Mohammed, and D. W. A. Sharp, *Spectrochim. Acta*, **21**, 1013 (1965).
56. T. L. Brown and G. L. Morgan, *Inorg. Chem.*, **2**, 736 (1963).
57. T. L. Brown and J. C. Puckett, *J. Chem. Phys.*, **44**, 2238 (1966).
58. T. L. Brown and K. Stark, *J. Phys. Chem.*, **69**, 2679 (1965).
59. M. P. Brown and D. E. Webster, *J. Phys. Chem.*, **64**, 698 (1960).
60. W. Brügel, T. Ankel, and F. Krückeberg, *Z. Elektrochem.*, **64**, 1121 (1960).
61. D. Bryce-Smith and N. A. Perkins, *Chem. Ind.*, 1022 (1959).
62. E. V. Bryuchova, G. K. Semin, V. I. Goldanskii, and V. V. Khrapov, *Chem. Commun.*, 491 (1968).
63. V. A. Bryukhanov, V. I. Goldanskii, N. N. Delyagin, L. A. Korylko, E. F. Makarov, I. P. Suzdalev, and V. S. Shpinel, *Zh. Eksp. Teor. Fiz.*, **43**, 448 (1962).
64. O. Buchman, M. Grosjean, and J. Nasielski, *Helv. Chim. Acta*, **47**, 1679 (1964).
65. O. Buchman, M. Grosjean, and J. Nasielski, *Helv. Chim. Acta*, **47**, 2037 (1964).
66. O. Buchman, M. Grosjean, J. Nasielski, and B. Wilmet-Devos, *Helv. Chim. Acta*, **47**, 1688 (1964).
67. G. B. Buckton, *Ann. Chem.*, **112**, 220 (1859).
68. R. H. Bullard, *J. Am. Chem. Soc.*, **51**, 3065 (1929).
69. R. H. Bullard and F. R. Holden, *J. Am. Chem. Soc.*, **53**, 3150 (1931).
70. R. H. Bullard and W. B. Robinson, *J. Am. Chem. Soc.*, **49**, 1368 (1927).
71. R. H. Bullard and R. A. Vingee, *J. Am. Chem. Soc.*, **51**, 892 (1929).
72. I. Burdon, P. L. Coc, and M. Fulton, *J. Chem. Soc.*, 2094 (1965).
73. J. J. Burke and P. C. Lauterbur, *J. Am. Chem. Soc.*, **83**, 326 (1960).
74. F. K. Butcher, W. Gerrard, E. F. Mooney, R. G. Raas, H. A. Willis, A. Anderson, and H. A. Gebbie, *J. Organometal. Chem.*, **1**, 431 (1964).
75. H. Buttgenbach, *Mem. Soc. Roy. Sci. Liège*, **12**, 3 (1924).
76. A. Cahours, *Ann. Chem.*, **114**, 227; 354 (1860).
77. A. Cahours, *Compt. Rend.*, **76**, 133 (1873).
78. A. Cahours, *Compt. Rend.*, **77**, 1403 (1873).
79. A. Cahours, *Compt. Rend.*, **88**, 725 (1879).
80. A. Cahours and E. Demarcey, *Compt. Rend.*, **89**, 68 (1879).
81. A. Cahours and E. Demarcey, *Bull. Soc. Chim. France*, **34**, 475 (1880).
82. A. Cassol, *Gazz. Chim. Ital.*, **96**, 1764 (1966).
83. A. Cassol and R. Barbieri, *Ann. Chim.*, **55**, 606 (1965).
84. A. Cassol and L. Magon, *Gazz. Chim. Ital.*, **96**, 172[illegible] (1966).
85. A. Cassol and L. Magon, *Gazz. Chim. Ital.*, **96**, 1752 (1966).
86. A. Cassol, L. Magon, and R. Barbieri, *J. Chromatog.*, **19**, 57 (1965).
87. A. Cassol, L. Magon, and R. Barbieri, *Inorg. Nucl. Chem. Lett.*, **3**, 25 (1967).
88. A. Cassol and R. Portanova, *Gazz. Chim. Ital.*, **96**, 1734 (1966).

89. A. Cassol, R. Portanova, and L. Magon, *Ricerca Sci.*, **36**, 1180 (1966).
90. C. J. Cattanach and E. F. Mooney, *Spectrochim. Acta*, **24**, 407 (1968).
91. F. Challenger and E. Rothstein, *J. Chem. Soc.*, 1258 (1934).
92. R. D. Chambers and T. Chivers, *J. Chem. Soc.*, 4782 (1964).
93. R. D. Chambers, H. C. Clark, and C. J. Willis, *Can. J. Chem.*, **39**, 131 (1961).
94. R. D. Chambers and P. C. Scherer, *J. Am. Chem. Soc.*, **48**, 1054 (1926).
94a. J. Chatt and A. A. Williams, *J. Chem. Soc.*, 4403 (1954).
95. H. C. Clark and R. G. Goel, *J. Organometal. Chem.*, **7**, 263 (1967).
96. H. C. Clark, J. T. Kwon, L. W. Reeves, and E. J. Wells, *Inorg. Chem.*, **3**, 907 (1964).
97. H. C. Clark and R. J. O'Brien, *Proc. Chem. Soc.*, 111 (1963).
98. H. C. Clark and R. J. O'Brien, *Inorg. Chem.*, **2**, 740 (1963).
99. H. C. Clark and R. J. O'Brien, *Inorg. Chem.*, **2**, 1020 (1963).
100. H. C. Clark, R. J. O'Brien, and A. L. Pickard, *J. Organometal. Chem.*, **4**, 43 (1965).
101. H. C. Clark, R. J. O'Brien, and J. Trotter, *Proc. Chem. Soc.*, 85 (1963).
102. H. C. Clark, R. J. O'Brien, and J. Trotter, *J. Chem. Soc.*, 2332 (1964).
103. J. P. Clark and C. J. Wilkins, *J. Chem. Soc. A*, 871 (1966).
104. R. J. H. Clark and C. S. Williams, *Spectrochim. Acta*, **21**, 1861 (1965).
105. G. E. Coates, *Organometallic Compounds*, Wiley, New York, 1960.
106. G. Cohen, C. M. Murphy, J. G. O'Rear, H. Ravner, and W. A. Zisman, *Ind. Engng. Chem.*, **45**, 1766 (1953).
107. D. Colaitis and M. Lesbre, *Bull. Soc. Chim. France*, 1069 (1952).
108. J. Combarieu, I. Raitzyn, and G. Wetroff, French Pat. 1 399, 552 (1965).
109. M. Cordey-Hayes, R. D. Peacock, and M. Vucelic, *J. Inorg. Nucl. Chem.*, **29**, 1177 (1967).
110. G. Costa, *Gazz. Chim. Ital.*, **80**, 42 (1950).
111. G. Costa, *Ann. Chim. Rome*, **41**, 207 (1951).
112. R. A. Cummins, *Australian J. Chem.*, **16**, 985 (1963).
113. R. A. Cummins, *Australian J. Chem.*, **18**, 985 (1965).
114. O. Danek, *Coll. Czech. Chem. Commun.*, **26**, 2035 (1961).
115. J. D'Ans and H. Zimmer, *Chem. Ber.*, **85**, 585 (1952).
116. J. D'Ans, H. Zimmer, E. Endrulat, and K. Lübke, *Naturwissenschaften*, **39**, 450 (1952).
117. A. G. Davies and P. G. Harrison, *J. Organometal. Chem.*, **7**, P 13 (1967).
118. H. Davies and F. S. Kipping, *J. Chem. Soc.*, 296 (1911).
119. N. W. G. Debye, E. Rosenberg, and J. J. Zuckerman, *J. Am. Chem. Soc.*, **90**, 3234 (1968).
119a. M. L. Delwaulle, M. B. Buisset, and M. Delhaye, *J. Am. Chem. Soc.*, **74**, 5768 (1952).
120. R. E. Dessy, W. Kitching, and T. Chivers, *J. Am. Chem. Soc.*, **88**, 453 (1966).
121. R. E. Dessy and F. Paulik, *J. Chem. Educ.*, **40**, 185 (1963).
122. M. Devaud, *Compt. Rend.*, **263**, 1269 (1966).
123. M. Devaud, *J. Chim. Phys.*, **64**, 791 (1967).
124. M. Devaud, P. Souchay, and M. Person, *J. Chim. Phys.*, **64**, 646 (1967).
125. J. Devooght, M. Gielen, and S. Lejeune, *J. Organometal. Chem.*, **21**, 333 (1970).
126. V. H. Dibeler, *J. Res. Natl. Bur. Standards*, **49**, 235 (1952).
127. R. C. Dillard, E. H. McNeill, D. E. Simmons, and J. B. Yeldell, *J. Am. Chem. Soc.*, **80**, 3607 (1968).
128. L. Domange and J. Guy, *Ann. Pharm. Franç.*, **16**, 161 (1958).
129. J. G. F. Druce, *Chem. News*, **120**, 229 (1920).
130. J. G. F. Druce, *J. Chem. Soc.*, 758 (1921).
131. J. G. F. Druce, *J. Chem. Soc.*, 1859 (1922).
132. J. G. F. Druce, *Chem. News*, **127**, 306 (1923).

133. M. Dub, *Organometallic Compounds*, Vol. II, Springer, Berlin, 1961.
134. T. Dupuis and C. Duval, *Anal. Chim. Acta*, **4**, 615 (1950).
135. I. R. Durig, C. W. Sink, and S. F. Bush, *J. Chem. Phys.*, **45**, 66 (1966).
136. S. A. Edgar and P. A. Teer, *Poultry Sci.*, **36**, 329 (1957); through *CA*, **51**, 15772 (1957).
137. W. F. Edgell and C. H. Ward, *J. Mol. Spectry.*, **8**, 343 (1962).
138. J. C. Egmont, M. J. Janssen, J. G. A. Luijten, G. J. M. Van der Kerk, and G. M. Van der Want, *J. Appl. Chem.*, **12**, 17 (1962).
139. Yu. P. Egorov, *Teor. Eksp. Khim.*, **1**, 30 (1965).
140. Yu. P. Egorov and V. A. Khranovskii, *Teor. Eksp. Khim.*, **2**, 175 (1966).
141. F. W. B. Einstein and B. R. Penfold, *Chem. Commun.*, 780 (1966).
142. I. T. Eskin, A. N. Nesmeyanov, and K. A. Kocheshkov, *Zh. Obshch. Khim.*, **8**, 35 (1938).
143. F. Fairbrother and B. Wright, *J. Chem. Soc.*, 1058 (1949).
144. W. McFarlane, *J. Chem. Soc. A*, 528 (1967).
145. H. N. Farrer, M. M. McGrady, and R. S. Tobias, *J. Am. Chem. Soc.*, **87**, 5019 (1965).
146. C. J. Faulkner, German Pat. 946,447 (1959); through *CA*, **53**, 7014 (1959).
147. A. F. Fentiman, R. E. Wyant, D. A. Jeffrey, and J. F. Kircher, *J. Organometal. Chem.*, **4**, 302 (1965).
148. A. F. Fentiman, R. E. Wyant, J. L. McFarling, and J. F. Kircher, *J. Organometal. Chem.*, **6**, 645 (1966).
149. J. E. Fergusson, W. R. Roper, and C. J. Wilkins, *J. Chem. Soc.*, 3716 (1965).
150. A. Finch, R. C. Poller, and D. Steele, *Trans. Faraday Soc.*, **61**, 2628 (1965).
151. B. W. Fitzsimmons, N. J. Sealey, and A. W. Smith, *Chem. Commun.*, 390 (1968).
152. E. A. Flood and L. Horvitz, *J. Am. Chem. Soc.*, **55**, 2534 (1933).
153. E. Fluck, *Fortschr. Chem. Forsch.*, **5**, 395 (1966).
154. E. Frankland, *Ann. Chem.*, **85**, 329 (1853).
155. O. Fuchs and H. W. Post, *Rec. Trav. Chim.*, **78**, 566 (1959).
156. V. Galasso, G. De Alti, and A. Bigotto, *Z. Phys. Chem.*, **57**, 132 (1968).
157. H. Geissler and H. Kriegsmann, *J. Organometal. Chem.*, **11**, 85 (1968).
158. W. Gerrard, J. B. Leane, E. F. Mooney, and R. G. Rees, *Spectrochim. Acta*, **19**, 1965 (1963).
158a. W. Gerrard, E. F. Mooney, and R. G. Rees, *J. Chem. Soc. A*, 740 (1964).
158b. A. J. Gibbons, A. K. Sawyer, and A. Ross, *J. Org. Chem.*, **26**, 2304 (1961).
159. M. Gielen and G. Mayence, *J. Organometal. Chem.*, **12**, 363 (1968).
160. M. Gielen and J. Nasielski, *J. Organometal. Chem.*, **1**, 173 (1963).
161. M. Gielen and J. Nasielski, *J. Organometal. Chem.*, **7**, 273 (1967).
162. M. Gielen and J. Nasielski, *Ind. Chim. Belge*, **26**, 1393 (1961).
163. M. Gielen and J. Nasielski, *Bull. Soc. Chim. Belges*, **71**, 32; 601 (1962).
164. M. Gielen and J. Nasielski, *Rec. Trav. Chim.*, **82**, 228 (1963).
165. M. Gielen and J. Nasielski, *Ind. Chim. Belge.*, **29**, 767 (1964).
166. M. Gielen, J. Nasielski, J. E. Dubois, and P. Fresnel, *Bull. Soc. Chim. Belges*, **73**, 293 (1964).
167. M. Gielen, J. Nasielski, and R. Yernaux, *Bull. Soc. Chim. Belges*, **72**, 594 (1963).
168. H. Gilman and C. E. Arntzen, *J. Org. Chem.*, **15**, 994 (1950).
169. H. Gilman and L. A. Gist, Jr., *J. Org. Chem.*, **22**, 250 (1957).
170. H. Gilman and L. A. Gist, Jr., *J. Org. Chem.*, **22**, 368 (1957).
171. H. Gilman and T. N. Goreau, *J. Org. Chem.*, **17**, 1470 (1952).
172. H. Gilman and R. W. Leeper, *J. Org. Chem.*, **16**, 466 (1951).
173. H. Gilman and H. W. Melvin, *J. Am. Chem. Soc.*, **71**, 4050 (1949).

174. H. Gilman and S. D. Rosenberg, (a) *J. Org. Chem.*, **18**, 680 (1953); (b) *J. Org. Chem.*, **18**, 1554 (1953).
175. K. Gingold, E. G. Rochow, D. Seyferth, A. Smith, and R. C. West, *J. Am. Chem. Soc.*, **74**, 6306 (1952).
176. A. E. Goddard, J. N. Ashley, and R. B. Evans, *J. Chem. Soc.*, **121**, 978 (1922).
177. D. Goddard and A. E. Goddard, *J. Chem. Soc.*, 256 (1922).
178. V. I. Goldanskii, Consultants Bureau Enterprises, Inc., New York, 1964.
179. V. I. Goldanskii, O. Yu. Okhlobystin, V. I. Rochev, and V. V. Khrapov, *J. Organometal. Chem.*, **4**, 160 (1965).
180. V. I. Goldanskii, V. I. Rochev, and V. V. Khrapov, *Dokl. Akad. Nauk S.S.S.R.*, **156**, 909 (1964).
181. I. P. Goldshtein, N. K. Faizi, N. A. Slovokhotova, E. N. Guryanova, I. M. Viktorova, and K. A. Kocheshkov, *Dokl. Akad. Nauk S.S.S.R.*, **138**, 839 (1961).
182. I. P. Goldshtein, E. N. Guryanova, E. D. Delinskaya, and K. A. Kocheshkov, *Dokl. Akad. Nauk S.S.S.R.*, **136**, 1079 (1961).
183. I. P. Goldshtein, E. N. Guryanova, N. N. Zelmlyanskii, O. P. Syutkina, E. M. Panov, and K. A. Kocheshkov, *Izv. Akad. Nauk S.S.S.R., Ser. Khim.*, **10**, 2201 (1967).
184. D. Grant and J. R. Van Wazer, *J. Organometal. Chem.*, **4**, 229 (1965).
185. P. J. Green and J. D. Craeybeal, *J. Am. Chem. Soc.*, **89**, 4305 (1967).
186. V. S. Griffiths and G. A. W. Derwish, *J. Mol. Spectry.*, **3**, 165 (1959).
187. V. S. Griffiths and G. A. W. Derwish, *J. Mol. Spectry.*, **5**, 148 (1960).
188. V. S. Griffiths and G. A. W. Derwish, *J. Mol. Spectry.*, **7**, 233 (1961).
189. V. S. Griffiths and G. A. W. Derwish, *J. Mol. Spectry.*, **9**, 83 (1962).
190. M. Grosjean, M. Gielen, and J. Nasielski, *Ind. Chim. Belge*, **28**, 721 (1963).
191. G. Grüttner, *Chem. Ber.*, **47**, 3257 (1914).
192. G. Grüttner and E. Krause, *Chem. Ber.*, **50**, 1802 (1917).
193. G. Grüttner, E. Krause, and M. Wiernick, *Chem. Ber.*, **50**, 1549 (1917).
194. V. Gutman, *Halogen Chemistry*, Vol. II, Academic, New York, 1967.
195. T. Harada, *Bull. Chem. Soc. Japan*, **4**, 266 (1929).
196. T. Harada, *Sci. Papers Inst. Phys. Chem. Res. Tokyo*, **35**, 290 (1939).
197. T. Harada, *Sci. Papers Inst. Phys. Chem. Res. Tokyo*, **36**, 504 (1939).
198. T. Harada, *Sci. Papers Inst. Phys. Chem. Res. Tokyo*, **42**, 57; 59; 62; 64 (1947).
199. T. Harada, *Rep. Sci. Res. Inst. Japan*, **24**, 177 (1948).
200. L. A. Harrah, M. T. Ryan, and C. Tamboski, *Spectrochim. Acta*, **18**, 21 (1962).
201. B. J. Hathaway and D. E. Webster, *Proc. Chem. Soc.*, 14 (1963).
202. T. Hayashi, S. Kikhawa, and S. Matsuda, *Kogyo Kagaku Zasshi*, **70**, 1389 (1967).
203. T. Hayashi, S. Kikhawa, S. Matsuda, and K. Fujita, *Kogyo Kagaku Zasshi*, **70**, 2298 (1967).
204. M. C. Hayes, *J. Inorg. Nucl. Chem.*, **26**, 915 (1964).
205. M. C. Hayes, *J. Inorg. Nucl. Chem.*, **26**, 2306 (1964).
206. R. Heap, B. C. Saunders, and G. J. Stacey, *J. Chem. Soc.*, 919 (1949).
207. R. H. Herber and H. I. Parisi, *Inorg. Chem.*, **5**, 769 (1966).
208. R. H. Herber and H. A. Stöckler, *Trans. N.Y. Acad. Sci.*, **26**, 929 (1964).
209. R. H. Herber, H. A. Stöckler, and W. T. Reichle, *J. Chem. Phys.*, **42**, 2447 (1965).
210. S. Hilpert and M. Ditmar, *Chem. Ber.*, **46**, 3738 (1913).
211. M. Hjortdahl, *Compt. Rend.*, **88**, 584 (1879).
212. B. G. Hobroek and R. W. Kiser, *J. Phys. Chem.*, **65**, 2186 (1961).
213. J. R. Holmes and H. D. Kaesz, *J. Am. Chem. Soc.*, *83*, 3903 (1961).
214. J. M. Holmes, R. D. Peacock, and J. C. Tatlow, *Proc. Chem. Soc.*, 108 (1963).
215. J. M. Holmes, R. D. Peacock, and J. C. Tatlow, *J. Chem. Soc. A*, 150 (1966).

216. G. Van Hooydonck, unpublished results.
217. H. H. Huang, K. M. Hui, and K. K. Chiu, *J. Organometal. Chem.*, **11**, 515 (1968).
218. G. Hügel, *Kolloid. Z.*, **131**, 4 (1953).
219. R. Hulme, *J. Chem. Soc.*, 1524 (1963).
220. B. K. Hunter and L. W. Reeves, *Can. J. Chem.*, **46**, 1399 (1968).
221. R. K. Ingham, S. D. Rosenberg, and H. Gilman, *Chem. Rev.*, **60**, 459 (1960).
222. C. K. Ingold, *Structure and Mechanism in Organic Chemistry*, G. Bell and Sons Ltd., London, 1953.
223. J. Ireland, British Pat. 713,727 (1955).
224. N. A. Irisova and E. M. Dianov, *Optika Spektrosk.*, **9**, 261 (1960).
225. R. Irmscher, W. Knöpke, and H. Kunze, German Pat. 1,050,336 (1959).
226. H. Jehring and H. Mehner, *Fresenius Z. Anal. Chem.*, **224**, 136 (1967).
227. H. Jenker and H. W. Shmidt, U.S. Pat. 3,027,393 (1957).
228. W. K. Johnson, Abstracts of papers, 136th Meeting American Chemical Society, 1959, p. 117P.
229. W. T. Johnson, *J. Org. Chem.*, **25**, 2253 (1960).
230. O. H. Johnson and H. E. Fritz, *J. Org. Chem.*, **19**, 74 (1954).
231. O. H. Johnson, H. E. Fritz, D. O. Halvorson, and R. L. Evans, *J. Am. Chem. Soc.*, **77**, 5857 (1957).
231a. O. H. Johnson, *J. Org. Chem.*, **25**, 2262 (1960).
232. W. J. Jones, W. C. Davies, S. T. Bowden, C. Edwards, V. E. Davis, and L. H. Thomas, *J. Chem. Soc.*, 1446 (1947).
233. W. J. Jones, D. P. Evans, T. Gulwell, and D. C. Griffith, *J. Chem. Soc.*, 39 (1935).
234. K. K. Joshi and P. A. Wyatt, *J. Chem. Soc.*, 3825 (1959).
235. J. C. Jungers, L. Sajus, I. De Aguirre, and D. De Croocq, *Rev. Inst. Franç. Petrole Ann. Combust. liquides*, **20**, 545 (1965); *ibid.* **21**, 342 (1966).
236. L. V. Kaabak and A. P. Tomilov, *Zh. Obshch. Khim.*, **33**, 2808 (1963).
237. T. Karantassis and K. Bassileiados, *Compt. Rend.*, **205**, 460 (1937).
238. G. N. Kartsev, Ya. K. Syrkin, A. L. Karvchenko, and V. F. Mironov, *J. Struct. Chem.*, **5**, 591 (1964); *CA*, **61**, 13988a (1964).
239. J. Katomtzeff, *Compt. Rend.*, **230**, 536 (1950).
240. K. Kawakami and R. Okawara, *J. Organometal. Chem.*, **6**, 249 (1966).
241. K. Kawakami, T. Saito, and R. Okawara, *J. Organometal. Chem.*, **8**, 377 (1967).
242. Y. Kawasaki, *J. Inorg. Nucl. Chem.*, **29**, 840 (1967).
243. Y. Kawasaki, *Mol. Phys.*, **12**, 287 (1967).
244. Y. Kawasaki and T. Tanaka, *J. Chem. Phys.*, **43**, 3396 (1965).
245. Y. Kawasaki, T. Tanaka, and R. Okawara, *Inorg. Nucl. Chem. Lett.*, **2**, 9 (1966).
246. Y. Kawasaki, T. Tanaka, and R. Okawara, *Spectrochim. Acta*, **22**, 1571 (1966).
247. Y. Kawasaki, T. Tanaka, and R. Okawara, *Bull. Chem. Soc. Japan*, **40**, 1562 (1967).
248. Y. Kawasaki, R. Ueeda, and T. Tanaka, *Int. Symp. NMR*, 2-M-16, Tokyo, September, 1965.
249. G. P. Van der Kelen, *Nature*, **193**, 1069 (1962).
250. G. J. M. Van der Kerk and J. G. A. Luijten, *J. Appl. Chem.*, **4**, 301 (1954).
251. G. J. M. Van der Kerk and J. G. A. Luijten, *J. Appl. Chem.*, **4**, 307 (1954).
252. G. J. M. Van der Kerk and J. G. A. Luijten, *J. Appl. Chem.*, **4**, 314 (1954).
253. G. J. M. Van der Kerk and J. G. A. Luijten, *J. Appl. Chem.*, **6**, 49 (1956).
254. G. J. M. Van der Kerk and J. G. A. Luijten, *J. Appl. Chem.*, **6**, 56 (1956).
255. G. J. M. Van der Kerk and J. G. A. Luijten, *J. Appl. Chem.*, **6**, 93 (1956).
256. G. J. M. Van der Kerk and J. G. Noltes, *J. Appl. Chem.*, **9**, 179 (1959).
257. G. J. M. Van der Kerk and J. G. Noltes, *J. Appl. Chem.*, **16**, 271 (1966).

258. G. J. M. Van der Kerk, J. G. Noltes, and J. G. A. Luijten, *J. Appl. Chem.*, **7**, 366 (1957).
259. V. V. Khrapov, V. I. Goldanskii, A. K. Prokofev, and R. O. Kostyanovskii, *Zh. Obshch. Khim.*, **37**, 3 (1967).
260. L. K. Van Kien and T. Tuony, *Compt. Rend. Soc. Biol.*, **149**, 2196 (1955).
261. F. B. Kipping, *J. Chem. Soc.*, **131**, 2365 (1928).
262. W. Kitching, *J. Organometal. Chem.*, **6**, 586 (1966).
263. W. Kitching, *Tetrahedron Letters*, **31**, 3689 (1966).
264. K. A. Kocheshkov, *J. Russ. Phys. Chem. Soc.*, **60**, 1191 (1928).
265. K. A. Kocheshkov, *Chem. Ber.*, **61**, 1659 (1928).
266. K. A. Kocheshkov, *Chem. Ber.*, **62**, 996 (1929).
267. K. A. Kocheshkov, *J. Russ. Phys. Chem. Soc.*, **61**, 1385 (1929).
268. K. A. Kocheshkov, *Chem. Ber.*, **66**, 1661 (1933).
269. K. A. Kocheshkov, *J. Gen. Chem. U.S.S.R.*, **4**, 1359 (1934).
270. K. A. Kocheshkov, *J. Gen. Chem. U.S.S.R.*, **5**, 211 (1935).
271. K. A. Kocheshkov and R. K. Freidlina, *Izv. Akad. Nauk S.S.S.R. Otdel. Khim. Nauk*, 203 (1950).
272. K. A. Kocheshkov and R. K. Freidlina, *Uchenye Zapiski Moskov. Gosudarst. Univ. M.V. Lomonosova*; No. 132 (1950); *Org. Khim.* No. 7, 144 (1956); through *CA*, **50**, 7728 (1956).
273. K. A. Kocheshkov and M. M. Nad, *Chem. Ber.*, **67**, 717 (1934).
274. K. A. Kocheshkov, M. M. Nad, and A. P. Alexandrov, *Chem. Ber.*, **67**, 1348 (1934).
275. K. A. Kocheshkov and A. N. Nesmeyanov, *J. Russ. Phys. Chem. Soc.*, **62**, 1795 (1930).
276. K. A. Kocheshkov and A. N. Nesmeyanov, *Chem. Ber.*, **64**, 628 (1931).
277. K. A. Kocheshkov and A. N. Nesmeyanov, *J. Gen. Chem. U.S.S.R.*, **4**, 1102 (1934).
278. K. A. Kocheshkov, A. N. Nesmeyanov, and W. A. Klimova, *J. Gen. Chem. U.S.S.R.*, **6**, 167 (1936).
279. K. A. Kocheshkov, A. N. Nesmeyanov, and W. P. Puzyreva, *Chem. Ber.*, **69**, 1639 (1936).
280. K. A. Kocheshkov, N. I. Sheverdina, and L. V. Abramova, *Ind. Chim. Belge, Suppl.*, **2**, 331 (1959).
281. M. Komura, Y. Kawasaki, T. Tanaka, and R. Okawara, *J. Organometal. Chem.*, **4**, 308 (1965).
282. M. M. Koton and T. M. Kiseleva, *Zh. Obshch. Khim.*, **27**, 2553 (1957).
283. C. A. Kraus and R. H. Bullard, *J. Am. Chem. Soc.*, **48**, 2131 (1926).
284. C. A. Kraus and C. C. Callis, *J. Am. Chem. Soc.*, **45**, 2624 (1923).
285. C. A. Kraus and W. N. Greer, *J. Am. Chem. Soc.*, **45**, 2946 (1923).
286. C. A. Kraus and W. N. Greer, *J. Am. Chem. Soc.*, **47**, 2361 (1925).
287. C. A. Kraus and W. N. Greer, *J. Am. Chem. Soc.*, **47**, 2568 (1925).
288. C. A. Kraus and A. M. Neal, *J. Am. Chem. Soc.*, **51**, 2403 (1929).
289. C. A. Kraus and A. M. Neal, *J. Am. Chem. Soc.*, **52**, 4426 (1930).
290. C. A. Kraus and W. V. Sessions, *J. Am. Chem. Soc.*, **47**, 2361 (1925).
291. E. Krause, *Chem. Ber.*, **51**, 912 (1918).
292. E. Krause, *Chem. Ber.*, **51**, 1447 (1918).
293. E. Krause and R. Becker, *Chem. Ber.*, **53**, 173 (1920).
294. E. Krause and A. von Grosse, *Die Chemie der metallorganischen Verbindungen*, Börnträger, Berlin, 1937.
295. E. Krause and R. Pohland, *Chem. Ber.*, **57**, 544 (1924).
296. E. Krause and R. Pohland, *Chem. Ber.*, **57**, 532 (1924).
297. E. Krause and O. Schlötting, *Chem. Ber.*, **63**, 1381 (1930).
298. E. Krause and G. Renwanz, *Chem. Ber.*, **62**, 710 (1929); *ibid.*, **9**, 179 (1959).

299. E. Krause and K. Weinberg, *Chem. Ber.*, **62**, 2235 (1929).
300. E. Krause and K. Weinberg, *Chem. Ber.*, **63**, 381 (1930).
301. H. Kriegsmann and H. Geissler, *Z. Anorg. Chem.*, **328**, 170 (1963).
302. H. Kriegsmann and H. Hoffmann, *Z. Anorg. Chem.*, **321**, 224 (1963).
303. H. Kriegsmann and S. Pauly, *Z. Anorg. Chem.*, **330**, 275 (1964).
304. H. Kriegsmann and S. Pischtshan, *Z. Anorg. Allgem. Chem.*, **308**, 212 (1961).
305. H. G. Kuivila and J. A. Verdone, *Tetrahedron Letters*, 119 (1964).
306. P. Kulmitz, *J. Prakt. Chem.*, **80**, 1860 (1860).
307. A. Ladenburg, *Chem. Ber.*, **3**, 358 (1870).
308. A. Ladenburg, *Chem. Ber.*, **4**, 17 (1871).
309. A. Ladenburg, *Chem. Ber.*, **4**, 19 (1871).
310. A. Ladenburg, *Ann. Chem.*, **159**, 251 (1871).
311. A. Ladenburg, *Ann. Chem. Suppl.*, **8**, 55 (1872).
312. L. V. Laine, M. F. Shotaskovskii, V. N. Kotrelev, and S. P. Kalina, *U.S.S.R.*, patent **137**, 519 (1960); through *CA*, **56**, 1481a (1962).
313. H. G. Langer and A. H. Blut, *J. Organometal. Chem.*, **5**, 288 (1966).
314. I. I. Lapkin and V. A. Dumler, *Uch. Zap., Permsk. Gas. Univ.*, **111**, 185 (1964); through *CA*, **64**, 12045e (1966).
315. R. Lecoq, *Compt. Rend.*, **239**, 678 (1954).
316. R. W. Leeper, *Iowa State Coll. J. Sci.*, **18**, 57 (1943); through *CA*, **38**, 727 (1944).
317. L. A. Leites, I. D. Pavlova, and Y. P. Egorov, *Teor. Eksp. Khim.*, **1**, 311 (1965).
318. M. Lesbre and R. Buisson, *Bull. Soc. Chim. France*, 1204 (1957).
319. M. Lesbre and J. Pocques, *Compt. Rend. 78e Congr. Soc. Savants Paris Dep. Sect. Sci.*, 423 (1953).
320. E. R. Lippincott, P. Mercier, and M. C. Tobin, *J. Phys. Chem.*, **57**, 939 (1953).
321. E. R. Lippincott and M. C. Tobin, *J. Am. Chem. Soc.*, **75**, 4141 (1953).
322. D. H. Lohmann, *J. Organometal. Chem.*, **4**, 382 (1965).
323. J. Lorberth and M. R. Kula, *Chem. Ber.*, **97**, 3444 (1964).
324. J. Lorberth and H. Nöth, *Chem. Ber.*, **98**, 969 (1965).
325. J. Lorberth and H. Vahrenkamp, *J. Organometal. Chem.*, **11**, 111 (1968).
326. C. Löwig, *Ann. Chem.*, **84**, 308 (1852).
327. H. Lutz, *Z. Naturforsch.*, **20b**, 1011 (1965).
328. J. G. A. _uijten and G. J. M. Van der Kerk, *Investigations in the Field of Organotin Chemistry*, Tin Research Institute, London, 1959.
329. J. G. A. Luijten and F. Rijkens, *Rec. Trav. Chim. Pays-Bas Belges*, **83**, 857 (1964).
330. G. P. Mack and E. Parker, U.S. Pat. 2,619,625 (1953).
331. G. P. Mack and E. Parker, U.S. Pat. 2,634,281 (1954).
332. J. C. Maire, *J. Organometal. Chem.*, **9**, 271 (1967).
333. J. C. Maire, J. Cassan, B. Leprêtre, and J. Marrot, *Compt. Rend.*, **260**, 5290 (1965).
334. J. C. Maire and F. Hemmert, *Bull. Sci. Chim. France*, 2785 (1963).
335. L. Malatesta, A. Sacco, and L. Ormerzano, *Gazz. Chim. Ital.*, **80**, 658 (1950).
336. Z. M. Manulkin, *J. Gen. Chem. U.S.S.R.*, **11**, 386 (1941).
337. Z. M. Manulkin, *J. Gen. Chem. U.S.S.R.*, **13**, 42 (1943).
338. Z. M. Manulkin, *J. Gen. Chem. U.S.S.R.*, **13**, 46 (1943).
339. Z. M. Manulkin, *J. Gen. Chem. U.S.S.R.*, **14**, 1047 (1944).
340. Z. M. Manulkin, *J. Gen. Chem. U.S.S.R.*, **16**, 235 (1946).
341. Z. M. Manulkin, *J. Gen. Chem. U.S.S.R.*, **18**, 299 (1948).
342. Z. M. Manulkin, *J. Gen. Chem. U.S.S.R.*, **20**, 2004 (1950).
343. Z. M. Manulkin, *Uzbek. Khim. Zh.*, **2**, 66 (1960).
344. J. Marrot, J. C. Maire, and J. Cassan, *Compt. Rend.*, **260**, 3931 (1965).

345. D. F. Martin, P. C. Maybury, and R. D. Walton, *J. Organometal. Chem.*, **7**, 362 (1967).
346. D. F. Martin and R. D. Walton, *J. Organometal. Chem.*, **5**, 57 (1966).
347. R. Mathis, M. Lesbre, and I. S. de Roche, *Compt. Rend.*, **243**, 257 (1956).
348. G. Matsubayashi, Y. Kawasaki, T. Tanaka, and R. Okawara, *J. Inorg. Nucl. Chem.*, **28**, 2937 (1966).
349. G. Matsubayashi, Y. Kawasaki, T. Tanaka, and R. Okawara, *Bull. Chem. Soc. Japan*, **40**, 1566 (1967).
350. H. Matsuda, Ph.D. Thesis, Osaka Univ., 1961.
351. S. Matsuda and H. Matsuda, *J. Chem. Soc. Japan*, **63**, 1658 (1961).
352. S. Matsuda and H. Matsuda, *Bull. Chem. Soc. Japan*, **35**, 208 (1962).
353. S. Matsuda, H. Matsuda, and J. Hayashi, *Kogyo Kagaku Zasshi*, **64**, 1951 (1961).
354. S. Matsuda, H. Matsuda, and M. Nakamara, *Kogyo Kagaku Zasshi*, **64**, 1948 (1961).
355. N. A. Matwiyoff and R. S. Drago, *Inorg. Chem.*, **3**, 337 (1964).
356. M. M. McGrady and R. S. Tobias, *Inorg. Chem.*, **3**, 1157 (1964).
357. R. N. Meals, *J. Org. Chem.*, **9**, 211 (1944).
358. J. Mendelsohn, A. Marchand, and J. Valade, *J. Organometal. Chem.*, **6**, 25 (1966).
359. Metal Thermit Corp., British Pat. 739,883 (1955).
360. K. Moedritzer and J. R. Van Wazer, *Inorg. Chem.*, **3**, 943 (1964).
361. A. Mohammad and D. P. N. Satchell, *Chem. Ind. London*, 2013 (1966).
362. P. W. Moore, *Dissertation Abs.*, **23**, 4550 (1963); through *CA*, **59**, 9460h (1963).
363. M. A. Mullins and C. Curran, *Inorg. Chem.*, **6**, 2107 (1967).
364. A. N. Murin and V. D. Nefedov, *Primenenie Mechenykh Atomov v. Anal. Khim. Akad. Nauk S.S.S.R. Inst. Geokhim. Anal. Khim.*, 75 (1955).
365. M. M. Nad and K. A. Kocheshkov, *Zh. Obshch. Khim.*, **8**, 42 (1938).
366. J. Nasielski, O. Buchman, M. Grosjean, and E. Hannaert, *Bull. Soc. Chim. Belges*, **77**, 15 (1968).
367. J. Nasielski, N. Sprecher, J. Devoogt, and S. Lejeune, *J. Organometal. Chem.*, **8**, 97 (1967).
368. S. N. Naumov and Z. M. Manulkin, *Zh. Obshch. Khim.*, **5**, 281 (1935).
368a. S. N. Naumov and Z. M. Manulkin, *Acta Univ. Asiae Mediae* (VII), No. 31, 12 (1937).
369. M. B. Neimann and V. A. Shushunov, *Dokl. Akad. Nauk SSSR*, **60**, 1347 (1950).
370. A. N. Nesmeyanov, A. E. Borisov, and A. N. Abramov, *Izv. Akad. Nauk SSSR, Otdel. Khim. Nauk*, 647 (1946).
371. A. N. Nesmeyanov, A. E. Borisov, and A. Abramov, *Izv. Akad. Nauk SSSR, Otdel. Khim. Nauk*, 570 (1949).
372. A. N. Nesmeyanov, A. E. Borisov, and N. V. Novikova, *Dokl. Akad. Nauk SSSR*, **119**, 504 (1958).
373. A. N. Nesmayanov, A. E. Borisov, and N. V. Novikova, *Izv. Akad. Nauk SSSR, Otdel. Khim. Nauk*, 259 (1959).
374. A. N. Nesmeyanov, A. E. Borisov, and N. V. Novikova, *Izv. Akad. Nauk SSSR, Otdel. Khim. Nauk*, 644 (1959).
375. A. N. Nesmeyanov, A. E. Borisov, N. V. Novikova, and M. A. Osipova, *Izv. Akad. Nauk SSSR, Otdel. Khim. Nauk*, 263 (1959).
376. A. N. Nesmeyanov, A. E. Borisov, I. S. Saeleva, and E. S. Golubeva, *Izv. Akad. Nauk SSSR, Otdel. Khim. Nauk*, 1490 (1958).
377. A. N. Nesmeyanov and K. A. Kocheshkov, *Chem. Ber.*, **63**, 2496 (1930).
378. A. N. Nesmeyanov and K. A. Kocheshkov, *Zh. Obshch. Khim.*, **1**, 219 (1931).
379. A. N. Nesmeyanov and K. A. Kocheshkov, *Chem. Ber.*, **67**, 317 (1934).

380. A. N. Nesmeyanov, K. A. Kocheshkov, and W. P. Puzyreva, *J. Gen. Chem. U.S.S.R.*, 7, 118 (1937).
381. A. N. Nesmeyanov, K. A. Kocheshkov, V. A. Klimova, and N. A. Gipp, *Chem. Ber.*, **68**, 1877 (1935).
382. A. N. Nesmeyanov and L. G. Makarova, *Dokl. Akad. Nauk S.S.S.R.*, **87**, 421 (1952).
383. A. N. Nesmeyanov, O. A. Reutov, T. P. Tolstaya, O. A. Ptitsyna, L. S. Isaeve, M. F. Turchinskii, and G. P. Bochkareva, *Dokl. Akad. Nauk SSSR*, **125**, 1265 (1959).
384. W. P. Neumann, *Angew. Chem.*, **75**, 225 (1963).
385. W. P. Neumann, *Ann. Chem.*, **653**, 157 (1962).
386. W. P. Neumann and G. Burkhardt, *Ann. Chem.*, **663**, 11 (1963).
387. W. P. Neumann, R. Schik, and R. Köster, *Angew. Chem.*, **76**, 380 (1964).
387a. W. P. Neumann and J. Pedain, *Tetrahedron Letters*, 2461 (1964).
388. J. F. Nobis, L. Moormeier, and R. E. Robinson, *Adv. Chem. Ser.*, **23**, 63 (1959).
389. J. G. Noltes, Dissertation, Utrecht, 1958.
390. J. G. Noltes and G. J. M. Van der Kerk, *Functionally Substituted Organotin-Compounds*, Tin Research Institute, London, 1958.
391. V. Oakes and R. E. Hutton, *J. Organometal. Chem.*, **9**, 133 (1966).
392. V. Oakes and R. E. Hutton, *J. Organometal. Chem.*, **3**, 472 (1965).
393. J. L. Occolowitz, *Tetrahedron Letters*, 5291 (1966).
394. P. F. Oesper and C. P. Smyth, *J. Am. Chem. Soc.*, **64**, 173 (1942).
395. R. Okawara, B. J. Hathaway, and D. E. Webster, *Proc. Chem. Soc.*, 13 (1963).
396. R. Okawara and E. G. Rochow, *J. Am. Chem. Soc.*, **82**, 3285 (1960).
397. R. Okawara and M. Wada, *J. Organometal. Chem.*, **1**, 81 (1963).
398. R. Okawara, D. E. Webster, and E. G. Rochow, *J. Am. Chem. Soc.*, **82**, 3287 (1960).
399. O. A. Osipov and O. E. Kashireninov, *Zh. Obshch. Khim.*, **32**, 1717 (1962).
400. J. Otera, Y. Kawasaki, and T. Tanaka, *Inorg. Chim. Acta.*, **1**, 294 (1967).
401. S. Papetti and H. W. Post, *J. Org. Chem.*, **22**, 526 (1957).
402. H. J. Passino and G. G. Lauer, U.S. Pat. 2,665,286 (1955); through *CA*, **49**, 367 (1955).
403. D. J. Patmore and W. A. G. Graham, *Inorg. Chem.*, **7**, 771 (1968).
404. L. Pauling, *The Nature of the Chemical Bond*, Cornell University Press, New York, 1960.
405. G. J. D. Peddle and G. Redl, *Chem. Commun.*, 626 (1968).
406. J. B. Pedley and H. A. Skinner, *Trans. Faraday Soc.*, **55**, 544 (1959).
407. J. B. Pedley, H. A. Skinner, and C. L. Chernick, *Trans. Faraday Soc.*, **53**, 1612 (1957).
408. E. J. Pikina, T. V. Talalaeva, and K. A. Kocheshkov, *Zh. Obshch. Khim.*, **8**, 1844 (1938).
409. P. Pfeiffer, *Chem. Ber.*, **35**, 3303 (1902).
410. P. Pfeiffer, *Chem. Ber.*, **44**, 1269 (1911).
411. P. Pfeiffer, *Chem. Ber.*, **44**, 1273 (1911).
412. P. Pfeiffer and I. Heller, *Chem. Ber.*, **37**, 4618 (1904).
413. P. Pfeiffer and R. Lehnhardt, *Chem. Ber.*, **36**, 1054 (1903).
414. P. Pfeiffer and R. Lehnhardt, *Chem. Ber.*, **36**, 3027 (1903).
415. P. Pfeiffer, R. Lehnhardt, H. Luftensteiner, R. Prade, K. Schurmann, and P. Truskier, *Z. Anorg. Chem.*, **68**, 102 (1910).
416. P. Pfeiffer and K. Schurmann, *Chem. Ber.*, **37**, 319 (1904).
417. A. Polis, *Chem. Ber.*, **22**, 2915 (1889).
418. R. C. Poller, *J. Inorg. Nucl. Chem.*, **24**, 593 (1962).
419. R. C. Poller, *Spectrochim. Acta*, **22**, 935 (1966).
420. R. C. Poller and D. L. B. Toley, *J. Chem. Soc. A*, **10**, 1578 (1967).

421. W. J. Pope and S. J. Peachy, *Proc. Chem. Soc.*, **16**, 42 (1900).
422. W. J. Pope and S. J. Peachy, *Proc. Chem. Soc.*, **16**, 116 (1900).
423. W. J. Pope and S. J. Peachy, *Proc. Chem. Soc.*, **19**, 290 (1903).
424. W. J. Pope and S. J. Peachy, *Chem. News*, **87**, 253 (1903).
425. W. J. Pope and S. J. Peachy, *Proc. R. Soc.*, **72**, 7 (1903).
426. W. J. Pope and S. J. Peachy, *Proc. Chem. Soc.*, **28**, 42; 116 (1912).
427. S. J. W. Price and A. F. Trotman-Dickenson, *Trans. Faraday Soc.*, **54**, 1630 (1958).
428. See Ref. 427.
429. R. H. Prince, *Trans. Faraday Soc.*, **54**, 838 (1958).
430. R. H. Prince, *J. Chem. Soc.*, 1783 (1959).
431. C. Quintin, *Ing. Chim. Milan*, **14**, 205 (1930).
432. H. E. Ramsden, U.S. Pat. 2,873,287 (1959).
433. G. A. Razuvaev, *Akad. Nauk SSSR, Inst. Org. Khim. Sintezy Org. Soedinenii Sbornik*, **1**, 41 (1950).
434. G. A. Razuvaev and E. S. Fedotova, *Zh. Obshch. Khim.*, **21**, 1219 (1951).
435. G. A. Razuvaev and V. Fetyukova, *Zh. Obshch. Khim.*, **21**, 1010 (1951).
436. G. A. Razuvaev and N. S. Vyazankin, *Zh. Obshch. Khim.*, **3**, 3762 (1961).
437. G. A. Razuvaev, N. S. Vyazankin, Y. I. Dergunov, and E. N. Gl'adyshev, *Bull. Acad. Sci. USSR*, **2**, 794 (1964).
438. G. A. Razuvaev, N. S. Vyazankin, Y. I. Dergunov, and O. S. D'yachkovskava, *Dokl. Akad. Nauk SSSR*, **132**, 364 (1960).
439. W. T. Reichle, *Inorg. Chem.*, **5**, 87 (1966).
440. G. F. Reifenberg and W. J. Considine, *J. Organometal. Chem.*, **9**, 505 (1967).
441. O. A. Reutov, O. A. Ptitsyna, and N. D. Tatrina, *Zh. Obshch. Khim.*, **28**, 588 (1958).
442. L. Riccoboni, *Gazz. Chim. Ital.*, **71**, 696 (1941).
443. L. Riccoboni and P. Popoff, *Atti Inst. Veneto Sci.*, **107**, 123 (1949).
444. E. G. Rochow and D. Seyferth, *J. Am. Chem. Soc.*, **75**, 2877 (1953).
445. S. D. Rosenberg, E. Debreczeni, and E. L. Weinberg, *J. Am. Chem. Soc.*, **81**, 972 (1959).
446. S. D. Rosenberg and A. J. Gibbons, *J. Am. Chem. Soc.*, **79**, 2138 (1957).
447. S. D. Rosenberg, A. J. Gibbons, and H. E. Ramsden, *J. Am. Chem. Soc.*, **79**, 2137 (1957).
448. G. F. Rubinchik and Z. M. Manulkin, *J. Gen. Chem. U.S.S.R.*, **36**, 271 (1966).
449. J. J. Rush and W. C. Hamilton, *Inorg. Chem.*, **5**, 2238 (1966).
449a. L. A. Rothmann and E. J. Becker, *J. Org. Chem.*, **24**, 294 (1959).
450. A. Sacco, *Atti Accad. Naz. Lincei. Rend. Chem. Sci.*, **11**, 101 (1951).
451. M. K. Saikina, *Uchen. Zap. Kazan. Gos. Univ. Khim.*, **116**, 129 (1956); through *CA*, **51**, 7191e (1957).
452. G. S. Sasin, *J. Org. Chem.*, **18**, 1142 (1953).
453. L. Savidan, *Bull. Soc. Chim. France*, 411 (1953).
454. A. K. Sawyer and J. E. Brown, *J. Organometal. Chem.*, **5**, 438 (1966).
455. A. K. Sawyer, J. E. Brown, and E. L. Hanson, *J. Organometal. Chem.*, **3**, 464 (1965).
456. A. K. Sawyer, J. E. Brown, and G. S. May, *J. Organometal. Chem.*, **11**, 192 (1968).
456a. A. K. Sawyer and H. G. Kuivila, *J. Am. Chem. Soc.*, **85**, 1010 (1963).
457. R. Sayre, *J. Chem. Eng. Data*, **6**, 560 (1961).
458. E. O. Schlemper and D. Britton, *Inorg. Chem.*, **5**, 507 (1966).
459. E. O. Schlemper and W. C. Hamilton, *Inorg. Chem.*, **5**, 995 (1966).
460. H. Schmidbauer and I. Ruidisch, *Inorg. Chem.*, **3**, 599 (1964).
461. G. K. Semin and E. V. Bryuchova, *Chem. Commun.*, 605 (1968).
462. D. Seyferth, *J. Org. Chem.*, **22**, 478 (1957).

463. D. Seyferth, *J. Am. Chem. Soc.*, **79**, 2133 (1957).
464. D. Seyferth, *J. Am. Chem. Soc.*, **79**, 5881 (1957).
465. D. Seyferth, *J. Org. Chem.*, **22**, 1599 (1957).
466. D. Seyferth, *Naturwissenschaften*, **44**, 34 (1957).
467. D. Seyferth and M. M. Cohen, *Inorg. Chem.*, **2**, 652 (1963).
468. D. Seyferth and S. O. Grim, *J. Am. Chem. Soc.*, **83**, 1610 (1961).
469. D. Seyferth and E. G. Rochow, *J. Am. Chem. Soc.*, **77**, 3102 (1955).
470. D. Seyferth and E. G. Rochow, *J. Am. Chem. Soc.*, **84**, 2050 (1962).
471. D. Seyferth and F. G. A. Stone, *J. Am. Chem. Soc.*, **79**, 515 (1957).
472. D. Seyferth, M. A. Weiner, S. O. Grim, and N. Kahlen, Abstracts of papers, 135th Meeting American Chemical Society, 1959, p.20.
472a. D. W. A. Sharp and J. M. Winfield, *J. Chem. Soc.*, 2278 (1965).
473. J. Shukoff, *Chem. Ber.*, **36**, 2691 (1905).
474. H. Siebert, *Z. Anorg. Allgem. Chem.*, **268**, 177 (1952).
475. H. Siebert, *Z. Anorg. Allgem. Chem.*, **275**, 210 (1954).
476. K. Sisido, S. Kozima, and T. Tuzi, *J. Organometal. Chem.*, **9**, 109 (1967).
477. K. Sisido, T. Miyanisi, K. Nabika, and S. Kozima, *J. Organometal. Chem.*, **11**, 281 (1968).
478. K. Sisido, S. Kozima, and T. Isibasi, *J. Organometal. Chem.*, **10**, 439 (1967).
479. K. Sisido, Y. Takeda, and Z. Kinugawa, *J. Am. Chem. Soc.*, **83**, 538 (1961).
480. K. Sisido and Y. Takeda, *J. Org. Chem.*, **26**, 2301 (1961).
481. H. A. Skinner and L. E. Sutton, *Trans. Faraday Soc.*, **40**, 164 (1944).
482. A. P. Skoldinov and K. A. Kocheshkov, *Zh. Obshch. Khim.*, **12**, 398 (1942).
483. A. C. Smith and E. G. Rochow, *J. Am. Chem. Soc.*, **75**, 4103 (1953).
484. A. C. Smith and E. G. Rochow, *J. Am. Chem. Soc.*, **75**, 4105 (1953).
485. A. L. Smith, *Spectrochim. Acta*, **24A**, 695 (1968).
486. T. A. Smith and F. S. Kipping, *J. Chem. Soc.*, **101**, 2553 (1912).
487. T. A. Smith and F. S. Kipping, *J. Chem. Soc.*, **103**, 2034 (1913).
488. C. P. Smyth, *J. Org. Chem.*, **6**, 421 (1941).
489. C. P. Smyth, *J. Am. Chem. Soc.*, **63**, 57 (1941).
489a. R. Sommer, B. Schneider, and W. P. Neumann, *Ann.*, **692**, 12 (1966).
490. A. Solerio, *Gazz. Chim. Ital.*, **85**, 61 (1955).
491. M. E. Spaght, F. Hein, and H. Pauling, *Phys. Z.*, **34**, 212 (1933).
492. N. Sprecher, Dissertation, Univ. Brussels, 1967.
493. T. S. Srivastava, *J. Organometal. Chem.*, **10**, 373 (1967).
494. T. S. Srivastava, *J. Organometal. Chem.*, **10**, 375 (1967).
495. H. A. Stöckler and H. Sano, *Trans. Faraday Soc.*, **64**, 577 (1968).
496. H. A. Stöckler, H. Sano, and R. H. Herber, *J. Chem. Phys.*, **47**, 1567 (1967).
497. A. Strecker, *Ann. Chem.*, **123**, 365 (1862).
498. E. D. Swiger and J. D. Graybeal, *J. Am. Chem. Soc.*, **87**, 1464 (1965).
499. G. Tagliavini and P. Zanella, *J. Organometal. Chem.*, **5**, 299 (1966).
500. G. Tagliavini and P. Zanella, *Anal. Chim. Acta*, **40**, 33 (1968).
501. C. Tamborski, F. E. Ford, W. L. Lehn, G. J. Moore, and E. I. Sokoloski, *J. Org. Chem.*, **27**, 619 (1962).
502. P. Taimsalu and J. L. Wood, *Spectrochim. Acta*, **20**, 1043 (1964).
503. P. Taimsalu and J. L. Wood, *Spectrochim. Acta*, **20**, 1357 (1964).
504. Y. Takeda, Y. Hayakawa, T. Fueno, and J. Furukawa, *Makromolek. Chem.*, **83**, 234 (1965).
505. T. Tanaka, *Inorg. Chim. Acta*, **1**, 217 (1967).
506. T. Tanaka and T. Kamitani, *Inorg. Chim. Acta*, **2**, 175 (1968).

507. T. Tanaka, M. Komura, Y. Kawasaki, and R. Okawara, *J. Organometal. Chem.*, **1**, 484 (1964).
508. A. Tchakirian, M. Lesbre, and M. Lewinsohn, *Compt. Rend.*, **202**, 138 (1936).
509. A. B. Thomas and E. G. Rochow, *J. Am. Chem. Soc.*, **79**, 1843 (1957).
510. A. B. Thomas and E. G. Rochow, *J. Inorg. Nucl. Chem.*, **4**, 205 (1957).
511. R. S. Tobias, *Organometal. Chem. Rev.*, **1**, 93 (1966).
512. V. F. Toropova and M. K. Saikina, *Obshch. Khim. Akad. Nauk SSSR*, **210**, 1 (1953); through *CA*, **48**, 12579 (1954).
512a. P. M. Treichel and P. A. Goodwich, *Inorg. Chem.*, **4**, 1424 (1965).
513. T. T. Tsai and W. L. Lehn, *J. Org. Chem.*, **3**, 2891 (1966).
514. L. Verdonck and G. P. Van der Kelen, *Bull. Soc. Chim. Belges*, **74**, 361 (1965).
515. L. Verdonck and G. P. Van der Kelen, *Ber. Bunsenges. Phys. Chem.*, **69**, 478 (1965).
516. L. Verdonck and G. P. Van der Kelen, *J. Organometal. Chem.*, **5**, 532 (1966).
517. L. Verdonck and G. P. Van der Kelen, *Bull. Soc. Chim. Belges*, **76**, 258 (1967).
518. L. Verdonck, G. P. Van der Kelen, and Z. Eeckhaut, *J. Organometal. Chem.*, **11**, 487 (1968).
519. L. Verdonck and G. P. Van der Kelen, *J. Organometal. Chem.*, **11**, 491 (1968).
520. T. Vladimiroff and E. R. Malinowski, *J. Chem. Phys.*, **42**, 440 (1965).
521. A. J. Vogel, W. T. Cresswell, and J. Leicester, *J. Phys. Chem.*, **58**, 174 (1954).
522. N. S. Vyazankin and V. T. Bychkov, *Zh. Obshch. Khim.*, **36**, 1648 (1966).
523. K. V. Vyayraghavan, *J. Indian Chem. Soc.*, **22**, 135 (1945).
524. N. N. Vyshinskii, T. V. Kozlova, and N. K. Rudnevskii, *Trudy Komissii Spektrosk. Akad. Nauk SSSR*, 451 (1964); through *CA*, **63**, 13027d (1965).
525. M. Wada, M. Shindo, and R. Okawara, *J. Organometal. Chem.*, **1**, 95 (1963).
526. M. Wada, K. Kawakami, and R. Okawara, *J. Organometal. Chem.*, **4**, 160 (1965).
527. J. L. Wardell, *J. Organometal. Chem.*, **9**, 89 (1967).
528. J. L. Wardell, *J. Organometal. Chem.*, **10**, 53 (1967).
529. E. L. Weinberg, U.S. Pat. 2,679,505 (1955); through *CA*, **53**, 4705d (1955).
530. E. L. Weinberg and E. G. Rochow, German Pat. 1,046,052 (1958); through *CA*, **56**, 2485a (1958).
531. A. Werner and P. Pfeiffer, *Z. Anorg. Chem.*, **17**, 82 (1898).
532. R. West and E. G. Rochow, *J. Am. Chem. Soc.*, **74**, 2490 (1952).
533. A. H. Westlake and D. F. Martin, *J. Inorg. Nucl. Chem.*, **27**, 1579 (1965).
534. C. J. Wilkins and H. M. Haendler, *J. Chem. Soc.*, 3174 (1965).
535. D. Wittenberg, *Ann. Chem.*, **654**, 23 (1962).
536. G. Wittig, F. J. Meyer, and G. Lange, *Ann. Chem.*, **571**, 167 (1951).
537. T. C. Wu, Doctoral Dissertation, Iowa State College, 1952.
538. R. E. Wyant, J. L. McFarling, J. F. Kircher, and E. J. Kahler, *U.S. Atomic Energy Comm. B.M.I.*, 1654 (1963).
539. A. Y. Yakubovich, S. P. Makarov, and G. S. Garilov, *Zh. Obshch. Khim.*, **22**, 1788 (1952).
540. A. Y. Yakubovich, S. P. Makarov, and V. A. Ginsburg, *Zh. Obshch. Khim.*, **28**, 1036 (1968).
541. A. Y. Yakubovich, S. P. Makarov, V. A. Ginsburg, G. J. Gavrilov, and E. N. Merkulova, *Dokl. Akad. Nauk SSSR*, **72**, 69 (1950).
542. K. Yasuda, Y. Kawasaki, N. Kasai, and T. Tanaka, *Bull. Chem. Soc. Japan*, **38**, 1216 (1965).
543. K. Yasuda, H. Matsumoto, and R. Okawara, *J. Organometal. Chem.*, **6**, 528 (1966).
544. A. L. Yergey and F. W. Lampe, *J. Am. Chem. Soc.*, **87**, 4204 (1965).

545. L. I. Zakharin, O. Y. Okhlobystin, and B. N. Strunin, *Zh. Prikl. Khim.*, **36**, 2034 (1963); through *CA*, **60**, 3002a (1964).
546. P. Zanella and G. Tagliavini, *J. Organometal. Chem.*, **12**, 355 (1968).
547. K. Ziegler, German Pat. 1,157,617 (1963); through *CA*, **59**, 12842d (1963).
548. K. Ziegler, British Pat. 923,179 (1963); through *CA*, **60**, 3008a (1964).
549. K. Ziegler and W. P. Neumann, Belgian Pat. 590,209; 590,210 (1960).
550. J. R. Zietz, S. M. Blitzer, H. E. Redman, and G. C. Robinson, *J. Org. Chem.*, **22**, 60 (1957).
551. H. Zimmer, I. Hechenbleikner, O. Hornberg, and M. Danzik, *J. Org. Chem.*, **29**, 2632 (1964).
552. H. Zimmer and K. Lubke, *Chem. Ber.*, **85**, 1119 (1952).
553. H. Zimmer and H. W. Sparman, *Chem. Ber.*, **87**, 645 (1954).
554. J. J. Zuckermann, *J. Inorg. Nucl. Chem.*, **29**, 2191 (1967).

4. ORGANOTIN COMPOUNDS WITH Sn—O BONDS

Organotin Alkoxides, Oxides, and Related Compounds

A. J. BLOODWORTH AND ALWYN G. DAVIES

University College
London, England

This chapter deals with the preparation and reactions of those compounds which contain a tin-oxygen bond, and which may be regarded as derivatives of the molecules H_2O, HOR, HO·OH, and HO·OR, that is, the hydroxides, oxides, alkoxides (and phenoxides), peroxides, and alkyl peroxides. Organotin derivatives of all the other inorganic and organic oxyacids are dealt with in the following chapter.

Because of their simplicity, the preparation of the alkoxides and alkyl peroxides is considered in the first section, which is subdivided according to the preparative method. The second section deals with the preparation of the various possible types of alkyltin hydroxides and oxides, in the sequence of the decreasing degree of alkylation of the tin. Thirdly, the reactions of all these types of tin-oxygen bonded compounds are discussed in terms of the various combinations of addition and substitution reactions which may be observed.

The discussion aims to be critical rather than comprehensive, but we have included in the tables all those compounds of which we were aware (in March 1968) and whose identity we regarded as being firmly established.

I. Preparation, Structure, and Physical Properties of Alkoxides, Phenoxides, and Alkyl Peroxides

A. Alkoxides and Phenoxides

When the last comprehensive review on organotin chemistry was published in 1960 (*137*), about forty organotin alkoxides* appeared in the literature, many of which were claimed only in patents and were inadequately characterized. The tables which accompany this chapter list over four hundred organotin alkoxides which the authors believe to have a justifiable claim to existence. Although some have not been characterized by melting or boiling points and indeed have not been isolated analytically pure, other evidence such as infrared and nuclear magnetic resonance spectra provide convincing evidence of their composition.

The dramatic rise in the number of alkoxides reflects the general growth rate in organotin chemistry and stems very largely from the development of addition reactions of Sn—X bonds. Thus the addition of any Sn—X bond to a carbonyl compound [Eq. (1)] provides an organotin alkoxide, and the tables consequently include many rather exotic examples [e.g., $Bu_3Sn \cdot OCH(CCl_3)NPh \cdot CO \cdot OMe$], although some very simple ones (e.g., $Me_3Sn \cdot OPr$) are still unknown:

$$R_3Sn{-}X + R'R''{\cdot}C{=}O \longrightarrow R_3Sn{\cdot}OCR'R''{\cdot}X \qquad (1)$$

Apart from additions with tin hydrides, simple alkoxides are prepared by nucleophilic substitution at tin, a process which has remained the principal route to these compounds.

1. *Preparation by Nucleophilic Substitution at Tin*

The basis of the method is generalized in Eq. (2) where alkoxide displaces some other substituent X (halide, alkoxide, oxide, amine, etc.) by nucleophilic

* The term "alkoxides" is used throughout to include also phenoxides.

attack at tin:

$$Y\text{—}OR' \quad R_3Sn\text{—}X \longrightarrow \overset{+}{Y} + R_3SnOR' + \overset{-}{X} \qquad (2)$$

a. Nucleophilic Displacement of Halide. The most popular method for preparing trialkyltin and dialkyltin alkoxides has been the reaction of an organotin halide (usually chloride) with the sodium salt of the appropriate alcohol, e.g.:

$$Bu_3SnCl + NaOMe \longrightarrow Bu_3Sn\cdot OMe + NaCl \quad (13) \qquad (3)$$

$$Et_2SnCl_2 + 2\,NaOBu \longrightarrow Et_2Sn(OBu)_2 + 2\,NaCl \quad (104) \qquad (4)$$

The anhydrous alcohol is frequently employed as a solvent to ensure homogeneous conditions. A very fine precipitate of sodium halide is produced which is removed by filtration or centrifugation and the alkoxide is purified by distillation, or recrystallization.

Disadvantages of the method are twofold; firstly, several transfers of the very moisture-sensitive alkoxide are necessary and, secondly, some sodium halide remains in solution, is precipitated as the solution concentrates, and can cause troublesome "bumping" during distillation. More rapid and cleaner methods, which are described later, have largely superseded this process.

Other metal alkoxides have been employed particularly for the preparation of chelated alkoxides such as acetylacetonates, where thallium is the usual choice (*211*):

$$Ph_2SnCl_2 + 2\,Tl\mathit{acac} \longrightarrow Ph_2Sn\mathit{acac}_2 + 2\,TlCl \qquad (5)$$

The transfer of alkoxy groups from silicon to tin provides the dialkoxytin dihalides [e.g., Eq. (6)], and magnesium derivatives of acetylenic alcohols have been used to prepare alkynyloxystannanes [e.g., Eq. (7)].

$$SnCl_4 + Si(OPr^i)_4 \longrightarrow (Pr^iO)_2SnCl_2 \quad (242) \qquad (6)$$

$$Et_3SnCl + BrMg\cdot O(CH_2)_2C{:}CMgBr \longrightarrow Et_3Sn\cdot O(CH_2)_2C{:}CH \quad (291) \qquad (7)$$

Tin alkoxides themselves can be used. Thus dialkyltin dihalides undergo a disproportionation reaction with dialkyltin dimethoxides to provide new types of methoxides of the composition $R_2Sn(OMe)X$

$$R_2SnX_2 + R_2Sn(OMe)_2 \longrightarrow 2\,R_2Sn(OMe)X \qquad (8)$$

Bipyridyl reverses the disproportionation, giving R_2SnX_2,bipy. and, presumably, the dialkyltin dimethoxide.

Kocheshkov and his co-workers (*107*), who prepared $Bu_2Sn(OMe)Cl$ and $Bu_2Sn(OMe)Br$ in this way, proposed the structure $Bu_2Sn(OMe)_2$, R_2SnX_2, principally because, when the product of the reaction between dibutyltin dimethoxide and diethyltin dichloride was treated with tin tetrachloride, the diethyltin dichloride could be recovered in 80% yield. It has been suggested (*72*), however, that the ligands X and OMe about tin are probably labile, and, in solution, take part in a series of mobile equilibria [Eq.(9)] similar to those which are usually considered for the Grignard reagent.

Molecular weights are in agreement with a monomer-dimer equilibrium as are preliminary ^{119}Sn NMR studies (*192*) which show two concentration-dependent resonances. Some evidence that the Kocheshkov structure (**1**) is, alone at least, not a complete description of the system is that, when the product of the interaction of dibutyltin dimethoxide and diethyltin dichloride is treated with bipyridyl, both the derivatives Bu_2SnCl_2, bipy. and Et_2SnCl_2, bipy. are obtained:

$$R_2Sn(\mu\text{-}OMe)_2SnR_2X_2 \;(1) \rightleftharpoons R_2Sn(OMe)_2 + R_2SnX_2$$

$$R_2Sn(\mu\text{-}OMe)_2SnR_2X_2 \;(1) \rightleftharpoons 2\,R_2Sn(X)(OMe) \;(2) \rightleftharpoons XR_2Sn(\mu\text{-}OMe)_2SnR_2X \;(3) \qquad (9)$$

The three products (**1**), (**2**), and (**3**), which might result, contain 4- and 6-, 4-, and 5- coordinate tin, respectively. Mössbauer spectra of $Me_2Sn(OMe)Cl$ and $Bu_2Sn(OMe)Cl$ each show only one doublet with high ρ factors (*70*). This is consistent with structure (**3**) for the solid state, but only an X-ray crystallographic analysis will be completely conclusive.

The only organotin alkoxide to have its structure determined by crystallography is tetrachloro-1,4-bis(triethylstannyloxy)benzene,

$$Et_3SnO\cdot C_6Cl_4\cdot OSnEt_3$$

which is hardly a typical example. The bulky chlorine and ethyl groups prevent free rotation about the C—O bond, and all molecules in the crystal exist in the *trans*-conformation, the Sn—O—C groups lying in a plane normal to the plane of the ring. No detectable changes were found in the angles formed by the tin atom which were assumed to be tetrahedral, and the Sn—O bond length and the angle at oxygen were found to be 2.08 ± 0.06 Å and 127°, respectively (*330*). It is unfortunate that the one compound examined by X-ray diffraction is not only atypical but also does not appear to have been characterized in any other way.

Structural information on the simple alkoxides is somewhat scanty, although a number of workers have made vibrational assignments in the infrared region. Butcher and his co-workers (*48*) and Lorberth and Kula (*185*) independently assigned bands near 1035 cm^{-1} and 1060 cm^{-1} to the symmetric and asymmetric C—O stretching modes, respectively, in dialkoxides. Mendelsohn et al. (*206*), however, have shown that the intensities of the bands in, for example, dibutyltin dimethoxide, are concentration dependent, the lower energy absorption decreasing with dilution and virtually disappearing at a concentration of 0.01 mole/liter. They suggest, therefore, that the bands arise from stretching vibrations of associated and free C—O groups, and together with molecular weight evidence (*107*), propose a dimeric structure for the alkoxide which is identical with (**3**) if X = OMe. Significantly the low energy band is completely absent in neat dibutyltin di-*t*-butoxide, steric hindrance apparently preventing the association.

b. Nucleophilic Displacement of Alkoxide. Organotin alkoxides can be prepared very conveniently by the exchange of alkoxy groups between a tin alkoxide derived from a low boiling alcohol, e.g., methoxide, and a higher boiling alcohol. Transalkoxylation is brought about by removing the lower-boiling alcohol by distillation, thus displacing the equilibrium in favor of the products (*146*):

$$Bu_2Sn(OPr^i)_2 + 2\,PhCH_2OH \rightleftharpoons Bu_2Sn(OCH_2Ph)_2 + 2\,Pr^iOH \qquad (10)$$

The method has received surprisingly little use for simple alkoxides, but it is a principal route to triethyltin derivatives of acetylenic alcohols [e.g., Eq. (11)]. Acetylacetonates [Eq. (12)] and dibutyltin derivatives of glycols [Eq. (13)] have also been prepared in this way:

$$Et_3Sn{\cdot}OMe + HO{\cdot}CHMe{\cdot}C{:}CH \longrightarrow Et_3Sn{\cdot}O{\cdot}CHMe{\cdot}C{:}CH + MeOH \qquad (292)\ (11)$$

$$Et_3Sn{\cdot}OMe + acac{\cdot}H \longrightarrow Et_3Sn{\cdot}acac + MeOH \qquad (204)\ (12)$$

$$Bu_2Sn(OEt)_2 + \begin{matrix} HO-CMe_2 \\ | \\ HO-CMe_2 \end{matrix} \longrightarrow Bu_2Sn\begin{matrix} \diagup O-CMe_2 \\ \quad\ | \\ \diagdown O-CMe_2 \end{matrix} + 2\,EtOH \qquad (203)\ (13)$$

In a variation of the method, trimethylsilyl or trimethylgermyl alkoxides undergo exchange with tributyltin methoxide (*238*):

$$Bu_3Sn{\cdot}OMe + Me_3Si{\cdot}OBu \longrightarrow Bu_3Sn{\cdot}OBu + Me_3Si{\cdot}OMe \qquad (14)$$

$$Bu_3Sn{\cdot}OMe + Me_3Ge{\cdot}O{-}C_6H_{11} \longrightarrow Bu_3Sn{\cdot}O{-}C_6H_{11} + Me_3Ge{\cdot}OMe \qquad (15)$$

However, reaction with esters is the most common method of transalkoxylation. As in the alkoxide-alcohol exchange, formation of the new organotin alkoxide is dependent upon displacing an equilibrium by removing the most volatile component. The reaction is successful only where a large difference exists between the boiling point of the starting ester and of the product. Thus tributyltin phenoxide has been prepared in 85% yield from the reaction of tributyltin methoxide and phenyl acetate or benzoate:

$$Bu_3Sn\cdot OMe + PhO\cdot CO\cdot R \longrightarrow Bu_3Sn\cdot OPh + MeO\cdot CO\cdot R \qquad (16)$$

where R = Me or Ph (*314*).

The most important application here has been to the preparation of vinyloxystannanes. Lutsenko and his co-workers (*254*) prepared a number of these compounds by the reaction of trialkyltin methoxide with enol acetates of aldehydes:

$$Pr_3Sn\cdot OMe + Me_2C{=}CH\cdot O\cdot CO\cdot Me \longrightarrow Pr_3Sn\cdot O\cdot CH{=}CMe_2 + MeO\cdot CO\cdot Me \qquad (17)$$

The vinyloxystannane structure was supported by NMR data which showed the presence of a vinyl proton, by infrared spectroscopy which showed a strong band at 1640–1660 cm^{-1} attributable to the C=C stretching vibration, and by ultraviolet spectroscopy which showed the complete absence of any carbonyl absorption in the 250–300 mμ region. A similar reaction with isopropenyl acetate, on the other hand, had previously been reported to give α-stannylacetone (*216*):

$$Bu_3Sn\cdot OMe + CH_2{=}CMe\cdot O\cdot CO\cdot Me \longrightarrow Bu_3Sn\cdot CH_2\cdot CO\cdot Me + MeO\cdot CO\cdot Me \qquad (18)$$

Recently, Valade and his co-workers (*237*) have carried out a more complete investigation, studying the reaction of tributyltin methoxide with enol acetates of a variety of aldehydes and ketones and assigning the structure of products on the basis of infrared and NMR data. They find that the products obtained depend upon the ester, the formation of stannylated enols (*O*-derivatives), α-stannylated ketones (*C*-derivatives) or even a mixture of these two isomeric structures being observed.

Enol acetates of ketones in general give a mixture, except for isopropenyl acetate which gives 100% of the *C*-derivative thus confirming the earlier report. By contrast, all enol acetates derived from aldehydes which were investigated give 100% of the *O*-derivative. If geometric isomers of the ester are available it is found that the stereochemical arrangement is not conserved during the stannylation, but that an identical mixture is isolated from reaction with either isomer:

$$\left.\begin{array}{l} \underset{trans}{(Et)(H)C{=}C(O\cdot CO\cdot Me)(Me)} + Bu_3Sn\cdot OMe \\ \underset{cis}{(Et)(H)C{=}C(Me)(O\cdot CO\cdot Me)} + Bu_3Sn\cdot OMe \end{array}\right] \longrightarrow \left\{\begin{array}{l} 75\%\ EtCH{=}CMe\cdot OSnBu_3 \\ (trans\text{: } 10\%;\ cis\text{: } 90\%) \\ 25\%\ EtCH(SnBu_3)CO\cdot Me \end{array}\right. \qquad (19)$$

These results suggest that the transalkoxylation proceeds via an ionic intermediate, the limiting structures of which are shown below:

$$\underset{\textbf{(4a)}}{RR'C{=}C(R'')O^{\ominus}} \longleftrightarrow \underset{\textbf{(5a)}}{RR'\overset{\ominus}{C}{-}C(R''){=}O} \rightleftharpoons \underset{\textbf{(5b)}}{RR'\overset{\ominus}{C}{-}C(=O)R''} \longleftrightarrow \underset{\textbf{(4b)}}{RR'C{=}C(O^{\ominus})R''}$$

The isomeric vinyloxystannanes are then formally derived from (**4a**) and (**4b**) and the α-stannyl ketones from (**5**). The final balance in the mixture which is isolated appears to be determined by steric control.

c. Nucleophilic Displacement of Hydroxide or Stannyloxide. The reaction of bis(triethyltin) oxide with phenols was first carried out to demonstrate the basicity of an organotin oxide (*278*). The equilibrium [Eq. (20)] is well over to the side of organotin phenoxide and if the water is removed this reaction provides an easy preparative route to trialkyltin phenoxides. It has been used widely for the preparation of 8-quinolinoxy compounds (oxinates). Dehydration is accomplished by distillation or by adding desiccants, and trimethyltin hydroxide can be estimated by titration with phenol in the presence of molecular sieves [Eq. (21)] (*202*).

$$(Et_3Sn)_2O + 2\,p\text{-}Me\cdot C_6H_4\cdot OH \rightleftharpoons 2\,Et_3Sn\cdot OC_6H_4p\text{-}Me + H_2O$$

$$Me_3Sn\cdot OH + PhO\cdot H \xrightarrow{-H_2O} Me_3Sn\cdot OPh \qquad (20)$$

Acetylenic alcohols also react readily, the hydroxylic proton being replaced more readily than its acetylenic competitor. Thus the reaction of bis(triethyltin) oxide with an excess of but-3-yn-2-ol [Eq. (21)] (*295*), at −10° goes rapidly with predominant formation of (**6**), but under more severe conditions (**7**) is obtained:

$$(Et_3Sn)_2O + HO\cdot CHMe\cdot C{\vdots}CH \longrightarrow \underset{\textbf{(6)}}{Et_3Sn\cdot OCHMe\cdot C{\vdots}CH} + H_2O$$

$$(Et_3Sn)_2O + HO\cdot CHMe\cdot C{\vdots}CH \longrightarrow \underset{\textbf{(7)}}{Et_3Sn\cdot O\cdot CHMe\cdot C{\vdots}C\cdot SnEt_3} + H_2O \qquad (21)$$

By azeotropic dehydration, the method may be extended to include much less acidic alcohols as long as they are reasonably high boiling [e.g., Eq. (22)] (*85*). The reagents are heated under reflux in benzene or toluene and the water collected in a Dean and Stark separator. Better than 90% yields are obtained in a maximum reaction time of 1 h.

$$(Bu_3Sn)_2O + 2\,HO\cdot CHMePh \xrightarrow{-H_2O} 2\,Bu_3Sn\cdot O\cdot CHMePh \qquad (22)$$

Because of the low boiling points of methanol and ethanol this does not present a very satisfactory route to methoxides and ethoxides, although these are just the compounds which are in greatest demand. However they may be obtained from the oxide by a new, alternative reaction which also rapidly gives high yields.

The preparation of trialkyltin alkoxides by transalkoxylation with esters has been described in Sec. I.A.1.*b*. Contrary to earlier reports (*20*), it has been shown that a similar reaction occurs with bis(trialkyltin) oxides (*83*):

$$(Et_3Sn)_2O + EtO{\cdot}CO{\cdot}Me \longrightarrow Et_3Sn{\cdot}OEt + Et_3Sn{\cdot}O{\cdot}CO{\cdot}Me \qquad (23)$$

This in itself does not present an attractive preparative route, as half the tin is wasted in the carboxylate and separation of the two products is often difficult. It is also known that alkyl trialkyltin carbonates decarboxylate when they are heated to give trialkyltin alkoxides (*33*, *39*). The new route to alkoxides is based upon these two principles. If a bis(trialkyltin) oxide is heated under reflux with a dialkyl carbonate, carbon dioxide is evolved and the trialkyltin alkoxide is formed (*85*):

$$\begin{aligned}(Oct_3Sn)_2O + MeO{\cdot}CO{\cdot}OMe &\longrightarrow Oct_3Sn{\cdot}OMe + [Oct_3Sn{\cdot}O{\cdot}CO{\cdot}OMe] \\ &\xrightarrow{-CO_2} 2\ Oct_3Sn{\cdot}OMe\end{aligned} \qquad (24)$$

Some evidence to support the carbonate intermediate comes from the reaction of bis(trialkyltin) oxides with pyrocarbonates from which alkyl trialkyltin carbonates can be isolated (*90*):

$$(Bu_3Sn)_2O + EtO{\cdot}CO{\cdot}O{\cdot}CO{\cdot}OEt \longrightarrow 2\ Bu_3Sn{\cdot}O{\cdot}CO{\cdot}OEt \qquad (25)$$

The traditional but tedious procedure of preparing trialkyltin alkoxides from the corresponding halide and sodium alkoxide has become obsolete. Methoxides and ethoxides, and indeed other alkoxides, are available by the dialkyl carbonate method; alkoxides derived from higher alcohols may be prepared by transalkoxylation, or from the oxide and alcohol, by azeotropic dehydration. In the new methods, the time needed is less, the yields are higher, the secondary products are volatile, and no transfers are involved.

Unfortunately, these improved techniques do not, in general, appear to be applicable to the preparation of dialkyltin dialkoxides. The report that dibutyltin diallyloxide may be prepared by refluxing dibutyltin oxide and allyl alcohol in benzene (*179*) is probably incorrect in the light of a more complete investigation of the reaction of dibutyltin oxide with alcohols (*84*). This latter work indicates that the initial products are tetrabutyl dialkoxydistannoxanes (**8**), although disproportionation to dialkoxide and oxide occurs during distillation:

$$4\,Bu_2SnO + 4\,ROH \longrightarrow \underset{(\mathbf{8})}{[(RO)Bu_2Sn\cdot O\cdot SnBu_2(OR)]_2} + 2\,H_2O \xrightarrow{\Delta} 2\,Bu_2Sn(OR)_2 + 2\,Bu_2SnO \qquad (26)$$

(R = allyl, butyl, 2,2,2-trifluoroethyl and benzyl)

It has previously been shown that phenols (*58*) and alkyl hydroperoxides (*38*) react similarly to give compounds analogous to (**8**) where (OR) is replaced by (OAr) and (OOR), respectively.*

By contrast, it is possible to prepare derivatives of 1,2-glycols directly from the oxide. For example, a quantitative yield of 2,2-dibutyl-*cis*-4,5-dimethyl-2-stanna-1,3-dioxolane (**9**) was obtained from the reaction of dibutyltin oxide and *meso*-2,3-butanediol in benzene [Eq. (27)] (*55*), and some related, though slightly more exotic, compounds are claimed in a patent (*135*).

$$Bu_2Sn\begin{matrix}/\,O-CH_2-CH_2-O\,\backslash \\ \backslash\,O-CH_2-CH_2-O\,/\end{matrix}SnBu_2 \quad (\mathbf{10})$$

The five-membered ring structure in (**9**), 2-stanna-1,3-dioxolane, is proposed for a number of other glycol derivatives and is also suggested for the products obtained when dialkyltin oxides are treated with catechol (*94*, *136*). An apparent exception is the compound prepared from dibutyltin oxide and ethylene glycol. This was originally formulated (*261*) as $HO(CH_2)_2\cdot[OSnBu_2\cdot O(CH_2)_2]_2\cdot OH$, but Bornstein et al. (*41*) subsequently proposed the ten-membered ring structure (**10**), which was later supported by Considine (*55*). These structures

$$Bu_2SnO + \begin{matrix}HO-CHMe \\ | \\ HO-CHMe\end{matrix} \longrightarrow Bu_2Sn\begin{matrix}/\,O-CHMe \\ \quad\;\, | \\ \backslash\,O-CHMe\end{matrix} + H_2O \qquad (27)$$

(**9**)

are based upon the evidence of molecular weights, and of infrared spectra which show the absence of hydroxyl groups.

Molecular weights of dibutyltin derivatives of glycols prepared from the diethoxide by Mehrotra and Gupta (*203*) show that they exhibit varying

* In a very recent publication (*265*) it has been shown that dibutyltin diphenoxides can be prepared by dehydrating the oxide and phenol in boiling tetralin. It therefore appears that the reaction can be taken beyond the distannoxane stage at a sufficiently high temperature.

tendencies to polymerize, a process which was represented by

$$2\ Bu_2Sn\begin{matrix}O\\ \diagup\ \diagdown\\ \\ \diagdown\ \diagup\\ O\end{matrix}(CH_2)_n \rightleftharpoons Bu_2Sn\begin{matrix}O-(CH_2)_n-O\\ \\ O-(CH_2)_n-O\end{matrix}SnBu_2 \qquad (28)$$

The possibility of association by intermolecular coordination similar to that already described for dibutyltin dimethoxide, does not appear to have been considered, yet molecular weights in pyridine, a good coordinating solvent, are generally lower than those in benzene.*

A compound identical to that described as (**10**) has been prepared from dibutyltin oxide and ethylene carbonate (*85*). However, alkoxides derived from 1,2-glycols are not typical of dialkyltin dialkoxides, although the reason is not obvious. They can be formed very readily and, once formed, are hydrolytically unusually stable. Thus the method could not be extended to the preparation of dibutyltin dimethoxide, an alkoxydistannoxane being the product (*90*).

d. Nucleophilic Displacement of Amine. Organotin amines are highly susceptible to protolysis (*145*) and their reaction with alcohols provides a route to organotin alkoxides under very mild conditions. The method, based upon an analogous reaction yielding tin tetraisopropoxide (*310*), found initial success in the preparation of trimethyltin methoxide, which was obtained by careful methanolysis of diethylaminotrimethyltin in ether at −50°C [Eq. (29)] (*19*). It had been shown earlier (*18*) that this compound cannot be prepared from trimethyltin chloride and sodium methoxide, since it disproportionates during distillation into tetramethyltin and dimethyltin dimethoxide [Eq. (30)], at temperatures below 80°C. This disproportionation would also appear to rule out any preparation from dimethyl carbonate and trimethyltin hydroxide, since this route would necessarily require elevated temperatures:

$$Me_3Sn{\cdot}NEt_2 + MeOH \longrightarrow Me_3Sn{\cdot}OMe + HNEt_2 \qquad (29)$$

$$2\ Me_3Sn{\cdot}OMe \xrightarrow{\Delta} Me_4Sn + Me_2Sn(OMe)_2 \qquad (30)$$

In a tetrahedral molecule, Me_3SnX, both symmetrical and asymmetrica- Sn—C_3 stretching vibrations associated with the trimethyltin group are infrared active. If the stereochemistry about tin is changed to trigonal bi-

* In a report (*249a*) which appeared after the completion of this chapter, such association has been considered for nine dibutyltin derivatives of 1,2- or 1,3-diols. These are dimeric in benzene. The molecular weights are temperature dependent, but the large ring structure (*cf.* **10**) is favored over the coordination complex on the basis of infrared and NMR data, although the evidence is not yet conclusive.

pyramidal by association of the type:

```
 Me Me    Me Me
  \ /      \ /
X—Sn----X—Sn---
   |        |
  Me       Me
```

the trimethyltin group becomes planar, hence no change of dipole moment accompanies the symmetrical vibration which consequently disappears from the infrared spectrum. Such a structure exists in trimethyltin hydroxide and the infrared spectrum of the solid shows only one (asymmetrical) Sn—C_3 vibration (see Sec. II). The appearance of a second (symmetrical) vibration in trimethyltin alkoxides, which increases in intensity and shifts to higher wave number in going from methoxide to ethoxide, is taken as an indication of decreasing pentacoordination (*185*).

Only eight simple alkyltin trialkoxides are known and these have all been prepared by alcoholysis of appropriate organotin diethylamines (*185*):

$$PhSn(NEt_2)_3 + 3\ EtOH \longrightarrow PhSn(OEt)_3 + 3\ HNEt_2 \quad (31)$$

The Sn—O stretching vibration is reported to occur at about 550–600 cm^{-1}, 640 and 605 cm^{-1}, and 650 cm^{-1} in mono-, di- and trialkoxymethyltins, respectively (*185*).

The only disadvantage of the method is that it requires the initial preparation and purification of highly reactive organotin amines. Its great value is as a complement to the more robust methods described earlier, in making available organotin alkoxides which cannot otherwise be prepared.

e. Nucleophilic Displacement of Groups Bonded through Carbon. Simple alkyl groups attached to tin are very resistant to nucleophilic attack, and even vinyl groups are not readily displaced. However, the introduction of electron-withdrawing substituents can increase the lability of these groups to such an extent that they become susceptible to alcoholysis. For example, dialkyltin diethoxides were prepared by refluxing the corresponding perfluorovinyltin compounds with ethanol [e.g., Eq. (32)], although use of the more powerful ethoxide nucleophile was necessary to obtain trialkyltin compounds (*286*).

$$Et_2Sn(CF{:}CF_2)_2 + 2\ EtOH \longrightarrow Et_2Sn(OEt)_2 + 2\ CHF{:}CF_2 \quad (32)$$

Lutsenko and his co-workers have shown that α-stannyl-ketones, -esters and -amides react with alcohols or enols by cleaving the tin-carbon bond (*250–253*):

$$Et_3Sn{\cdot}CH_2{\cdot}CO{\cdot}Me + MeOH \longrightarrow Et_3Sn{\cdot}OMe + Me_2CO \quad (33)$$

$$Et_2Sn(CH_2{\cdot}CO{\cdot}OMe)_2 + 2\ MeOH \longrightarrow Et_2Sn(OMe)_2 + 2\ Me{\cdot}CO{\cdot}OMe \quad (34)$$

$$Pr_3Sn\cdot CH_2\cdot CO\cdot Me + Me\cdot CO\cdot CH_2\cdot CO\cdot Me \longrightarrow \underset{\mathbf{(11)}}{Pr_3Sn\cdot OCMe{=}CH\cdot CO\cdot Me} + Me_2CO \quad (35)$$

$$Et_3Sn\cdot CH_2CO\cdot NMe_2 + Me\cdot CO\cdot CH_2\cdot CO\cdot OEt \longrightarrow \underset{\mathbf{(12)}}{Et_3Sn\cdot OCMe{=}CH\cdot CO\cdot OEt} + Me\cdot CO\cdot NMe_2 \quad (36)$$

Infrared spectra of the functionally substituted vinyloxides (**11**) and (**12**), indicate that intramolecular coordination from the carbonyl to tin occurs to create a conjugated chelate system (**13**).

Me—C(=CH—C(—X)=O→)—O—SnR_3 (chelate ring) **(13)**

R = Et, Pr
X = Me, OEt

Related inorganic chelates have been prepared by cleaving phenyl groups from diphenyltin dichloride at 200°C with chelating agents such as benzoylacetone (*212*); a similar process provided tetrakis-(8-quinolinato)tin, an eight coordinate tin compound (*260*). These reactions do not constitute an alternative general preparative route to organotin alkoxides.

2. *Preparation by Addition of Tin Hydrides, Alkoxides, Amines, etc. to Carbonyl Compounds*

An organotin alkoxide is formed when any Sn—X bond adds to a carbonyl compound [Eq. (1)] and this forms the basis of a very versatile preparative route to these compounds. Additions usually proceed under mild conditions, indeed they are frequently exothermic, and by variation of the group X, a wide selection of simple and substituted alkoxides is made available.

a. Addition of the Tin-Hydrogen Bond. In the absence of catalysts, triphenyltin hydride reacts with aldehydes and ketones by hydrogenation of the carbonyl group (*174*, *315*):

$$2\,Ph_3Sn\cdot H + R'\cdot CO\cdot R'' \longrightarrow R'R''CH\cdot OH + Ph_3Sn\cdot SnPh_3 \quad (37)$$

Triphenyltin alkoxides, formed by hydride addition, are possible intermediates in this reduction and the corresponding trialkyltin adducts were successfully isolated by treating carbonyl compounds with trialkyltin hydrides in the presence of catalysts.

Valade and his co-workers (*51*) obtained tributyltin alkoxides in 30–70% yields from the prolonged exposure of ketone-tributyltin hydride mixtures to ultraviolet light:

$$Bu_3Sn\cdot H + Ph\cdot CO\cdot Me \xrightarrow[20h]{uv} Bu_3Sn\cdot OCH(Ph)Me \quad (38)$$

As the reaction temperature was increased to 150°C, the competing reduction to alcohol [c.f. Eq. (37)] became predominant. Neumann and Heymann (*217*) obtained 60–90% yields of alkoxides by adding triethyltin hydride to aldehydes and ketones in the presence of radical-generating catalysts such as azobisisobutyronitrile (AIBN) or the Lewis acid zinc chloride:

$$Et_3Sn\cdot H + Pr^i\cdot CHO \xrightarrow[\text{AIBN}]{40\text{–}90^\circ C} Et_3Sn\cdot OBu^i \tag{39}$$

In a more complete report later (*218*) it was shown that the technique may be extended to the preparation of dialkyltin dialkoxides.

The mechanistic aspects of the hydrostannation of aldehydes and ketones have received considerable attention, and it appears that hydrides can add both by radical and polar mechanisms. Additions which are brought about under free radical-generating conditions [e.g., Eqs. (38) and (39)] evidently proceed by a free radical mechanism (*218*, *248*). On the other hand, hydrostannations under the influence of zinc chloride were thought to follow a polar course (*218*). The catalytic activity of zinc chloride probably arises from an enhancement of the electrophilic character of the carbon atom of the carbonyl bond. In agreement with this, aldehydes and ketones, in which such electrophilic enhancement is achieved by electron-withdrawing substituents, readily undergo uncatalyzed addition with tin hydrides. Thus trimethyltin hydride adds exothermically to hexafluoroacetone (*63*):

$$Me_3Sn\cdot H + CF_3\cdot CO\cdot CF_3 \longrightarrow Me_3Sn\cdot OCH(CF_3)_2 \tag{40}$$

to chloral and pentafluorobenzaldehyde:

$$R_3Sn\cdot H + C_6F_5\cdot CHO \longrightarrow R_3Sn\cdot OCH_2\cdot C_6F_5 \qquad R = Me, Et \tag{41}$$

and to 2,2,2-trifluoroacetophenone (*183*).

These reactions are usually complicated by further addition of the generated alkoxide to more aldehyde or ketone:

$$Et_3Sn\cdot OCH_2\cdot C_6F_5 + C_6F_5\cdot CHO \rightleftharpoons Et_3Sn\cdot O\cdot CH(C_6F_5)\cdot OCH_2\cdot C_6F_5 \tag{42}$$

a general process which is discussed later.

One novel complication in the reaction of triethyltin hydride with chloral is that, under appropriate conditions, the 2,2,2-trichloroethoxytriethylstannane is partially reduced by an excess of triethyltin hydride (*183*):

$$Et_3Sn\cdot OCH_2\cdot CCl_3 + Et_3Sn\cdot H \longrightarrow Et_3Sn\cdot OCH_2\cdot CHCl_2 + Et_3Sn\cdot Cl \tag{43}$$

Leusink (*183*) has studied, in detail, the mechanism of hydrostannation of 2,2,2-trifluoroacetophenone, where no subsequent reactions occur to complicate the picture. The addition is first order in each reactant. The rate of reaction increases with increasing polarity of the solvent, and the reactivity order $Ph_3SnH \ll Bu_3SnH < Me_3SnH < Et_3SnH$ indicates that electron-releasing substituents at tin accelerate the hydrostannation, the reversed order $Et_3SnH > Bu_3SnH$ being attributed to steric effects. Finally, the free-radical scavenger galvinoxyl does not affect the rate of reaction. An ionic mechanism is clearly indicated and the results suggest that the rate-determining step is a nucleophilic attack of hydridic hydrogen on carbon [Eq.(44)], rather than nucleophilic attack on tin as previously suggested for the hydrostannation of hexafluoroacetone (*63*):

$$R_3Sn{-}H + Ph(F_3C)C{=}O \xrightarrow{\text{slow}} R_3Sn^{\oplus} + H{-}C(Ph)(F_3C){-}O^{\ominus} \xrightarrow{\text{fast}} R_3SnO{\cdot}CH(CF_3)Ph \qquad (44)$$

The reported catalytic activity of Lewis acids like zinc chloride and weak acids such as phenol and methanol (*218*) can now be explained in terms of electrophilic assistance in the rate-determining stage of the ionic mechanism:

$$R_3Sn{\cdots}H{\cdots}C(R')(R''){=}O{\cdots}ZnCl_2 \qquad (45)$$

The high reactivity of salicylaldehyde as compared with benzaldehyde probably arises, at least partly, from electrophilic assistance by the phenolic proton. The addition of triethyltin hydride to salicylaldehyde [Eq. (46)] was first described by Neumann and Heymann (*218*), who reported a complete rearrangement of the 1:1 adduct (**14**) to an organotin phenoxide (**15**), and who suggested that the overall reaction proceeds by a free radical mechanism. The reaction was reinvestigated by Leusink et al. (*183*) who suggest that, in solution, an equilibrium exists between the species (**14**) and (**15**) which is strongly dependent upon conditions of solvent and temperature:

$$Et_3Sn{\cdot}H + o\text{-}HO{-}C_6H_4{-}HC{=}O \longrightarrow \underset{(\mathbf{14})}{o\text{-}HO{-}C_6H_4{-}CH_2{\cdot}O{\cdot}SnEt_3} \rightleftharpoons \underset{(\mathbf{15})}{o\text{-}Et_3Sn{\cdot}O{-}C_6H_4{-}CH_2{\cdot}OH} \qquad (46)$$

It was also confirmed that the hydrostannation is subject to autocatalysis, but kinetic studies indicate that the autocatalytic reaction involves a polar mechanism. It was concluded that the overall reaction does not proceed by a free radical mechanism, but rather that the polar autocatalytic mechanism becomes dominant once a sufficient amount of adduct has been generated. These first amounts of adduct, however, may be formed by an ionic reaction in polar solvents, but by a free radical reaction in nonpolar solvents in the presence of free radical generators.

In the hydrostannation of γ-ethylenic ketones, addition at the olefinic bond competes with addition at, and reduction of, the carbonyl group (*239*). With the conjugated α-ethylenic ketones, the primary reaction is that of 1,4-addition (*183, 240*). Whereas the reaction stops at this stage with tributyltin hydride and yields tributyltin vinyloxides:

$$Bu_3Sn \cdot H + PhCH{=}CH \cdot CO \cdot Ph \longrightarrow Bu_3Sn \cdot O \cdot CPh{=}CH \cdot CH_2Ph \quad (47)$$

subsequent hydrogenolysis occurs in the case of triphenyltin hydride, and rearrangement of the resulting vinyl alcohols yields saturated ketones:

$$Ph_3Sn \cdot H + CH_2{=}CH \cdot CO \cdot Ph \longrightarrow Ph_3Sn \cdot O \cdot CPh{=}CH \cdot Me \xrightarrow{Ph_3Sn \cdot H} Ph_3Sn \cdot SnPh_3 + Ph \cdot CO \cdot Et \quad (48)$$

A more complicated series of substituted trialkyltin vinyloxides of the type $R_3Sn \cdot OC(OEt){=}CX \cdot CH_2R'$ (where $X = CN$ or $O \cdot CO \cdot Et$) and $R_3Sn \cdot OCR''{=}C(CN) \cdot CH_2R'$ from similar 1,4-hydrostannations are also reported (*223*).

Thus the additions of trialkyl- and dialkyl-tin hydrides to carbonyl compounds provide a route to the corresponding primary alkoxides (from aldehydes), secondary alkoxides (from ketones), and vinyloxides (from α-ethylenic ketones). The method is not generally applicable to the synthesis of phenyltin alkoxides and obviously tertiary alkoxides cannot be prepared in this way.

b. Addition of the Tin-Oxygen Bond. As will be discussed in Sec. III, organotin alkoxides and oxides add across multiply bonded groups of a variety of acceptors. With aldehydes and ketones, the adducts which are formed, are themselves organotin alkoxides with 1-alkoxy- or 1-stannyloxy substituents in their alkoxide skeleton:

$$R_3Sn \cdot OR' + R'' \cdot CO \cdot R''' \rightleftharpoons R_3Sn \cdot O \cdot CR''R''' \cdot OR' \quad (49)$$

(R' = alkyl, aryl or R_3Sn; R'' = alkyl or aryl; R''' = alkyl or H)

The first reported reactions of this class were the additions of tributyltin methoxide and of bis(tributyltin) oxide to chloral, which are exothermic at room temperature (*32, 33*). The majority of the additions, which were subsequently described, also involved chloral, or other aldehydes and ketones which are activated by halogen substitution, although it has been shown that ordinary aldehydes such as acetaldehyde are also reactive (*89*). Ordinary ketones like acetone and cyclohexanone do not appear to undergo addition at room temperature.

The alkoxides obtained by these reactions cannot be purified by distillation. If distillation is attempted, the initial addition is reversed. For example, 1,1,1-trichloro-2-methoxy-2-tributylstannyloxyethane dissociates into tributyltin methoxide and chloral to the extent of 25% if distilled at 90°C/0.8

mm, and the adduct between butyraldehyde and tributyltin methoxide is in equilibrium with its precursors at 33°C. the mixture containing some 15–20% of free aldehyde (*89*). Alkoxides derived from chloral and trimethyltin or triethyltin 2,2,2-trichloroethoxides dissociate to the extent of 88 and 100% respectively during distillation (*183*). By contrast it is reported that triethyltin alkoxides obtained by the addition of acetylenic alkoxystannanes to chloral are distillable (*290*).

Some alkoxides derived from trihalogenomethyl aldehydes or ketones decompose by an alternative route in which trihalogenomethyltrialkylstannanes are formed (*88*):

$$(Bu_3Sn)_2O + CBr_3{\cdot}CHO \longrightarrow Bu_3Sn{\cdot}O{\cdot}CH(CBr_3)O{\cdot}SnBu_3 \longrightarrow Bu_3Sn{\cdot}CBr_3 + Bu_3Sn{\cdot}O{\cdot}CHO \quad (50)$$

$$Bu_3Sn{\cdot}OMe + CCl_3{\cdot}CO{\cdot}CCl_3 \longrightarrow Bu_3Sn{\cdot}O{\cdot}C(CCl_3)_2OMe \longrightarrow Bu_3Sn{\cdot}CCl_3 + CCl_3{\cdot}CO{\cdot}OMe \quad (51)$$

Such alkoxides provide a useful source of the trihalogenomethyltin compounds.

Attempted distillation of $Bu_3Sn{\cdot}OCH(CCl_3)OCH_2{\cdot}CCl_3$ yielded not only tributyltin 2,2,2-trichloroethoxide, but also tributyltin chloride (*183*). The formation of chloride was attributed to decomposition of the 2,2,2-trichloroethoxide, yet this compound is thermally stable when prepared by dehydration of the oxide-alcohol mixture (*85*). It seems likely then, that the tributyltin chloride was formed by thermal decomposition of trichloromethyltributyltin derived from a reaction analogous to Eq. (50).

A detailed mechanistic study of the addition of trialkyltin alkoxides to carbonyl compounds has not yet been made. However, the relative power of tin-oxygen bonded compounds to act as addenda, which decreases in the series $Bu_3Sn{\cdot}OMe > (Bu_3Sn)_2O > Bu_3Sn{\cdot}OPh > Bu_3Sn{\cdot}OCH(CCl_3)OMe$, and the reactivity sequence for carbonyl compounds acting as acceptors, which is $Cl_3C{\cdot}CHO > Br_3C{\cdot}CHO, CF_2Cl{\cdot}CO{\cdot}CFCl_2 > (Cl_3C)_2CO > C_6F_5{\cdot}CHO > MeCHO > PrCHO, Bu^iCHO \gg PhCHO$, suggests that the principal process governing reactivity is the nucleophilic attack of the oxygen bonded to tin upon the carbonyl carbon atom.

All data so far in hand are in agreement with this. There is further evidence that the introduction of electronegative substituents into the alkoxide group depresses its reactivity (tributyltin alkoxides add to chloral in the reactivity sequence, $Bu_3Sn{\cdot}OC_6H_{11} > Bu_3Sn{\cdot}O{\cdot}CH_2{\cdot}CMe_3 > Bu_3Sn{\cdot}O{\cdot}CH_2{\cdot}CF_2{\cdot}CHF_2 > Bu_3Sn{\cdot}O{\cdot}CH_2{\cdot}CF_3$) (*84*), while similar substitution in the carbonyl compound increases its acceptor power ($Et_3Sn{\cdot}OCH_2{\cdot}C_6F_5$, when treated with a mixture of benzaldehyde and pentafluorobenzaldehyde, reacts exclusively with the latter) (*183*). The process appears, then, to be analogous to the polar hydrostannations discussed in the preceding section.

Triphenyltin alkoxides have been prepared by adding methoxide and oxide to chloral and bromal; since the products are solids they may be purified by recrystallization (*84*). By using triphenyltin alkyl peroxides, analogous alkoxides with 1-alkylperoxy substituents have been isolated (*31*):

$$Ph_3Sn{\cdot}O{\cdot}OBu^t + CCl_3{\cdot}CHO \longrightarrow Ph_3Sn{\cdot}O{\cdot}CH(CCl_3)O{\cdot}OBu^t \quad (52)$$

Dibutyltin dimethoxide reacts exothermically with acetaldehyde or chloral and products from the addition of one or both methoxytin bonds can be prepared (*71*):

$$Bu_2Sn(OMe)_2 + MeCHO \longrightarrow Bu_3Sn(OMe)O{\cdot}CHMe{\cdot}OMe \xrightarrow{MeCHO} Bu_2Sn(O{\cdot}CHMe{\cdot}OMe)_2 \quad (53)$$

Compounds of the type $Bu_2Sn(OMe)X$, (where X = Br, OAc, SCN), add to chloral and the alkoxides so formed may also be prepared by a disproportionation, analogous to Eq. (8), of a mixture of Bu_2SnX_2 and $Bu_2Sn[O{\cdot}CH(CCl_3)OMe]_2$:

$$Bu_2Sn(OAc)OMe + Cl_3C{\cdot}CHO \longrightarrow Bu_2Sn(OAc)O{\cdot}CH(CCl_3)OMe \longleftarrow Bu_2Sn(OAc)_2 + Bu_2Sn[O{\cdot}CH(CCl_3)OMe]_2 \quad (54)$$

We have seen previously (Sec. I.A.1.*b*) that organotin alkoxides can be converted into other alkoxides by substitution reactions. When the substitution involves a cyclic compound, the process resembles an addition. Thus β-substituted organotin alkoxides have been prepared by the reaction of trialkyltin methoxides with β-propionolactone [Eq. (55a)]:

$$R_3Sn{\cdot}OMe + \begin{matrix} H_2C - CH_2 \\ | \quad\quad | \\ O{=}C - O \end{matrix} \begin{matrix} \xrightarrow{(a)} R_3Sn{\cdot}OCH_2{\cdot}CH_2{\cdot}CO{\cdot}OMe \\ \xrightarrow{(b)} MeO{\cdot}CH_2{\cdot}CH_2{\cdot}CO{\cdot}OSnR_3 \end{matrix} \quad (55)$$

With trimethyltin methoxide (R = Me), Ishii and his co-workers (*138*) have shown that this acyl-oxygen fission predominates over the alternative ring-opening by alkyl-oxygen fission [Eq. (55b)] in several solvents, although of the solvents investigated it is only obtained exclusively in carbon tetrachloride. Acyl-oxygen fission was similarly proposed in a preliminary study with tributyltin methoxide (R = Bu) (*36*).

These alkoxides exhibit a carbonyl absorption at 1740 cm^{-1}, as for ordinary esters, suggesting that there is no intramolecular coordination of the type present in acetylacetonates, etc. (**13**). An S_N2 mechanism is indicated from kinetic studies and, as with additions, the important step is probably nucleophilic attack of alkoxide at carbonyl carbon.

Ketene dimer also has a β-lactone structure and one might anticipate an analogous reaction. However, the product with triethyltin ethoxide is the chelated triethylstannyl enolate of ethyl acetoacetate. This is said to arise from the rearrangement of initially formed ethyl 4-triethylstannylacetoacetate (*253*):

$$Et_3Sn{\cdot}OEt + \begin{array}{c} H_2C-C{=}CH_2 \\ | \quad\ \ | \\ O{=}C-O \end{array} \longrightarrow [Et_3Sn{\cdot}CH_2{\cdot}CO{\cdot}CH_2{\cdot}CO{\cdot}OEt]$$

$$\begin{array}{c} H \\ Me-C{\overset{C}{=}}\ \ C-OEt \\ | \qquad\quad \| \\ O \quad\ \ O \\ \searrow Sn \swarrow \\ Et_3 \end{array} \longleftarrow [Et_3Sn{\cdot}CH_2{\cdot}C(OH){:}CH{\cdot}CO{\cdot}OEt] \qquad (56)$$

c. *Addition of the Tin-Nitrogen Bond.* *N*-Stannyl derivatives of amines (*90*), amides (*82*), carbamates (*36, 84*), sulphonamides (*82*), and ureas (*90*) all undergo carbonyl additions with chloral or bromal, to provide organotin alkoxides:

$$Ph_3Sn{\cdot}NHBu + CCl_3{\cdot}CHO \longrightarrow Ph_3Sn{\cdot}O{\cdot}CH(CCl_3)NHBu \qquad (57)$$

$$Bu_3Sn{\cdot}NMe{\cdot}CHO + CBr_3CHO \longrightarrow Bu_3Sn{\cdot}O{\cdot}CH(CBr_3)NMe{\cdot}CHO \qquad (58)$$

$$Bu_3Sn{\cdot}NPh{\cdot}CO{\cdot}OMe + CCl_3{\cdot}CHO \longrightarrow Bu_3Sn{\cdot}O{\cdot}CH(CCl_3)NPh{\cdot}CO{\cdot}OMe \qquad (59)$$

$$Bu_3Sn{\cdot}NPr{\cdot}SO_2{\cdot}Ph + CCl_3{\cdot}CHO \longrightarrow Bu_3Sn{\cdot}O{\cdot}CH(CCl_3)NPr{\cdot}SO_2{\cdot}Ph \qquad (60)$$

$$Bu_3Sn{\cdot}NMe{\cdot}CO{\cdot}NMe{\cdot}SnBu_3 + CBr_3{\cdot}CHO \longrightarrow Bu_3Sn{\cdot}O{\cdot}CH(CBr_3)NMe{\cdot}CO{\cdot}NMe{\cdot}SnBu_3 \qquad (61)$$

Very few of these compounds have been isolated but infrared and ^1H NMR spectra establish their existence.

As with alkoxystannanes, metathetical reactions of aminostannanes with cyclic compounds can provide alkoxides. Thus dimethylaminotrimethyltin gives a chelated alkoxide with ketene dimer [Eq. (62); cf. Eq. (56)] (*133*), and diethylaminotributyltin reacts with epoxides in the presence of catalytic amounts of lithium diethylamine [e.g., Eq. (63)] (*312*):

$$Me_3Sn{\cdot}NMe_2 + \begin{array}{c} H_2C-C{=}CH_2 \\ | \quad\ \ | \\ O{=}C-O \end{array} \longrightarrow \begin{array}{c} O-CMe \\ Me_3Sn \qquad \overset{\|}{C}H \\ \nwarrow O{=}C \\ | \\ NMe_2 \end{array} \qquad (62)$$

$$Bu_3Sn{\cdot}NEt_2 + \begin{array}{c} Ph{\cdot}HC-CH_2 \\ \diagdown O \diagup \end{array} \xrightarrow{LiNEt_2} Bu_3Sn{\cdot}O{\cdot}CHPh{\cdot}CH_2{\cdot}NEt_2 \qquad (63)$$

d. Addition of the Tin-Carbon Bond. The addition of alkyltins to aldehydes and ketones would provide, in principle, a route to organotin alkoxides derived from secondary and tertiary alcohols respectively:

$$R_3Sn{\cdot}R' + R''{\cdot}CO{\cdot}R''' \longrightarrow R_3Sn{\cdot}OCR'R''R''' \quad \text{(where } R''' = \text{alkyl or H)} \qquad (64)$$

Ordinary alkyl groups are not reactive in this sense, but if they contain α-substituents (x) with a $-M$ effect, carbonyl additions proceed to completion under very mild conditions, and may even be exothermic (*225*):

$$R_3Sn{\cdot}CH_2X + R''{\cdot}CO{\cdot}R''' \longrightarrow R_3Sn{\cdot}O{\cdot}CR''R'''{\cdot}CH_2X \qquad (65)$$

(where $R''' =$ alkyl or H; $X =$ CN, CO·Me, CO·OEt, $CO{\cdot}NEt_2$)

By this reaction trialkyltin alkoxides with β-cyano, -acetyl, -carboethoxy, and -diethylamido substituents in the alkoxide skeleton have been prepared:

$$Bu_3Sn{\cdot}CH_2{\cdot}CN + C_6F_5{\cdot}CHO \longrightarrow Bu_3Sn{\cdot}O{\cdot}CH(C_6F_5)CH_2{\cdot}CN \qquad (66)$$

$$Et_3Sn{\cdot}CH_2{\cdot}CO{\cdot}OEt + Ph{\cdot}CO{\cdot}CF_3 \longrightarrow Et_3Sn{\cdot}OCPh(CF_3)CH_2{\cdot}CO{\cdot}OEt \qquad (67)$$

It is suggested that heterolytic cleavage of the Sn—C bond in compounds $R_3Sn{\cdot}CH_2X$ is facilitated by the ability of the group X to stabilize negative charge on the adjacent carbon atom, e.g.,

$$\bar{C}H_2{-}C{\equiv}N \longleftrightarrow CH_2{=}C{=}\bar{N}$$

Furthermore the reactivity sequence for carbonyl compounds follows the same pattern as that observed for polar additions of Sn—H and Sn—O bonds, and the reaction can be catalyzed by Lewis acids. All this suggests, once again, that nucleophilic attack at the carbonyl carbon is the rate-determining step in these reactions.

Acidolysis of the product alkoxides yields β-hydroxy-nitriles, -ketones, -esters, and -amides, which are not readily available via Grignard reagents or Reformatsky reactions. This provides a good example of the synthetic applications of organotin compounds in organic chemistry, a topic which will be discussed in more detail in Sec. III.

Unsaturation in R′ can similarly activate the Sn—C bond toward additions. Both triethyl(phenylethynyl)tin (*208*) and triethylallyltin (*160*) have been added to aldehydes, although rather more vigorous conditions are generally required:

$$Et_3Sn{\cdot}C{\vdots}CPh + CCl_3{\cdot}CHO \longrightarrow Et_3Sn{\cdot}O{\cdot}CH(CCl_3){\cdot}C{\vdots}CPh \qquad (68)$$

$$Et_3Sn{\cdot}CH_2{\cdot}CH{:}CH_2 + PhCHO \longrightarrow Et_3Sn{\cdot}O{\cdot}CHPh{\cdot}CH_2{\cdot}CH{:}CH_2 \qquad (69)$$

3. *Miscellaneous Preparations*

The reactions discussed in this section do not represent practical synthetic routes to organotin alkoxides which can compete with the methods described so far. These reactions generally require long reaction times at high temperatures and the organotin alkoxide is one component of a complex mixture of products from which isolation is difficult. Nevertheless there is considerable academic interest in synthesizing alkoxides directly from alkyltins.

a. Homolytic Substitution at Tin. Triethyltin *t*-butoxide is formed when *t*-butyl peroxides are heated in tetraethyltin at temperatures of 125°–145°C, where thermal homolysis of the peroxide linkage can be expected to occur. For example di-*t*-butyl peroxide, when heated in tetraethyltin at 145°C for 16 h, gave the following products (yields in moles per mole of peroxide): $Et_3Sn \cdot OBu^t$ (1.12), $HOBu^t$ (0.82), C_2H_4 (1.25), C_2H_6 (0.50), CH_4 (0.04), C_4H_{10} (0.01) and traces of Et_6Sn_2 and Sn (*263*). All the products can be accounted for by free radical reactions:

$$Bu^tO{-}OBu^t \longrightarrow 2\,Bu^tO^{\bullet} \tag{70}$$

$$Bu^tO^{\bullet} \longrightarrow Me^{\bullet} + Me_2CO \tag{71}$$

$$Bu^tO^{\bullet} + Et_4Sn \begin{cases} \xrightarrow{a} Et_3Sn{-}OBu^t + Et^{\bullet} \\ \xrightarrow{b} Et_3Sn\dot{C}_2H_4 + HOBu^t \end{cases} \tag{72}$$

$$Me^{\bullet} + Et_4Sn \longrightarrow Et_3Sn\dot{C}_2H_4 + CH_4 \tag{73}$$

$$Et_3Sn\dot{C}_2H_4 \longrightarrow Et_3Sn^{\bullet} + C_2H_4 \tag{74}$$

$$Et_3Sn^{\bullet} + Et_4Sn \longrightarrow Et_3Sn{-}SnEt_3 + \dot{Et} \tag{75}$$

$$2\dot{Et} \begin{cases} \xrightarrow{a} C_2H_6 + C_2H_4 \\ \xrightarrow{b} C_4H_{10} \end{cases} \tag{76}$$

However, to account for the almost equimolar amounts of the three main products it was suggested that they are formed by homolytic decomposition of the intermediate complex (**16**):

$$\left[\begin{matrix} & CHMe & \\ Et_3Sn & & H \\ Bu^tO & & OBu^t \end{matrix}\right] \longrightarrow Et_3Sn{-}OBu^t + HOBu^t + [Me\ddot{C}H] \longrightarrow CH_2{=}CH_2 \tag{77}$$

(**16**)

On the other hand, similar results obtained by decomposing *t*-butyl peroxybenzoate in tetraethyltin were accounted for only in terms of the free radical process (*320*).

Hexaethylditin similarly reacts with di-*t*-butyl peroxide at 130–135°C for 16 h to yield triethyltin *t*-butoxide (*323*). The alkoxide is also obtained under milder conditions from *β*-trimethylsilylperoxypropionic acid where a "crypto-radical" process is again suggested (*321*):

$$Et_3Sn{\cdot}SnEt_3 + Me_3SiCH_2{\cdot}CH_2{\cdot}CO{\cdot}O{\cdot}OBu^t \longrightarrow \left[\begin{array}{c} Et_3Sn \text{---} SnEt_3 \\ O \quad\quad\quad \vdots \\ \| \quad\quad\quad \vdots \\ Me_3Si{\cdot}CH_2{\cdot}CH_2{\cdot}C\text{---}O\text{---}OBu^t \end{array} \right] \longrightarrow Et_3Sn{\cdot}OBu^t + Et_3Sn{\cdot}O{\cdot}CO{\cdot}CH_2{\cdot}CH_2{\cdot}SiMe_3 \quad (78)$$

A similar intermediate may be involved in the reaction of dicyclohexylperoxydicarbonate and hexaethylditin in benzene, from which a 76% yield of cyclohexyloxytriethyltin was obtained (*264*):

$$Et_6Sn_2 + (C_6H_{11}O{\cdot}CO{\cdot}O)_2 \longrightarrow 2\,Et_3Sn{\cdot}O{\cdot}C_6H_{11} + 2\,CO_2 \quad (79)$$

Neumann and Lind (*219*) have recently shown that dibenzyloxydiimide undergoes a free radical decomposition induced by triethyltin hydride and triethyltin benzyloxide is a major product. Finally, Aleksandrov and his co-workers (*4*) have shown that triethyltin ethoxide is a product, formed via organotin peroxides, in the oxidation of hexaethylditin.

b. Nucleophilic Rearrangement of Tin Peroxides. Attempted preparations of trialkyltin peroxycarboxylates by nucleophilic substitution at tin lead to nonperoxidic compounds which appear to be dialkyltin alkoxide-carboxylates, although no pure samples have been isolated (*14*, *296*, *322*). By analogy with *t*-alkyl peroxycarboxylates (*69*), it was proposed that the products are formed by a redox rearrangement [Eq. (80)], and the kinetics of the reaction have recently been studied for some triphenyltin compounds (*42*):

$$R_3SnX + HOO{\cdot}CO{\cdot}R' \longrightarrow \left[\begin{array}{c} R \\ | \\ R_2Sn\text{---}O \\ | \\ O{\cdot}CO{\cdot}R' \end{array} \right] \longrightarrow R_2Sn \begin{array}{l} \diagup OR \\ \diagdown O{\cdot}CO{\cdot}R' \end{array} \quad (80)$$

Since dialkyltin alkoxide-carboxylates can be prepared readily by the disproportionation of dialkyltin dialkoxides and dicarboxylates [cf. Eq. (54)], the reaction has little synthetic appeal.

Lists of tri-, di-, and mono-alkyltin alkoxides are given in Tables 1–3, respectively. In each table the compounds are arranged in order of increasing carbon number, primarily of the alkyl groups on tin and secondly of the

TABLE 1

TRIALKYLTIN ALKOXIDES

Compound[a]	mp, °C	bp, °C/mm	Preparations[b]	References
$Me_3Sn\cdot OMe$	75	0/58	X	*(18)*
			N	*(19, 185)*
$Me_3Sn\cdot OEt$	—	81/0.1	N	*(185)*
			M	*(178)*
$Me_3Sn\cdot OCH_2\cdot CCl_3$	—	—	H	*(183)*
$Me_3Sn\cdot OCH(CCl_3)OCH_2\cdot CCl_3$	—	—	Σ	*(183)*
$Me_3Sn\cdot OCH(CCl_3)O\cdot OBu^t$	—	—	Σ	*(31)*
$Me_3Sn\cdot OCH_2\cdot CH_2\cdot CO\cdot OMe$	—	—	N	*(138)*
			M	*(138)*
$Me_3Sn\cdot OCH(CF_3)_2$	—	76/58	H	*(63)*
$Me_3Sn\cdot OC(CF_3)_2OC(CF_3)_2H$	23	—	Σ	*(63)*
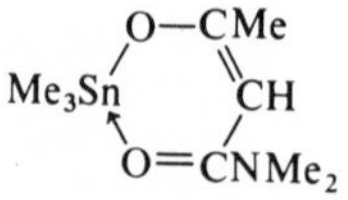 Me_3Sn (O—CMe=CH—CNMe_2=O→Sn, ring)	68–70	—	M	*(133)*
$Me_3Sn\cdot OPh$	—	222–224/760	X	*(165)*
		109/8	O	*(202)*
$Me_3Sn\cdot OCH_2\cdot C_6F_5$	—	—	H	*(183)*
$Me_3Sn\cdot OCH(C_6F_5)OCH_2\cdot C_6F_5$	—	—	Σ	*(183)*
$Me_3Sn\cdot OCH(CF_3)Ph$	—	62–64/0.3	H	*(183)*
$Me_3Sn\cdot$*Ox*	—	108.5–109.5 /0.2	X	*(99, 147)*
			O	*(99)*
$Et_3Sn\cdot OMe$	—	79–80/17	X	*(216, 224)*
		73–74/13	N	*(185)*
		37/0.1		
$Et_3Sn\cdot OEt$	—	190–192/760	X	*(137)*
		82–84/11	N	*(185)*
		45/0.1		

[a] Me = methyl; Et = ethyl; Pr = propyl; Bu = butyl; Ph = phenyl; C_6H_{11} = cyclohexyl; Oct = n-C_8H_{17}; Np = 1-naphthyl; 2-Np = 2-naphthyl;

acac = (O=C(Me)—CH=C(Me)—O, chelating); *bzac* = (O=C(Ph)—CH=C(Me)—O, chelating); *dbzm* = (O=C(Ph)—CH=C(Ph)—O, chelating); *Ox* = (8-quinolinolato, N→, O—)

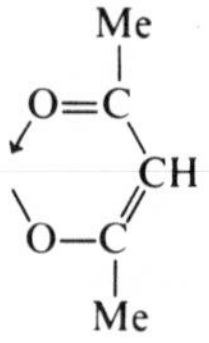

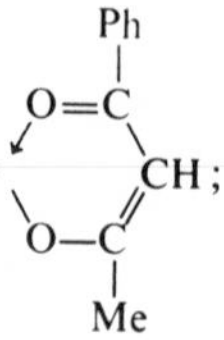

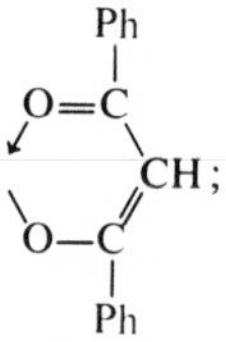

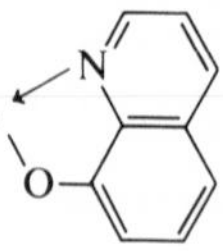

[b] X = preparation by substitution at an organotin halide. A = preparation by substitution at an organotin alkoxide. O = preparation by substitution at an organotin oxide or hydroxide. N = preparation by substitution at an organotin amine. H = preparation by addition of an Sn—H bond. Σ = preparation by addition of any other Sn—X bond. M = preparation by any other method or method not reported.

TABLE 1 (*continued*)

Compound[a]	mp, °C	bp, °C/mm	Preparations[b]	References
$Et_3Sn·OCH_2·CCl_3$	—	79/0.3	H	(*183*)
$Et_3Sn·OCH_2·CHCl_2$	—	—	M	(*183*)
$Et_3Sn·OCH(CCl_3)OCH_2·CCl_3$	—	—	Σ	(*183*)
$Et_3Sn·OCH_2·CH_2·OSnEt_3$	—	184–185/4.5	O	(*289*)
$Et_3Sn·OCH(CCl_3)OCH_2·C⋮CH$	—	85/2	Σ	(*290*)
$Et_3Sn·OCH(CCl_3)OCH_2·CH_2·C⋮CH$	—	69/1	Σ	(*290*)
$Et_3Sn·OCH(CBr_3)·NPh·CHO$	—	—	Σ	(*82*)
$Et_3Sn·OCH(CCl_3)OCH(CCl_3)CH_2·CO·OEt$	—	—	Σ	(*225*)
$Et_3Sn·OPr$	—	83–84/8	X	(*190*)
$Et_3Sn·OPr^i$	—	83/12	H	(*218*)
$Et_3Sn·OCH_2—CHO—CMe_2$ (ring: CHO / CH_2O bridged by CMe_2)	—	116/2	X	(*24*)
$Et_3Sn·OCH_2—CHO—CMeEt$ (ring: CHO / CH_2O bridged by CMeEt)	—	130/3	X	(*24*)
$Et_3Sn·OCH_2—CHO—CMePr$ (ring: CHO / CH_2O bridged by CMePr)	—	143/4	X	(*24*)
$Et_3Sn·O(CH_2)_3SiMe_2Et$	—	131–133/5	O	(*289*)
$Et_3Sn·OCH_2·C⋮CH$	—	80–81/5	A	(*292*)
		46–48/1	O	(*288, 295*)
$Et_3Sn·OCH_2·C⋮C·SnEt_3$	—	160/5	A	(*292*)
	—	124–125/1	O	(*288, 295*)
$Et_3Sn·OBu$	—	98–101/11	H	(*218*)
		69/1	M	(*319*)
$Et_3Sn·OBu^i$	—	95–96/11	H	(*217, 218*)
$Et_3Sn·OBu^s$	—	93–96/12	H	(*217*)
$Et_3Sn·OBu^t$	—	132–135/112	X	(*263*)
		52–59/0.03	M	(*61*)
$Et_3Sn·OCH(CCl_3)CH_2·CO·OEt$	—	108–114/0.25	Σ	(*225*)
$Et_3Sn·OCH(CCl_3)CH_2·CO·NEt_2$	—	—	Σ	(*225*)
$Et_3Sn·OCH{:}CMe_2$	—	62–63/1	A	(*254*)
Et_3Sn(O—CMe=CH—C(OMe)=O→) (chelate ring)	—	94–96/3	M	(*253*)
Et_3Sn(O—CMe=CH—C(OEt)=O→) (chelate ring)	—	91–93/2	M	(*253*)
Et_3Sn(O—CMe=CH—C(OPr)=O→) (chelate ring)	—	120–123/3	M	(*253*)

TABLE 1 (*continued*)

Compound[a]	mp, °C	bp, °C/mm	Preparations[b]	References
$Et_3Sn \cdot O(CH_2)_2C \vdots CH$	—	157–158/3	X	(*291*)
		60–62/1	A	(*292*)
			O	(*295*)
$Et_3Sn \cdot O(CH_2)_2C \vdots C \cdot SnEt_3$	—	134–135/1	A	(*292*)
			O	(*295*)
$Et_3Sn \cdot OCHMe \cdot C \vdots CH$	—	73/3	A	(*292*)
		57–58/1	O	(*288, 295*)
$Et_3Sn \cdot OCHMe \cdot C \vdots C \cdot SnEt_3$	—	138/2	A	(*292*)
		111–112/0.5	O	(*288, 295*)
$Et_3Sn \cdot OCHEt_2$	—	110/13	H	(*218*)
$Et_3Sn \cdot OCH_2 \cdot CH(Me)Et(+)$	—	80.5–80.9/1.5	X	(*304*)
$Et_3Sn \cdot OCMe_2 \cdot C \vdots CH$	—	83/2	A	(*292*)
		44/0.5	O	(*288, 295*)
$Et_3Sn \cdot OCMe_2 \cdot C \vdots C \cdot SnEt_3$	—	113–114/0.5	A	(*292*)
			O	(*295*)
$Et_3Sn \cdot OCH_2$–(2-furyl, O)	—	130/12	H	(*217, 218*)
$Et_3Sn \cdot OCH{:}CHPr^i$	—	63–64/0.5	A	(*254*)
$Et_3Sn \cdot$*acac*	41–43	94–95/1	A	(*204*)
			M	(*253*)
$Et_3Sn \cdot OC_6H_{11}$	—	129/11	X	(*264*)
		121–123/5	H	(*217, 218*)
		69/0.1	M	(*264*)
$Et_3Sn \cdot OC(CH_2Cl)_2CH_2 \cdot CO \cdot Me$	—	—	Σ	(*225*)
$Et_3Sn \cdot OC(Me)Et \cdot C \vdots CH$	—	70/1	O	(*111*)
$Et_3Sn \cdot OC(Me)Et \cdot C \vdots C \cdot SnEt_3$	—	131/2	X	(*111*)
			O	(*111*)
$Et_3Sn \cdot O$–(cyclopentenyl), $O{=}C \cdot OEt$ (O→Sn)	—	123–126/1.5	M	(*253*)
$Et_3Sn \cdot OPh$	—	147.5–148/12	X	(*137*)
		115/1	O	(*110, 137, 202*)
$Et_3Sn \cdot OC_6H_4$-4-Br	—	186–188/12	X	(*139*)
$Et_3Sn \cdot OC_6H_4$-4-Cl	—	174–175/2	X	(*139*)
		115–118/0.1	M	(*61*)
$Et_3Sn \cdot OC_6H_4$-4-NO_2	31–32	—	X	(*43*)
			O	(*316*)
$Et_3Sn \cdot OC_6H_4$-4-OH	103–104	—	X	(*139*)
$Et_3Sn \cdot OC_6H_4$-4-OMe	—	111–113/0.09	M	(*61*)
$Et_3Sn \cdot OC_6Cl_4$-4-$OSnEt_3$	—	—	M	(*330*)
$Et_3Sn \cdot OC_6H_4$-4-Me	—	124–127/1	O	(*278*)
			M	(*279*)

TABLE 1 (*continued*)

Compound[a]	mp, °C	bp, °C/mm	Preparations[b]	References
$Et_3Sn \cdot OC_6H_4\text{-2-}CH_2 \cdot OH$	64	—	H O	(*218*) (*218*)
$Et_3Sn \cdot OC_6H_4\text{-3-}CH_2 \cdot OH$	63	—	H	(*218*)
$Et_3Sn \cdot OCH_2 \cdot Ph$	—	150–155/14 99–101/0.5	H	(*217, 218*)
$Et_3Sn \cdot OCH_2 \cdot C_6F_5$	—	86–87/0.1	H	(*183*)
$Et_3Sn \cdot OCH(C_6F_5)OCH_2 \cdot C_6F_5$	—	—	Σ	(*183*)
$Et_3Sn \cdot OCHPr \cdot CH_2 \cdot CO \cdot Me$	—	—	Σ	(*225*)
O (furan ring) $Et_3Sn \cdot OCH \cdot CH_2 \cdot CO \cdot NEt_2$	—	—	Σ	(*225*)
$Et_3Sn \cdot O$ (cyclopentane ring) $HC⋮C$	—	99/2	O	(*111*)
$Et_3Sn \cdot O$ (cyclopentane ring) $Et_3Sn \cdot C⋮C$	—	157/2	X O	(*111*) (*111*)
$Et_3Sn \cdot OCH_2 \cdot C_6H_4\text{-4-OMe}$	—	132/0.04	H	(*217, 218*)
$Et_3Sn \cdot OCH(CF_3)Ph$	—	80–82/0.1	H	(*183*)
O (furan ring) $Et_3Sn \cdot OCH \cdot CH_2 \cdot CO \cdot Me$	—	—	Σ	(*225*)
$Et_3Sn \cdot O$ (cyclohexane ring) $HC⋮C$	—	110/5	O	(*111*)
$Et_3Sn \cdot O$ (cyclohexane ring) $Et_3Sn \cdot C⋮C$	—	152/1	X O	(*111*) (*111*)
$Et_3Sn \cdot OCHMe \cdot C_6H_4\text{-4-OMe}$	—	$90–93/10^{-4}$	H	(*217*)
$Et_3Sn \cdot OC_6H_2\text{-1,3,5-}(CH_2 \cdot NMe_2)_3$	—	—	A	(*97*)
$Et_3Sn \cdot OCHPh \cdot CH_2 \cdot CO \cdot OEt$	—	—	Σ	(*225*)
$Et_3Sn \cdot OCH(C_6F_5)CH_2 \cdot CO \cdot OEt$	—	107–108/0.07	Σ	(*225*)
$Et_3Sn \cdot OCH(C_6F_5)CH_2 \cdot CO \cdot NEt_2$	—	—	Σ	(*225*)
$Et_3Sn \cdot O$ (cyclohexane ring) $Me \cdot CO \cdot H_2C$	—	—	Σ	(*225*)
$Et_3Sn \cdot$*Ox*	—	152–154/2 132–134/0.05	X O	(*147*) (*129*)

TABLE 1 (*continued*)

Compound[a]	mp, °C	bp, °C/mm	Preparations[b]	References
$Et_3Sn{-}O$-(5-chloro-8-quinolyl)	—	168/2.2	O	(*129*)
$Et_3Sn{\cdot}O{\cdot}2\text{-}Np$	—	200–201/7	O	(*156*)
$Et_3Sn{\cdot}O$-(2-Pr^i-5-Me-cyclohexyl) (−)	—	125.8–125.9/0.9	X	(*304*)
$Et_3Sn{\cdot}OCHPh{\cdot}CH_2CH{:}CH_2$	—	96/0.003	Σ	(*160*)
$Et_3Sn{\cdot}OCH(C_6H_4\text{-}4\text{-}Cl)CH_2{\cdot}CH{:}CH_2$	—	128/0.02	Σ	(*160*)
$Et_3Sn{\cdot}OCH(C_6H_4\text{-}4\text{-}NO_2)CH_2{\cdot}CH{:}CH_2$	—	166/0.01	Σ	(*160*)
$Et_3Sn{\cdot}OCH(CCl_3)C{\vdots}CPh$	—	160–161/1	Σ	(*208*)
$Et_3Sn{\cdot}OCHPh{\cdot}CH_2{\cdot}CO{\cdot}Me$	—	124–126/2	Σ	(*225, 252*)
$Et_3Sn{\cdot}OCH(C_6H_4\text{-}2\text{-}Cl)CH_2{\cdot}CO{\cdot}Me$	—	—	Σ	(*225*)
$Et_3Sn{\cdot}OCH(C_6H_4\text{-}4\text{-}Cl)CH_2{\cdot}CO{\cdot}Me$	—	—	Σ	(*225*)
$Et_3Sn{\cdot}OCH(C_6H_4\text{-}3\text{-}NO_2)CH_2{\cdot}CO{\cdot}Me$	—	—	Σ	(*225*)
$Et_3Sn{\cdot}OCH(C_6H_4\text{-}4\text{-}NO_2)CH_2{\cdot}CO{\cdot}Me$	—	—	Σ	(*225*)
$Et_3Sn{\cdot}OCH(C_6F_5)CH_2{\cdot}CO{\cdot}Me$	—	—	Σ	(*225*)
$Et_3Sn{\cdot}OCPh(CF_3)CH_2{\cdot}CO{\cdot}OEt$	—	—	Σ	(*225*)
$Et_3Sn{\cdot}OCH(C_7H_{15})CH_2{\cdot}CH{:}CH_2$	27	122/0.35	Σ	(*160*)
$Et_3Sn{\cdot}OCH(C_6H_4\text{-}4\text{-}CCl_3)CH_2{\cdot}CH{:}CH_2$	—	98/0.2	Σ	(*160*)
$Et_3Sn{\cdot}OCPh(CF_3)CH_2{\cdot}CO{\cdot}Me$	—	—	Σ	(*225*)
$Et_3Sn{\cdot}OCH(CH{:}CHPh)CH_2{\cdot}CH{:}CH_2$	—	117/0.001	Σ	(*160*)
$Pr_3Sn{\cdot}OMe$	—	87–88/3	X	(*216*)
$Pr_3Sn{\cdot}OEt$	—	89–90/3	X	(*190*)
$Pr_3Sn{\cdot}OCHMe{\cdot}C{\vdots}CH$	—	92/1	O	(*288*)
$Pr_3Sn{\cdot}OCHMe{\cdot}C{\vdots}C{\cdot}SnPr_3$	—	165/1	O	(*288*)
$Pr_3Sn{\cdot}OCH{:}CMe_2$	—	84–85/1	A	(*254*)
Pr_3Sn chelate: O—CMe=CH—C(OMe)=O→Sn	—	128–131/4	M	(*253*)
Pr_3Sn chelate: O—CMe=CH—C(OEt)=O→Sn	—	114–116/2	M	(*252*)
Pr_3Sn chelate: O—CMe=CH—C(OPr^i)=O→Sn	—	134–136/5	M	(*253*)
$Pr_3Sn{\cdot}OCMe_2{\cdot}C{\vdots}CH$	—	109/4	O	(*288*)
$Pr_3Sn{\cdot}OCMe_2{\cdot}C{\vdots}C{\cdot}SnPr_3$	—	172/4	O	(*288*)

TABLE 1 (*continued*)

Compound[a]	mp, °C	bp, °C/mm	Preparations[b]	References
$Pr_3Sn\cdot$*acac*	51–53	106–109/1.5	M	*(253)*
$Pr_3Sn\cdot OC(Me)Et\cdot C\vdots CH$	—	101/2	O	*(111)*
$Pr_3Sn\cdot OC(Me)Et\cdot C\vdots C\cdot SnPr_3$	—	168–170/2	X	*(111)*
			O	*(111)*
$Pr_3Sn\cdot OPh$	—	145–147/1	M	*(279)*
$Pr_3Sn\cdot O$ / $HC\vdots C$ (cyclohexane ring)	—	144/2	O	*(111)*
$Pr_3Sn\cdot O$ / $Pr_3Sn\cdot C\vdots C$ (cyclohexane ring)	—	200/2	X	*(111)*
			O	*(111)*
$Pr^i_3Sn\cdot OMe$	—	66/1.2	X	*(97)*
$Pr^i_3Sn\cdot OC_6H_2$-1,3,5-$(CH_2\cdot NMe_2)_3$	—	—	A	*(97)*
$Bu_3Sn\cdot OMe$	—	101–102/2	X	*(13, 137, 205)*
		90/0.1	O	*(85)*
			N	*(185)*
$Bu_3Sn\cdot OEt$	—	115/0.1	X	*(205)*
			O	*(85)*
			N	*(185)*
			M	*(255, 286)*
$Bu_3Sn\cdot OCH_2\cdot CCl_3$	—	116/0.1	O	*(85)*
			H	*(183)*
$Bu_3Sn\cdot OCH_2\cdot CF_3$	—	79/0.05	O	*(84)*
$Bu_3Sn\cdot OCH_2\cdot CH_2O\cdot SnBu_3$	—	170/0.1	O	*(85)*
$Bu_3Sn\cdot OCHMe\cdot OMe$	—	—	Σ	*(89)*
$Bu_3Sn\cdot OCH(CCl_3)OMe$	—	91–92/0.8 with dissociation	Σ	*(89)*
$Bu_3Sn\cdot OCH(CBr_3)OMe$	—	—	Σ	*(89)*
$Bu_3Sn\cdot OCH(CCl_3)OCH_2\cdot CCl_3$	—	—	Σ	*(90, 183)*
$Bu_3Sn\cdot OCH(CCl_3)OCH_2\cdot CMe_3$	—	—	Σ	*(84)*
$Bu_3Sn\cdot OCH(CCl_3)OC_6H_{11}$	—	—	Σ	*(84)*
$Bu_3Sn\cdot OCH(CCl_3)OCHPh_2$	—	—	Σ	*(84)*
$Bu_3Sn\cdot OCH(CCl_3)OCH_2\cdot CF_3$	—	—	Σ	*(84)*
$Bu_3Sn\cdot OCH(CCl_3)OCH_2\cdot CF_2\cdot CHF_2$	—	—	Σ	*(84)*
$Bu_3Sn\cdot OCH(CCl_3)OPh$	—	—	Σ	*(89)*
$Bu_3Sn\cdot OCH(CBr_3)OPh$	—	—	Σ	*(89)*
$Bu_3Sn\cdot OCH(CCl_3)OSnBu_3$	—	90–120/0.2 decomp.	Σ	*(89)*
$Bu_3Sn\cdot OCH(CBr_3)O\cdot SnBu_3$	—	—	Σ	*(89)*
$Bu_3Sn\cdot OCH(CCl_3)O\cdot OBu^t$	—	—	Σ	*(31)*
$Bu_3Sn\cdot OCHMe\cdot NEt_2$	—	—	Σ	*(90)*
$Bu_3Sn\cdot OCH(CCl_3)NHPh$	—	128/0.15	O	*(80)*
		decomp.	Σ	*(80)*
$Bu_3Sn\cdot OCH(CCl_3)NH\cdot C_6H_4$-4-Me	—	130/0.2	O	*(80)*
		decomp.	Σ	*(80)*

TABLE 1 (*continued*)

Compound[a]	mp, °C	bp, °C/mm	Preparations[b]	References
$Bu_3Sn{\cdot}OCH(CBr_3)NH{\cdot}C_6H_4$-4-Me	—	—	A	(*80*)
$Bu_3Sn{\cdot}OCH(CCl_3)NH{\cdot}C_6H_4$-4-$NO_2$	—	—	O	(*80*)
$Bu_3Sn{\cdot}OCH(CCl_3)NH{\cdot}CO{\cdot}Me$	—	—	Σ	(*82*)
$Bu_3Sn{\cdot}OCH(CCl_3)NMe{\cdot}CHO$	—	—	Σ	(*82*)
$Bu_3Sn{\cdot}OCH(CBr_3)NMe{\cdot}CHO$	—	—	Σ	(*82*)
$Bu_3Sn{\cdot}OCH(CCl_3)NMe{\cdot}CO{\cdot}OMe$	—	—	Σ	(*84*)
$Bu_3Sn{\cdot}OCH(CBr_3)NMe{\cdot}CO{\cdot}OMe$	—	—	Σ	(*82*)
$Bu_3Sn{\cdot}OCH(CCl_3)NEt{\cdot}CO{\cdot}OMe$	—	—	Σ	(*84*)
$Bu_3Sn{\cdot}OCH(CCl_3)NPh{\cdot}CO{\cdot}OMe$	—	—	Σ	(*36*)
$Bu_3Sn{\cdot}OCH(CBr_3)NPh{\cdot}CO{\cdot}OMe$	—	—	Σ	(*84*)
$Bu_3Sn{\cdot}OCH(CCl_3)NPh{\cdot}CO{\cdot}OEt$	—	—	Σ	(*84*)
$Bu_3Sn{\cdot}OCH(CCl_3)NNp{\cdot}CO{\cdot}OMe$	—	—	Σ	(*84*)
$Bu_3Sn{\cdot}OCH(CCl_3)NH{\cdot}SO_2{\cdot}Ph$	—	—	Σ	(*82*)
$Bu_3Sn{\cdot}OCH(CCl_3)NH{\cdot}SO_2{\cdot}C_6H_4$-4-Me	—	—	O	(*80*)
			Σ	(*80*)
			M	(*80*)
$Bu_3Sn{\cdot}OCH(CCl_3)N(SnBu_3){\cdot}SO_2{\cdot}C_6H_4$-4-Me	—	—	Σ	(*80*)
$Bu_3Sn{\cdot}OCH(CCl_3)NPr{\cdot}SO_2{\cdot}Me$	—	—	Σ	(*82*)
$Bu_3Sn{\cdot}OCH(CCl_3)NPr{\cdot}SO_2{\cdot}Ph$	—	—	Σ	(*82*)
$Bu_3Sn{\cdot}OCH(CCl_3)NPr{\cdot}SO_2{\cdot}C_6H_4$-4-Me	—	—	Σ	(*82*)
$Bu_3Sn{\cdot}OCH(CCl_3)NPh{\cdot}SO_2{\cdot}Me$	—	—	Σ	(*82*)
$Bu_3Sn{\cdot}OCH(CCl_3)NMe{\cdot}CO{\cdot}NBu{\cdot}SnBu_3$	—	105/0.1	Σ	(*90*)
$Bu_3Sn{\cdot}OCH(CBr_3)NMe{\cdot}CO{\cdot}NMe{\cdot}SnBu_3$	—	—	Σ	(*90*)
$Bu_3Sn{\cdot}OPr$	—	100/0.1	X	(*205*)
			O	(*90*)
$Bu_3Sn{\cdot}OPr^i$	—	—	X	(*205*)
$Bu_3Sn{\cdot}OCH_2{\cdot}CF_2{\cdot}CHF_2$	—	86/0.05	O	(*84*)
$Bu_3Sn{\cdot}OC(CCl_3)_2OMe$	—	—	Σ	(*89*)
$Bu_3Sn{\cdot}OC(CCl_2F)(CClF_2)OMe$	—	—	Σ	(*89*)
$Bu_3Sn{\cdot}OC(CCl_3)_2O{\cdot}SnBu_3$	—	—	Σ	(*89*)
$Bu_3Sn{\cdot}OC(CCl_2F)(CClF_2)O{\cdot}SnBu_3$	—	—	Σ	(*89*)
$Bu_3Sn{\cdot}OBu$	—	124–128/3	X	(*60*)
		137–138/1.4	A	(*205, 314*)
		116/0.1	O	(*90*)
$Bu_3Sn{\cdot}OBu^i$	—	110/0.1	O	(*90*)
$Bu_3Sn{\cdot}OBu^s$	—	102/0.35	H	*218*)
$Bu_3Sn{\cdot}OBu^t$	—	96–97/1.2	X	(*194, 222*)
		82–83/0.5	A	(*205*)
$Bu_3Sn{\cdot}OCH{:}CHEt$	—	119–121/0.6	A	(*237*)
$Bu_3Sn{\cdot}OCH{:}CMe_2$	—	104–106/0.7	A	(*237*)
$Bu_3Sn{\cdot}OCMe{:}CHMe$ (+77% $Bu_3Sn{\cdot}CHMe{\cdot}CO{\cdot}Me$)	—	115/0.8	A	(*237*)
$Bu_3Sn{\cdot}OCHMe{\cdot}C{\vdots}CH$	—	117/3	O	(*288*)
$Bu_3Sn{\cdot}OCHMe{\cdot}C{\vdots}C{\cdot}SnBu_3$	—	200/3	O	(*288*)
$Bu_3Sn{\cdot}OCHEt{\cdot}CH_2{\cdot}NEt_2$	—	160–162/1.5	X	(*312*)
		122/0.1	N	(*312*)
$Bu_3Sn{\cdot}OCH(CCl_3){\cdot}CH_2{\cdot}CN$	—	130–133/0.08	Σ	(*225*)

TABLE 1 (*continued*)

Compound[a]	mp, °C	bp, °C/mm	Preparations[b]	References
$Bu_3Sn \cdot OCHPr \cdot OMe$	—	—	Σ	(*89*)
$Bu_3Sn \cdot OCHPr \cdot NEt_2$	—	—	Σ	(*90*)
$Bu_3Sn \cdot OCHEt_2$	—	119/1.0	H	(*51*)
$Bu_3Sn \cdot OCH_2 \cdot CMe_3$	—	122/0.3	O	(*85*)
$Bu_3Sn \cdot OCH{:}CHPr^i$	—	143–144/3	A	(*237*)
$Bu_3Sn \cdot OCMe{:}CHEt$ (+25% $Bu_3Sn \cdot CHEt \cdot CO \cdot Me$)	—	117–118/0.3	A	(*237*)
$Bu_3Sn \cdot COEt{:}CHMe$ (+30% $Bu_3Sn \cdot CHMe \cdot CO \cdot Et$)	—	121–123/0.8	A	(*237*)
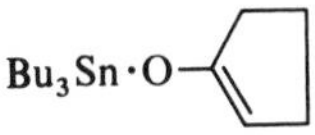 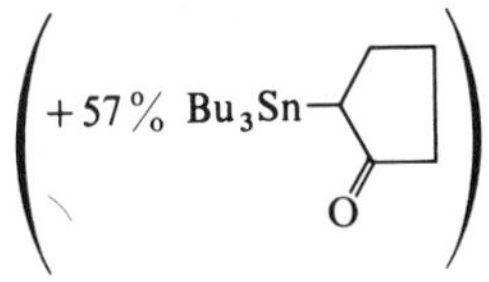 $Bu_3Sn \cdot O$— (+57% Bu_3Sn— ; O)	—	144–146/1.4	A	(*237*)
$Bu_3Sn \cdot OCMe_2 \cdot C{\vdots}CH$	—	118/2	O	(*288*)
$Bu_3Sn \cdot OCMe_2 \cdot C{\vdots}C \cdot SnBu_3$	—	198/2	O	(*288*)
$Bu_3Sn \cdot OCH(Me)Bu^t$	—	—	M	(*205*)
$Bu_3Sn \cdot OC_6H_{11}$	—	139–140/0.7	X	(*194*)
		129–130/0.3	A	(*205, 314*)
			O	(*84*)
			H	(*51*)
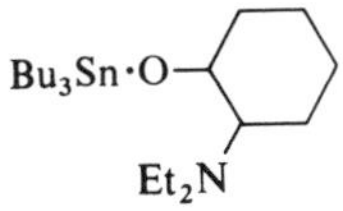 $Bu_3Sn \cdot O$— ; Et_2N	—	144/0.1	X	(*312*)
			N	(*312*)
$Bu_3Sn \cdot OCMe{:}CHPr^i$	—	154–155/5.5	A	(*237*)
		122–123/0.4	H	(*240*)
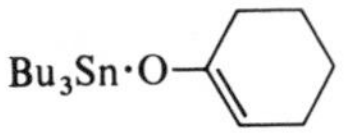 $Bu_3Sn \cdot O$—	—	146–148/1.3	A	(*237*)
$Bu_3Sn \cdot OCBu^t{:}CH_2$ (+75% $Bu_3Sn \cdot CH_2 \cdot CO \cdot Bu^t$)	—	118/0.3	A	(*237*)
$Bu_3 \cdot SnOPh$	—	152/1	X	(*33*)
		124/0.01	A	(*238, 314*)
			O	(*90*)
$Bu_3Sn \cdot OC_6H_4$-4-Cl	—	214–215/12	X	(*139*)
$Bu_3Sn \cdot OC_6F_5$	—	120/0.1	O	(*90*)
$Bu_3Sn \cdot O(CH_2)_6Me$	—	132/0.3	O	(*85*)
		90–95/1		

TABLE 1 (*continued*)

Compound[a]	mp, °C	bp, °C/mm	Preparations[b]	References
$Bu_3Sn\cdot OCH_2\cdot Ph$	—	128–130/0.1	X	(*194*)
			O	(*84*)
$Bu_3Sn\cdot OCHMe\cdot (CH_2)_2CH{:}CHMe$	—	136–137/0.6	H	(*239*)
$Bu_3Sn\cdot OC_6H_4$-2-$CH_2\cdot NMe_2$	—	—	A	(*97*)
$Bu_3Sn\cdot OC_6H_4$-3-$CH_2\cdot OH$	—	—	H	(*218*)
$Bu_3Sn\cdot O$–(cyclohexyl ring), Me (2-position)	—	157/0.5	H	(*248*)
$Bu_3Sn\cdot O$–(cyclohexyl ring), Me (3-position)	—	155/0.5	H	(*248*)
$Bu_3Sn\cdot O$–(cyclohexyl ring)–Me (4-position)	—	160/0.5	H	(*248*)
$Bu_3Sn\cdot OCH{:}CH\cdot C_5H_{11}$	—	142/0.8	A	(*237, 238*)
$Bu_3Sn\cdot OCH(C_6F_5)OMe$	—	—	Σ	(*89*)
$Bu_3Sn\cdot OCH(C_6F_5)O\cdot SnBu_3$	—	—	Σ	(*89*)
$Bu_3Sn\cdot OOct$	—	—	A	(*205*)
$Bu_3Sn\cdot OCHMe\cdot C_6H_{13}$	—	130–132/0.2	O	(*84*)
$Bu_3Sn\cdot OCH(Me)Ph$	—	145/0.9	O	(*85*)
		138/0.4	H	(*51*)
$Bu_3Sn\cdot OCPh{:}CH_2$ (+78% $Bu_3Sn\cdot CH_2\cdot CO\cdot Ph$)	—	138/0.7	A	(*237*)
$Bu_3Sn\cdot OCHMe\cdot (CH_2)_2CH{:}CMe_2$	—	151–152/1.2	H	(*239*)
$Bu_3Sn\cdot OCHPh\cdot CH_2\cdot NEt_2$	—	154/0.2	X	(*312*)
		132/0.05	N	(*312*)
$Bu_3Sn\cdot OCH(C_6F_5)CH_2\cdot CN$	—	146–156/0.1	Σ	(*225*)
$Bu_3Sn\cdot OC_6H_2$-1,3,5-$(CH_2\cdot NMe_2)_3$	—	—	A	(*97*)
$Bu_3Sn\cdot$*Ox*	126	149–151/0.007	X	(*99, 147*)
			O	(*99*)
$Bu_3Sn\cdot OCBu^i{:}CHPr^i$	—	145–146/2	A	(*237*)
$Bu_3Sn\cdot O$–(cyclohexyl ring), Me, Pr^i	—	113–114/0.05	O	(*84*)
$Bu_3Sn\cdot OCMe_2$–(cyclohexenyl ring)–Me	—	160/0.05	O	(*84*)
$Bu_3Sn\cdot OCHPh_2$	—	140/0.3	O	(*85*)
$Bu_3Sn\cdot OCPh_3$	—	180/0.3	O	(*85*)
$Bu^i_3Sn\cdot OBu^i$	—	122–124/2	H	(*218*)
$Bu^i_3Sn\cdot OBu^s$	—	88/0.4	H	(*218*)
$Bu^i_3Sn\cdot OC_6H_4$-2-$CH_2\cdot OH$	33–34	—	H	(*218*)

TABLE 1 (*continued*)

Compound[a]	mp, °C	bp, °C/mm	Preparations[b]	References
$(C_5H_{11})_3Sn\cdot OMe$	—	120–125/0.1	X	(*97*)
$Ph_2(Bu)Sn\cdot \mathit{Ox}$	206–221	—	M	(*329*)
$Ph_3Sn\cdot OMe$	69–70	—	X	(*13, 43*)
			N	(*18, 185*)
$Ph_3Sn\cdot OEt$	112	—	N	(*185*)
$Ph_3Sn\cdot OCH(CCl_3)OMe$	98–99	—	Σ	(*84*)
$Ph_3Sn\cdot OCH(CBr_3)OMe$	108–110	—	Σ	(*84*)
$Ph_3Sn\cdot OCH(CCl_3)O\cdot SnPh_3$	131–132	—	Σ	(*84*)
$Ph_3Sn\cdot OCH(CBr_3)O\cdot SnPh_3$	108–109	—	Σ	(*84*)
$Ph_3Sn\cdot OCH(CCl_3)O\cdot OBu^t$	77–79	—	Σ	(*31*)
$Ph_3Sn\cdot OCH(CCl_3)O\cdot OCMe_2Ph$	—	—	Σ	(*31*)
$Ph_3Sn\cdot OCH(CCl_3)NHBu$	89–91	—	Σ	(*90*)
$Ph_3Sn\cdot OCH(CCl_3)NH\cdot C_6H_4\text{-}4\text{-}Me$	84	—	O	(*80*)
$Ph_3Sn\cdot OCH(CBr_3)NH\cdot C_6H_4\text{-}4\text{-}Me$	—	—	A	(*80*)
$Ph_3Sn\cdot OPh$	84–85	—	X	(*43*)
			O	(*84, 90*)
$Ph_3Sn\cdot OC_6H_4\text{-}2\text{-}Br$	60–61	—	X	(*68*)
$Ph_3Sn\cdot OC_6H_4\text{-}4\text{-}Br$	85–86	—	X	(*68*)
$Ph_3Sn\cdot OC_6H_4\text{-}2\text{-}Cl$	59–60	—	X	(*68*)
$Ph_3Sn\cdot OC_6H_4\text{-}4\text{-}Cl$	77–78	—	X	(*68*)
$Ph_3Sn\cdot OC_6H_4\text{-}2\text{-}NO_2$	104–105	—	X	(*68*)
$Ph_3Sn\cdot OC_6H_4\text{-}3\text{-}NO_2$	170–171	—	X	(*68*)
$Ph_3Sn\cdot OC_6H_4\text{-}4\text{-}NO_2$	106–108	—	X	(*43, 68*)
			O	(*244*)
$Ph_3Sn\cdot OC_6H_2\text{-}2,4,6\text{-}Br_3$	123–124	—	X	(*68*)
$Ph_3Sn\cdot OC_6H_2\text{-}2,4,6\text{-}Cl_3$	103–104	—	X	(*68*)
$Ph_3Sn\cdot OC_6Br_5$	164–165	—	X	(*68*)
$Ph_3Sn\cdot OC_6H_4\text{-}2\text{-}CH_2\cdot OH$	195–196	—	H	(*218*)
$Ph_3Sn\cdot OCH_2Ph$	—	128–130/0.05	O	(*84*)
$Ph_3Sn\cdot OC_6H_4\text{-}4\text{-}CO\cdot Et$	111–112	—	X	(*68*)
$Ph_3Sn\cdot \mathit{Ox}$	145–146.5	—	X	(*147, 275*)
$Ph_3Sn\cdot O$ (structure: Br, Br, N)	165–166	—	X	(*68*)
$Ph_3Sn\cdot O$ (structure: Ph, Ph, Ph, Ph, $Mn(CO)_3$)	206–207	—	X	(*109*)
			Σ	(*108*)
$(4\text{-}Cl\cdot C_6H_4\cdot CH_2)_3Sn\cdot \mathit{Ox}$	260	—	X	(*99*)
			O	(*99*)
$Oct_3Sn\cdot OMe$	—	160/0.1	X	(*60*)
			O	(*85*)
$Oct_3Sn\cdot OCH_2Ph$	—	198–200/0.05	O	(*84*)

alkoxy groups, i.e., all methyltin compounds precede all ethyltin compounds etc., and within each set methoxide is followed by ethoxide, and so on. In Tables 2 and 3 alkoxides of the type $R_2Sn(OR')X$ and $RSn(OR')_2X$ are placed at the end of each R-set; each set of dialkyltin derivatives of glycols appears between $R_2Sn(OR')_2$ and $R_2Sn(OR')X$. In these tables as well as in Table 4 the method of preparation according to the discussion in this section is noted for each compound in the column preceding the reference.

TABLE 2

DIALKYLTIN ALKOXIDES

Compound[a]	mp, °C	bp, °C/mm	Preparations[b]	References
$Me_2Sn(OMe)_2$	86–87	—	X	(*18, 48, 72, 104*)
			N	(*185*)
$Me_2Sn(OEt)_2$	—	82/0.1	X	(*48, 104*)
			N	(*185*)
$Me_2Sn(OCH_2{\cdot}CH_2Cl)_2$	—	98/1.0	A	(*103*)
		66/0.05		
$Me_2Sn(OCH_2{\cdot}CH_2Br)_2$	—	126/32	A	(*103*)
$Me_2Sn(OPr)_2$	—	100/0.1	X	(*48, 104*)
$Me_2Sn[OCH(CF_3)_2]_2$	—	92/25	H	(*63*)
$Me_2Sn[OC(CF_3)_2OCH(CF_3)_2]_2$	—	—	Σ	(*63*)
$Me_2Sn(OBu)_2$	—	130–131/0.5	X	(*48, 104*)
$Me_2Sn{\cdot}$*acac*$_2$ (*trans*)	>300	—	X	(*148, 313*)
	177–178	—	O	(*193*)
$Me_2Sn{\cdot}$*Ox*$_2$	236.5–238	—	X	(*104, 306, 329*)
			M	(*134, 150*)
$Me_2Sn{\cdot}$*bzac*$_2$ (*trans*)	134–135	—	O	(*193*)
$Me_2Sn{\cdot}$*dbzm*$_2$ (*trans*)	189–191	—	O	(*193*)
Me_2Sn (O—CH·CO·OEt / O—CH·CO·OEt, cyclic)	220 decomp.	—	O	(*76*)
Me_2Sn (O, O-phenylene, cyclic)	>360	—	X	(*94*)
Me_2Sn (O, O-biphenylene, cyclic)	sublimes 260/ *in vacuo*	—	X	(*94*)
$Me_2Sn(OMe)Cl$	178–1[illegible]	—	A	(*72*)

[a] See footnote to TABLE 1.
[b] See footnote to TABLE 1.

TABLE 2 (*continued*)

Compound[a]	mp, °C	bp, °C/mm	Preparations[b]	References
$Me_2Sn(OMe)Br$	211–212	—	A	(*76*)
$Me_2Sn(OMe)I$	183–184	—	A	(*72*)
$Me_2Sn(Ox)Cl$	147–148	—	X	(*134, 147, 329*)
			A	(*147*)
			M	(*150*)
$Me_2Sn(Ox)I$	134–135	—	X	(*329*)
$Me_2Sn(Ox)NCS$	124	—	M	(*147*)
	decomp.			
$Me_2Sn(Ox)dbzm$	152–230	—	X	(*329*)
$Et_2Sn(OMe)_2$	—	124–126/3	X	(*48, 104, 195, 250*)
		102–104/0.05	N	(*185*)
$Et_2Sn(OEt)_2$	—	101–103/2	X	(*48, 104, 250*)
		78/0.1	N	(*185*)
			M	(*286*)
$Et_2Sn(OCH_2 \cdot CH_2Cl)_2$	—	96/0.7	A	(*103*)
$Et_2Sn(OCH_2 \cdot CH_2Br)_2$	—	135/27	A	(*103*)
$Et_2Sn(OPr)_2$	—	110–112/2	X	(*48, 104, 250*)
		88/0.01		
$Et_2Sn(OCHMe \cdot CO \cdot OBu)_2$	—	—	A	(*196*)
$Et_2Sn(OBu)_2$	—	112/0.03	X	(*48, 104*)
$Et_2Sn \cdot acac_2$ (*trans*)	86.5–87	—	X	(*313*)
$Et_2Sn \cdot Ox_2$	177–178	—	X	(*104, 306*)
			O	(*129*)
O—CH·CO·OEt Et_2Sn | O—CH·CO·OEt	171–172	—	O	(*76*)
$Et_2Sn(OMe)Cl$	135–136	—	A	(*72*)
$Et_2Sn(OMe)Br$	138–140	—	A	(*72*)
$Et_2Sn(OEt)F$	90–91	—	A	(*232*)
$Et_2Sn[O \cdot CH(CCl_3)OMe]Br$	—	—	Σ	(*73*)
$Et_2Sn(OC_6H_{11})Cl$	42	—	H	(*220*)
$Et_2Sn(OPh)GePh_3$	—	—	N	(*62*)
$Et_2Sn(Ox)Cl$	119–120.5	—	X	(*147*)
			A	(*147*)
$(CH_2{:}CH)_2Sn \cdot acac_2$ (*trans*)	87–88	—	X	(*313*)
$Pr_2Sn(OMe)_2$	—	121–125/0.1	X	(*48, 72, 104*)
		105–107/0.1		
$Pr_2Sn(OEt)_2$	—	104–106/1.5	X	(*48, 104, 250*)
		84/0.03		
$Pr_2Sn(OPr)_2$	—	90–92/0.05	X	(*48, 104*)
$Pr_2Sn(OBu)_2$	—	100/0.05	X	(*48, 104*)
$Pr_2Sn \cdot Ox_2$	162–163	—	X	(*104, 306*)
$Pr_2Sn(OMe)F$	180–185	—	A	(*72*)
$Pr_2Sn(Ox)Cl$	90–91	—	X	(*147*)
			A	(*147*)

TABLE 2 (*continued*)

Compound[a]	mp, °C	bp, °C/mm	Preparations[b]	References
$Pr_2Sn(Ox)NCS$	134	—	M	(*147*)
$Pr^i_2Sn(OMe)_2$	—	—	M	(*65*)
$Bu_2Sn(OMe)_2$	—	136–139/1.2	X	(*13, 195*)
		126–128/0.05	N	(*185*)
			M	(*206*)
$Bu_2Sn(OEt)_2$	—	96–97/0.2	X	(*48, 104*)
			N	(*185*)
			M	(*286*)
$Bu_2Sn(OCH_2\cdot CH_2Cl)_2$	—	106/0.3	A	(*103*)
		82.5/0.1		
$Bu_2Sn(OCH_2\cdot CH_2Br)_2$	—	160–162/14	A	(*103*)
$Bu_2Sn(OCHMe\cdot OMe)_2$	—	—	Σ	(*73*)
$Bu_2Sn[OCH(CCl_3)OMe]_2$	—	—	Σ	(*73*)
$Bu_2Sn(OPr)_2$	—	99–100/0.1	X	(*104*)
			M	(*206*)
$Bu_2Sn(OPr^i)_2$	—	131–132/10	X	(*146*)
$Bu_2Sn(OCH_2\cdot CH_2\cdot CH_2Cl)_2$	—	91/0.1	A	(*103*)
$Bu_2Sn[OCH(CF_3)_2]_2$	—	$75/10^{-3}$	H	(*63*)
$Bu_2Sn(O\cdot CH_2\cdot CH{:}CH_2)_2$	—	220/0.3	X	(*195*)
			O	(*84, 179*)
$Bu_2Sn(OBu)_2$	—	160/3	X	(*104, 340*)
		136–138/0.05	O	(*84*)
			M	(*206*)
$Bu_2Sn(OBu^t)_2$	—	102/1	X	(*103*)
			A	(*206*)
Bu_2Sn chelate: Sn bonded to two Bu (axial) and two ligands MeC(–O)=CH–C(OMe)=O (MeC—O, HC, MeO·C=O → Sn)	—	136–138/0.2–0.3	A	(*204*)
Bu_2Sn chelate: Sn bonded to two Bu (axial) and two ligands MeC(–O)=CH–C(OEt)=O (MeC—O, HC, EtO·C=O → Sn)	—	154/0.3	A	(*204*)
$Bu_2Sn(OC_5H_{11})_2$	—	—	A	(*205*)
$Bu_2Sn\cdot acac_2$ (*trans*)	—	132/0.4	A	(*204*)
$Bu_2Sn(OC_6H_{11})_2$	—	—	A	(*205*)
$Bu_2Sn[O(CH_2)_6Cl]_2$	—	114/0.1	A	(*103*)
$Bu_2Sn(OPh)_2$	51–53	161/0.35	X	(*57*)
		146–148/0.1	A	(*103*)
			O	(*265*)
$Bu_2Sn(OC_6H_4\text{-}4\text{-}Cl)_2$	80–85	178–180/0.07	O	(*265*)

TABLE 2 (*continued*)

Compound[a]	mp, °C	bp, °C/mm	Preparations[b]	References
$Bu_2Sn(OC_6H_4\text{-}4\text{-}Me)_2$	65–71	166.5–168.5 /0.05	O	(*265*)
$Bu_2Sn(OCH_2Ph)_2$	—	210/0.4	A	(*146*)
		146–150/0.006	O	(*84*)
$Bu_2Sn(O\cdot Oct)_2$	—	—	A	(*206*)
$Bu_2Sn(OCH_2\cdot CH_2Ph)_2$	—	144/0.1	X	(*103*)
			A	(*103*)
$Bu_2Sn\cdot Ox_2$	155–156	—	X	(*29, 104, 306*)
$Bu_2Sn(OC_{12}H_{23})_2$	—	—	A	(*196*)
$Bu_2Sn(O{-}CH_2{-}CH_2{-}O)_2SnBu_2$ (cyclic: Bu_2Sn joined to $SnBu_2$ by two $O{-}CH_2{-}CH_2{-}O$ bridges)	223–229	—	X	(*41*)
			A	(*203, 249a*)
			O	(*41, 55, 90*)
$Bu_2Sn\langle O{-}CH\cdot Me \,/\, O{-}CH_2\rangle$ (ring)	182 sublimes 210–230/0.1	—	X	(*261*)
			A	(*203, 249a*)
$Bu_2Sn\langle O{-}CH\cdot CH_2Cl \,/\, O{-}CH_2\rangle$ (ring)	—	—	A	(*203*)
$Bu_2Sn\langle O, O\rangle(CH_2)_3$ (ring)	88	200/1.5 184–185/0.3	A	(*203, 249a*)
$Bu_2Sn\langle O{-}CH\cdot Me \,/\, O{-}(CH_2)_2\rangle$ (ring)	134	185/1.5 165/0.25	A	(*203, 249a*)
$Bu_2Sn\langle O, O\rangle(CH_2)_4$ (ring)	—	175–178/0.4	A	(*203*)
$Bu_2Sn\langle O{-}CH\cdot Me \,/\, O{-}CH\cdot Me\rangle$ (ring)	124–126	171–173/0.2	A	(*203, 249a*)
			O	(*55*)
$Bu_2Sn\langle O{-}CH\cdot CO\cdot OEt \,/\, O{-}CH\cdot CO\cdot OEt\rangle$ (ring)	132–134	—	O	(*76*)
$Bu_2Sn\langle O, O\rangle(CH_2)_5$ (ring)	—	185–187/0.15	A	(*203*)
$Bu_2Sn\langle O{-}CHMe{-}CH_2{-}CHMe{-}O\rangle$ (ring)	155	159/0.7	A	(*249a*)

TABLE 2 (*continued*)

Compound[a]	mp, °C	bp, °C/mm	Preparations[b]	References
$Bu_2Sn(O)_2(CH_2)_6$ (cyclic)	—	—	A	(*203*)
Bu_2Sn ring: O—CMe₂—CH₂—CHMe—O	80	143/1.5 163–167/0.7–0.8	A	(*203*, *249a*)
Bu_2Sn ring: O—CMe₂—CMe₂—O	45	204–205/1.6 195/0.4	A	(*203*, *249a*)
Bu_2Sn ring: O—CMe₂—CH₂—CMe₂—O	70	119/0.8	A	(*249a*)
Bu_2Sn catecholate (O,O-benzo ring)	272	—	X O	(*94*) (*136*)
Bu_2Sn cyclohexane-1,2-diolate (*trans*)	234–236	—	O	(*55*)
Bu_2Sn cyclohexane-1,2-diolate (*cis*)	164–165	—	O	(*55*)
Bu_2Sn 4-Bu^t-catecholate	262–264	—	O	(*136*)
$Bu_2Sn(OMe)F$	95–97	—	A	(*72*)
$Bu_2Sn(OMe)Cl$	25–26 5	166–167/1.5	A	(*72*, *107*)
$Bu_2Sn(OMe)Br$	49–50 35.5–37.5	169–170/3 159/1.5	A	(*72*, *107*, *249*)
$Bu_2Sn(OMe)I$	73	—	A	(*72*)
$Bu_2Sn(OMe)SCN$	81–84	—	A	(*72*)
$Bu_2Sn(OMe)NEt\cdot CO\cdot OMe$	—	—	Σ	(*73*)
$Bu_2Sn(OMe)NPh\cdot CO\cdot OMe$	—	—	Σ	(*73*)
$Bu_2Sn(OMe)N{:}C(CCl_3)OMe$	93–94	—	Σ	(*73*)
$Bu_2Sn(OMe)NNp\cdot C({:}NNp)OMe$	—	—	Σ	(*73*)
$Bu_2Sn(OMe)OCHMe\cdot OMe$	—	—	Σ	(*73*)
$Bu_2Sn(OMe)OCH(CCl_3)OMe$	—	—	Σ	(*73*)
$Bu_2Sn(OMe)O\cdot CO\cdot Me$	94–96	—	A	(*72*)
$Bu_2Sn(OMe)O\cdot CO\cdot R$[c]	—	—	A	(*72*)

[c] O·CO·R = laurate.

TABLE 2 (*continued*)

Compound[a]	mp, °C	bp, °C/mm	Preparations[b]	References
Bu$_2$Sn(OMe)O·SO$_2$·R[d]	—	—	A	(*72*)
Bu$_2$Sn(OMe)O·SO·OMe	95.5–96.5	—	Σ	(*73*)
Bu$_2$Sn(OMe)S·C(:NPh)OMe	—	—	Σ	(*73*)
Bu$_2$Sn(OMe)SC(:NCH$_2$·CH:CH$_2$)OMe	—	—	Σ	(*73*)
Bu$_2$Sn(OMe)S·CS·OMe	64–67	—	Σ	(*73*)
Bu$_2$Sn(OEt)H	—	105–111/6	A	(*128*)
Bu$_2$Sn[O·CH(CCl$_3$)OMe]O·CO·Me	—	—	Σ	(*73*)
Bu$_2$Sn[O·CH(CCl$_3$)OMe]SCN	—	—	Σ	(*73*)
Bu$_2$Sn(*Ox*)Cl	Decomposes >48	—	X	(*134*)
Bu$^i_2$Sn(OMe)$_2$	—	106–108/0.2	X	(*103*)
Bu$^i_2$Sn(OBu^t)$_2$	~20	96/0.002	H	(*218*)
Bu$^i_2$Sn(OC$_6$H$_4$-2-CH$_2$·OH)$_2$	146	—	H	(*218*)
Bu$^i_2$Sn·*Ox*$_2$	188–189	—	X	(*104*)
Bu$^i_2$Sn(OBu^t)Cl	65	120/0.5	H	(*220*)
Ph$_2$Sn(OMe)$_2$	270 decomp.	—	X	(*195*)
	186		N	(*185*)
Ph$_2$Sn(OEt)$_2$	124 decomp.	—	X	(*23*)
	95		N	(*185*)
Ph$_2$Sn·*acac*$_2$ (*trans*)	125–126	—	X	(*193, 211, 313*)
Ph$_2$Sn·*Ox*$_2$ (*trans*)	252–253	—	X	(*29, 193, 211, 275, 329*)
Ph$_2$Sn·*bzac*$_2$ (*trans*)	181–182	—	X	(*211*)
Ph$_2$Sn·*dbzm*$_2$ (*trans*)	239–241	—	X	(*211*)
Ph$_2$Sn(*Ox*)Cl	167–168	—	X	(*134*)
			A	(*329*)
Ph$_2$Sn(*Ox*)NO$_3$	Decomposes >175	—	X	(*134*)
Ph$_2$Sn(*Ox*) *bzac*	182–200	—	M	(*329*)
Ph$_2$Sn(*Ox*) *dbzm*	182–186	—	M	(*329*)
(4-Cl·C$_6$H$_4$)$_2$Sn·*Ox*$_2$	255.7–256.3	—	X	(*99*)
			O	(*99*)
(Ph·CH$_2$)$_2$Sn·*Ox*$_2$	118–120	—	X	(*104*)
(4-Cl·C$_6$H$_4$·CH$_2$)$_2$Sn·*Ox*$_2$	155	—	X	(*99*)
			O	(*99*)
(4-EtO·CO·C$_6$H$_4$)$_2$Sn·*Ox*$_2$	216–217	—	X	(*95*)
(4-Et·C$_6$H$_4$·CH$_2$)$_2$Sn·*Ox*$_2$	158	—	X	(*99*)
			O	(*99*)
Oct$_2$Sn(OMe)$_2$	—	167–168/0.05	X	(*48, 104*)
Oct$_2$Sn(OEt)$_2$	—	170–0.07 to 175/0.1	X	(*48, 104*)
Oct$_2$Sn(OPr)$_2$	—	174/0.02 to 180/0.4	X	(*48, 104*)
Oct$_2$Sn(OBu)$_2$	—	182–188/0.05	X	(*48, 104*)
Oct$_2$Sn·*Ox*$_2$	78–79	—	X	(*104*)

[d] O·SO$_2$·R = camphorsulfonate.

TABLE 3

MONOALKYLTIN ALKOXIDES

Compound[a]	mp, °C	bp, °C/mm	Preparations[b]	References
$MeSn(OMe)_3$	195	—	N	(*185*)
$MeSn(OEt)_3$	—	110/0.1	N	(*185*)
$MeSn(acac)_2Cl$	135–136	—	X	(*147, 313*)
$MeSn(acac)_2Br$	129	—	X	(*313*)
$MeSn(acac)_2I$	115–116	—	X	(*313*)
$MeSn(Ox)_2Cl$	246.5–247.5	—	X	(*147*)
$MeSn(Ox)_2Br$	258–260	—	X	(*96*)
$EtSn(OMe)_3$	165	—	N	(*185*)
$EtSn(OEt)_3$	—	127–130/0.1	N	(*185*)
$EtSn(acac)_2Br$	94.5	—	X	(*313*)
$BuSn(OMe)_3$	250 decomp.	—	N	(*185*)
$BuSn(OEt)_3$	—	130/0.1	N	(*185*)
$BuSn \cdot Ox_3$	223	—	X	(*99*)
			O	(*99*)
$BuSn(Ox)_2Cl$	182–183	—	X	(*147*)
$PhSn(OMe)_3$	212	—	N	(*185*)
$PhSn(OEt)_3$	119	—	N	(*185*)
$PhSn(acac)_2Cl$	149–152	—	X	(*313*)
$PhSn(Ox)_2Cl$	218–219	—	X	(*96*)

[a] See footnote to TABLE 1.
[b] See footnote to TABLE 1.

B. ALKYL PEROXIDES

Alkyl trialkyltin peroxides are prepared by nucleophilic substitution at tin [Eq. (81)], a reaction which is analogous to that used to make the corresponding alkoxides [Eq. (2)].

$$Y\text{—}O\cdot OR' \quad R_3Sn\text{—}X \longrightarrow \overset{+}{Y} + R_3Sn\cdot O\cdot OR' + \overset{-}{X} \tag{81}$$

As with the alkoxides, the leaving group, X, can be a halide (*14*, *199*, *273*), an alkoxide (*14*, *273*), or an oxide (*38*) and in addition cyanides or hydrides (*14*) have been used.

The sodium salt of an alkyl hydroperoxide reacts with trialkyltin chlorides in absolute methanol to yield the corresponding alkyl trialkyltin peroxides (*269*):

$$Et_3Sn\cdot Cl + NaO\cdot OBu^t \longrightarrow Et_3Sn\cdot O\cdot OBu^t + NaCl \tag{82}$$

The disadvantages of the method were described for alkoxide preparations and a further problem here is that the peroxides are susceptible to thermal decomposition so that, for liquids, distillation should be rapid and at the lowest temperature possible.

A variation of the halide method, which has not been used for alkoxides, involves condensing the trialkyltin halide with the alkyl hydroperoxide in the presence of a base such as ammonia (*199*) or sodamide (*273*):

$$Bu_3Sn{\cdot}Cl + HO{\cdot}O{\cdot}CMe_2Ph \xrightarrow{NH_3} Bu_3Sn{\cdot}O{\cdot}O{\cdot}CMe_2Ph + NH_4Cl \quad (83)$$

Of course the same problems arise here as with the sodium salt method.

Tributyltin hydride reacts with an excess of *t*-butyl hydroperoxide to give *t*-butyl tributyltin peroxide, but two moles of hydroperoxide are required to yield one mole of tin peroxide and it is believed that the initial stage of the reaction is the formation of tributyltin *t*-butoxide:

$$Bu_3Sn{\cdot}H + HO{\cdot}OBu^t \xrightarrow{-H_2O} [Bu_3Sn{\cdot}OBu^t] \xrightarrow{HO\cdot OBu^t} Bu_3Sn{\cdot}O{\cdot}OBu^t + HOBu^t \quad (84)$$

The reaction of trialkyltin alkoxides with alkyl hydroperoxides does provide a very convenient route to the alkyl trialkyltin peroxides, since a volatile by-product is formed:

$$Pr_3Sn{\cdot}OMe + HO{\cdot}O{-}C_6H_9 \longrightarrow Pr_3Sn{\cdot}O{\cdot}O{-}C_6H_9 + MeOH \quad (85)$$

This method is more attractive now that pure alkoxides have become readily available from the corresponding oxides and hydroxides. However, it has recently been shown that the alkyl trialkyltin peroxides themselves can be prepared directly from oxides and hydroxides by azeotropic dehydration at room temperature (*38*):

$$(Ph_3Sn)_2O + 2\ HO{\cdot}OBu^t \xrightarrow{-H_2O} 2\ Ph_3Sn{\cdot}O{\cdot}OBu^t \quad (86)$$

Trimethyltin hydroxide reacts with an equimolar amount of *t*-butyl hydroperoxide in the presence of magnesium sulfate as a dehydrating agent, to give *t*-butyl trimethyltin peroxide, but if an excess of hydroperoxide is used, a considerable amount of a complex with the composition $Me_3Sn{\cdot}O{\cdot}OBu^t$, $HO{\cdot}OBu^t$ is formed. The free peroxide is a mobile liquid whereas the complex, which can also be synthesized from its components, is a crystalline solid, mp 28°C, which can be sublimed *in vacuo* and recrystallized from pentane although it is dissociated in solution.

From a study of the infrared spectrum, Walling and Heaton (*326*) suggested that *t*-butyl hydroperoxide forms a hydrogen-bonded dimer (**17**), and it has been shown that *t*-butyl trimethyltin peroxide is also partially associated (in benzene), presumably by coordinative bonding into dimers of the structure (**19**). The complex $Me_3Sn \cdot O \cdot OBu^t$, $HO \cdot OBu^t$ might then involve both types of bonding as illustrated in (**18**):

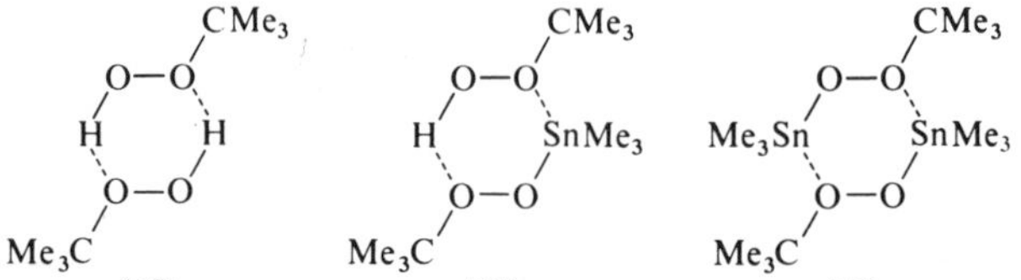

```
      CMe3                  CMe3                        CMe3
     /                     /                           /
 O—O                   O—O                         O—O
/   \                 /   \                       /   \
H    H               H    SnMe3           Me3Sn    SnMe3
 \  /                 \   /                     \   /
 O—O                   O—O                       O—O
/                     /                         /
Me3C                  Me3C                      Me3C
   (17)                  (18)                      (19)
```

The only comparable complexes which appear to have been reported are the compounds $[PhCMe_2O \cdot ONa, HO \cdot O \cdot CMe_2Ph]_n$, where $n = 1$ or 2 (*28*), and $[Et_3Sn \cdot O \cdot OH]_2HO \cdot OH$ (*8*), although it seems likely that further examples will come to light in the future.

In the infrared spectra the alkyl trialkyltin peroxides show a band at 820–835 cm^{-1}, which can probably be ascribed to the O—O stretching vibration (*38*). They are stable at room temperature for at least some months, but are readily hydrolyzed in air. The thermal decomposition of 2-phenyl-2-propyl triethyltin peroxide in octane (*11*), and of *t*-butyl triethyltin peroxide in tetraethyltin and hexaethylditin (*9*) have been studied. The nature of the products indicates that complex free radical processes are involved.

Attempts have been made to prepare di-, tri-, and tetra-alkylperoxides of tin. It was found that, as the number of alkylperoxy substituents is increased, the isolation of a peroxy derivative becomes progressively more difficult. The compounds are much less volatile and more thermally labile than the mono-peroxides and usually cannot be distilled without decomposition. They are more readily hydrolyzed and are difficult to purify by crystallization. The dialkylperoxy compounds cannot be prepared by dehydrating a mixture of dialkyltin oxide and hydroperoxide as, under the necessarily mild conditions employed, the reaction stops at a stage of partial dehydration and yields dialkylperoxydistannoxanes [cf. Eq. (26)] (*38*).

II. Preparation, Structure, and Physical Properties of Hydroxides, Oxides, and Peroxides

A. Hexaalkyldistannoxanes and Related Compounds

Organotin hydroxides and oxides are usually prepared by hydrolyzing the appropriate organotin halide under alkaline conditions.

$$\geq SnX + \bar{O}H \longrightarrow \geq SnOH \xrightarrow{-H_2O} \geq Sn \cdot O \cdot Sn \leq \tag{87}$$

TABLE 4

ORGANOTIN ALKYL PEROXIDES

Compound[a]	mp, °C	bp, °C/mm	Preparations[a]	References
$Me_3Sn\cdot O\cdot OEt$	95–97	—	X	(*273*)
$Me_3Sn\cdot O\cdot OBu^t$	—	56/12	X	(*269*, *270*, *273*)
			O	(*38*)
$Me_3Sn\cdot O\cdot OBu^t$, $HO\cdot OBu^t$	28	—	O	(*38*)
$Et_3Sn\cdot O\cdot OEt$	—	Room temp./ 0.2–0.5	M	(*2*)
$Et_3Sn\cdot O\cdot OBu^t$	—	62–62.5/1	X	(*14*, *270*)
		55/0.2	A	(*10*, *270*, *273*)
			O	(*10*)
$Et_3Sn\cdot O\cdot OCMe_2Ph$	—	105–110/2	X	(*269*, *270*)
		86–87/0.04	A	(*10*, *273*)
			O	(*10*)
$Pr_3Sn\cdot O\cdot O$— 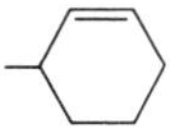	—	109/0.1–0.2 71/0.04	A	(*270*, *273*)
$Pr_3Sn\cdot O\cdot O$— 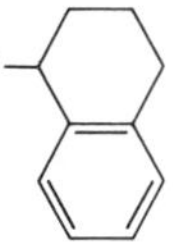	100–110 (decomp.)	—	A	(*270*, *273*)
$Bu_3Sn\cdot O\cdot OBu^t$	—	87/0.04	X	(*14*, *199*)
		71–72/0.001	A	(*273*)
			O	(*38*)
			M	(*14*)
$Bu_3Sn\cdot O\cdot OCMe_2Ph$	—	110–120(bath) /0.01	X	(*199*)
			A	(*14*)
$Bu_3Sn\cdot O\cdot O$— 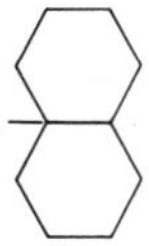	—	—	A	(*14*)
$(C_6H_{11})_3Sn\cdot O\cdot OBu^t$	144–146	—	X	(*270*, *273*)
$(C_6H_{11})_3Sn\cdot O\cdot OCMe_2Ph$	—	—	X	(*273*)
$Ph_3Sn\cdot O\cdot OBu^t$	63–65	—	X	(*273*)
			O	(*38*)
$Ph_3Sn\cdot O\cdot OCMe_2Ph$	111–114	—	X	(*273*)
			O	(*38*)
$Ph_3Sn\cdot O\cdot OCPh_3$	123–125	—	X	(*273*)

[a] See footnote to TABLE 1.

TABLE 4 (*continued*)

Compound[a]	mp, °C	bp, °C/mm Hg	Preparations[a]	References
$Ph_3Sn \cdot O \cdot O$– 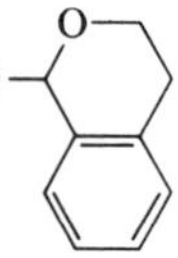 (2-chromanyl-type ring structure)	123–125	—	X A	(*273*) (*273*)
$Ph_3Sn \cdot O \cdot O$– 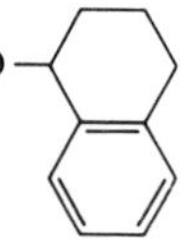 (tetrahydronaphthyl ring structure)	81.5–83	—	X	(*273*)
$Ph_3Sn \cdot O \cdot O$–  (decalyl ring structure)	102–105	—	A	(*14*)
$Me_2Sn(O \cdot OBu^t)_2$	160–161 (decomp.)	—	X	(*199*)
$Bu_2Sn(O \cdot OBu^t)_2$	—	100/<0.001	X	(*14*)
$MeSn(O \cdot OBu^t)_3$	—	—	X	(*14*)

Whether an hydroxide or oxide is isolated depends largely on the nature of the organic groups carried by the tin. Frequently the hydroxides are unstable and dehydrate spontaneously to the bis(trialkyltin) oxides, and sometimes it is difficult to distinguish between these two classes of compounds.

1. *Trialkyltin Hydroxides*, $R_3Sn \cdot OH$

Trimethyltin hydroxide is obtained by hydrolyzing trimethyltin chloride with aqueous sodium, potassium, or ammonium hydroxide (*188*). It is remarkably difficult to dehydrate, and can be purified by sublimation at 80°C, but at 100°C it disproportionates into tetramethyltin, dimethyltin oxide, and water (*162*). The melting point is 118°C; the value of 95°C which has been reported (*155*), probably refers to the complex $(Me_3Sn)_3OBr$ (mp 88°C) which can be isolated from partial hydrolysis of trimethyltin bromide (see Sec. II.A.3).

Trimethyltin hydroxide dissolves readily in water to give the basic species $Me_3Sn(OH_2)_nOH$ (*131*, *140*, *311*). It is almost insoluble in alkanes, but is slightly soluble in benzene, carbon tetrachloride, and chloroform, in which it exists as a dimer, with an Sn—O stretching vibration at 500–580 cm^{-1}; both the symmetric and asymmetric stretching vibrations of the Me_3Sn group are active in the infrared spectrum. In solution, therefore, the hydroxide is thought to have the structure (**20**) (*172*, *235*).

```
            H
            |
   Me       O       Me
     \    /   \   /
Me — Sn         Sn — Me
     /    \   /   \
   Me       O       Me
            |
            H
          (20)

Me  H    Me  H    Me  H    Me  H
|   |    |   |    |   |    |   |
Sn— O ···Sn— O ···Sn— O ···Sn— O
/ \      / \      / \      / \
Me Me    Me Me    Me Me    Me Me
              (21)
```

In the crystal the oxygen bridges between trimethyltin groups, which are in planes inclined at 75° to the axis of the Sn—O chain (**21**); the Sn—O bond shows a vibration frequency of 370 cm^{-1}, and the asymmetric stretching frequency of the Me_3Sn group is at 540 cm^{-1}, but the symmetric stretch of the planar group is infrared inactive (*172*, *235*).

Triethyltin hydroxide is more readily dehydrated, but it can be isolated by adding an equimolar quantity of water to bis(triethyltin) oxide, and allowing it to stand in a desiccator over 50% aqueous sodium hydroxide (*188*, *202*). As the size of the alkyl group is increased further, the hydroxides become less stable with respect to the corresponding bis(trialkyltin) oxides. For example, tributyltin hydroxide has probably not been characterized, and reports that it is an infusible solid (*200*), or a liquid of bp 186–190°C/5 mm (*137*), probably refer to dibutyltin oxide and bis(tributyltin) oxide, respectively

Branching in the alkyl group, however, can stabilize the hydroxide, perhaps by sterically destabilizing the oxide. For example tri-*t*-butyltin hydroxide can be isolated as a solid with mp 153°C (*168*) (although the corresponding oxide has not been reported to permit a clear distinction).

Usually, both the triaryltin hydroxides and the corresponding bis(triaryltin) oxides are available; they can often be readily interconverted, and may be difficult to distinguish. This has been carried out using the Karl Fischer titration (*175*), and the infrared spectra of the solid state. The OH group shows only a weak absorption associated with the stretching frequency, but triphenyltin hydroxide shows a strong doublet at 910 and 897 cm^{-1}, which may be associated with OH deformation, and is absent in bis(triphenyltin oxide) (*176*).

In solution, triphenyltin hydroxide does not dimerize like trimethyltin hydroxide, but is in equilibrium with bis(triphenyltin) oxide and water, obscuring any possible hydrogen-bonding interaction with the solvent (*101*, *169*, *202*, *327*, *328*).

The trialkyltin hydroxides which have been characterized are listed in Table 5.

TABLE 5

TRIALKYLTIN HYDROXIDES

$R_3Sn \cdot OH$ R	mp, °C	bp, °C/mm	Reference
Me	118		(*162, 188*)
$Me_2EtSn \cdot OH$			(*47*)
$CH_2{=}CH$	67.5–69		(*276*)
Et	49–50	272/760	(*188, 202*)
Pr	33–35		(*49, 188*)
$(N{:}C \cdot CH_2 \cdot CH_2)_3Sn \cdot OH, H_2O$	103–104		(*268*)
$Bu_2(CH_2{:}CH)Sn \cdot OH$		liquid	(*59*)
Bu^t	153		(*168*)
Pe^t		335–338/760	(*59*)
$Bu(Ph)(PhCH_2)Sn \cdot OH$	135–137		(*149*)
C_6F_5	150(decomp)		(*132*)
Ph	119–120		(*101, 167, 169, 188, 202, 327, 328*)
cyclo-C_6H_{11}	220–221		(*167, 188*)
$PhCH_2$	120		(*189*)
$3\text{-}Me \cdot C_6H_4$		syrup	(*189*)
$4\text{-}Me \cdot C_6H_4$	67.5–69		(*189*)
$Ph_2(4\text{-}CH_2{:}CH \cdot CH_2 \cdot C_6H_4)Sn \cdot OH$	—		(*258*)
$Ph(4\text{-}CH_2{:}CH \cdot CH_2 \cdot C_6H_4)_2Sn \cdot OH$	—		(*258*)
$4\text{-}CH_2{:}CH \cdot CH_2 \cdot C_6H_4$	—		(*258*)
$Me_2PhC \cdot CH_2$	145–146		(*267*)
$1\text{-}C_{10}H_7$	340–393		(*189*)
$C_{16}H_{33}$	75–76		(*226*)
$4\text{-}Pr^i \cdot C_6H_4$	95		(*66*)

2. *Hexaalkyldistannoxanes*, $R_3Sn \cdot O \cdot SnR_3$

If the trialkyltin hydroxide does not dehydrate spontaneously, the corresponding bis(trialkyltin) oxide can usually be obtained by desiccation of the hydroxide under reduced pressure over calcium chloride or phosphoric oxide, or by azeotropic distillation in benzene [e.g., triphenyltin hydroxide (*202*)]. Trimethyltin hydroxide is particularly difficult to dehydrate, but this can be brought about by sodium in dry benzene (*119, 172*). The oxide fumes in air, rehydrating to the hydroxide; it is also anomalous in that it reacts exothermically with trimethyltin halides, Me_3SnX, to give oxonium salts $(Me_3Sn)_3OX$ (see Sec. II.A.3).

The bis(trialkyltin) oxides (Table 6) are characterized in the infrared spectrum by a strong band at 735–785 cm^{-1}, associated with the asymmetric stretch of the Sn—O—Sn groups [e.g. $(Me_3Sn)_2O$—737 cm^{-1}, $(Et_3Sn)_2O$—779 cm^{-1}, $(Bu_3Sn)_2O$—783 cm^{-1}, $(Ph_3Sn)_2O$—775 cm^{-1}] (*171, 172*).

TABLE 6

BISTRIALKYLTIN OXIDES

$(R_3Sn)_2O$ R	mp, °C	bp °C/mm	Reference
Me		86/24	(*119, 172*)
Et		100/1	(*62, 115, 117, 120, 137, 173*)
Pr		142–144/1	(*137, 316*)
Pr^i		oil	(*188, 316*)
$(Et_2BuSn)_2O$		oil	(*317*)
Bu		220–230/10	(*137, 188, 316*)
Bu^i		142–146/0.001 197–198/12	(*222, 335, 336*)
Bu^s		138–142/0.0001	(*222*)
Pe		oil	(*188*)
Pe^i		182/2	(*157*)
Pe^{neo}		215–217	(*342*)
C_6F_5		—	(*132*)
$4\text{-F}\cdot C_6H_4$	>400		(*102*)
Ph	124		(*101, 137, 169, 175, 176, 202, 327, 328*)
cyclo-C_6H_{11}			(*166*)
C_6H_{13}		oil	(*188*)
$PhCH_2$	120		(*25, 189*)
$2\text{-Br}\cdot C_6H_4\cdot CH_2$	158–159		(*25*)
$2\text{-Cl}\cdot C_6H_4\cdot CH_2$	133.5		(*25*)
$2\text{-F}\cdot C_6H_4\cdot CH_2$	113		(*25*)
$(Ph_3Ge\cdot SnEt_2)_2O$	85–95		(*62*)
C_8H_{17}		268/10	(*144, 188*)
$2\text{-PhO}\cdot C_6H_4$	143–144		(*245*)

3. *Hexamethyldistannoxonium, Nonamethyltristannoxonium, and Nonaethyltristannoxonium Salts*

Trimethyl- or triethyl-tin hydroxide, or bis(trimethyltin) oxide, form a series of ionic complexes with the corresponding trialkyltin halides, as shown in Table 7.

Bis(trimethyltin) oxide reacts with trimethyltin bromide or iodide in benzene or light petroleum to give the complexes $(Me_3Sn)_2O,Me_3SnX$ (*121*). The Raman and infrared spectra suggest that these are oxonium salts (**22**) (*171*). In most solvents they are converted into the compounds $(Me_3SnOH)_2,Me_3SnX$ (*119*), which can also be prepared directly from the tin hydroxide and tin halide (*112*, *163*, *164*), or by autoxidation of a mixture of trimethyltin bromide

TABLE 7

Complexes Derived from R_3SnOH or $(R_3Sn)_2O$, and R_3SnX

Composition	mp, °C	Reference
$(Me_3Sn)_2O,Me_3SnBr$	88(decomp)	(*116, 170*)
$(Me_3Sn)_2O,Me_3SnI$	94	(*116*)
$(Me_3Sn \cdot OH)_2,Me_3SnBr$	116	(*163, 164, 170*)
$(Me_3Sn \cdot OH)_2,Me_3SnCl$	85–91	(*118, 164*)
$(Me_3Sn \cdot OH)_2,Me_3SnI$	143–153(decomp)	(*118, 164*)
$Me_3Sn \cdot OH,Me_3SnBr,H_2O$	210–211	(*164, 170*)
$Me_3Sn \cdot OH,Me_3SnCl,H_2O$	81–95(decomp)	(*115, 118, 164*)
$Me_3Sn \cdot OH,Me_3SnI,H_2O$	221(decomp)	(*164*)
$(Et_3SnOH)_2, Et_3SnBr,H_2O$	103	(*115*)
$(Et_3SnOH)_2, Et_3SnCl,H_2O$	77	(*115*)
$(Et_3SnOH)_2, EtSnI,H_2O$	108–110	(*115*)

or iodide and hexamethylditin in benzene (*112, 164*). These again appear to have an ionic structure (**23**) which involves hydrogen bonding (*170*):

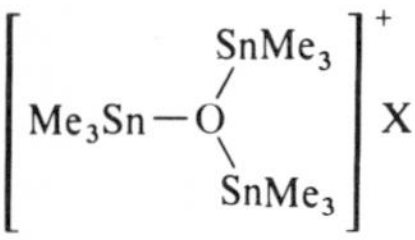

$$\left[Me_3Sn{-}O(SnMe_3)_2 \right]^+ X^- \qquad \left[Me_3Sn{\cdot}O(H){\rightarrow}SnMe_3{\leftarrow}O(H){\cdot}SnMe_3 \right]^+ X^- \qquad \left[Me_3Sn{\cdot}O(H){\rightarrow}SnMe_3{\leftarrow}OH_2 \right]^+ X^-$$

(22) 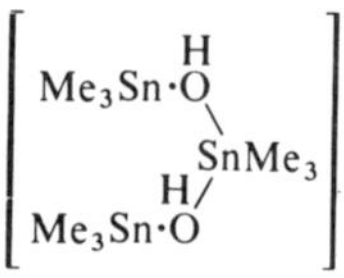 **(23)** 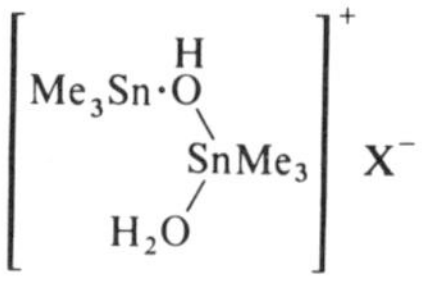 **(24)**

If these complexes are recrystallized from water (*118*), they are further hydrolyzed to the compounds $Me_3Sn \cdot OH,Me_3SnBr,H_2O$, which can also be prepared from the hydroxide and halide in a moist solvent (*112, 115, 118, 164*). The Raman and infrared spectra again indicate a hydrogen-bonded ionic structure (**24**).

Relatively few "complexes" derived from the corresponding triethyltin compounds are known. The compounds of composition $(Et_3Sn \cdot OH)_2,Et_3SnX,H_2O$ ($X = Cl$, Br I) can be synthesized from the components in benzene, and the iodo compound has also been obtained by exposing a mixture of triethyltin iodide and tetraethyltin to moist air and sunlight (*115, 120*).

Harada suggested that these were also ionic compounds, with the structure:

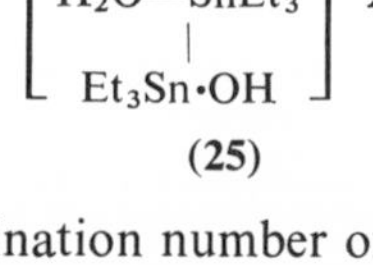

$$\left[\begin{array}{c} Et_3Sn{\cdot}OH \\ | \\ H_2O{-}SnEt_3 \\ | \\ Et_3Sn{\cdot}OH \end{array} \right]^+ X^-$$

(25)

where the tin has the coordination number of six (*120*).

B. Pentaalkyldistannoxanes, $R_3Sn \cdot O \cdot SnR_2X$

If a polymeric dialkyltin oxide, $(R_2SnO)_n$, is heated with a compound in which tin is bonded to an electronegative group X (e.g., halide or carboxylate), a telomerization reaction takes place.

$$\geq SnX + O \cdot SnR_2 \longrightarrow \geq Sn \cdot O \cdot SnR_2X \tag{88}$$

By this process, trialkyltin compounds, R_3SnX, provide a route to the functionally substituted pentaalkyldistannoxanes.

$$R_3SnX + OSnR_2 \longrightarrow R_3Sn \cdot O \cdot SnR_2X \tag{89}$$

1,1,1-Triethyl-3,3-dibutyl-3-chlorodistannoxane, $Et_3Sn \cdot O \cdot SnBu_2Cl$, and 1,1,1,3,3-pentabutyl-3-chloro, -3-bromo-, and -3-acetoxy-distannoxane have been isolated as very thick greases which revert to the parent oxide and halide above 150°C. In the infrared spectra they show a band associated with the asymmetric stretch of the Sn—O—Sn group at 680 cm^{-1} and 675 cm^{-1}, respectively [cf. 783 cm^{-1} in $(Bu_3Sn)_2O$], and in benzene they are only slightly associated (*74*, *75*).

C. 1,1,3,3-Tetraalkyldistannoxanes and Related Compounds

If a difunctional tin (IV) compound is hydrolyzed, a series of oligomeric dialkylstannoxanes can be isolated en route to the ultimate polymeric dialkyltin oxide. The principal members of this series are the 1,1,3,3-tetraalkyldistannoxanes as follows:

$$R_2Sn\begin{matrix} \diagup OH \\ \diagdown X \end{matrix} \tag{90}$$

The same distannoxanes can frequently be obtained by treating the (polymeric) dialkyltin oxide with the appropriate amount of the compound R_2SnX_2, in the presence of water if necessary:

$$R_2SnO + R_2SnX_2 \longrightarrow XR_2Sn \cdot O \cdot SnR_2X \tag{91}$$

$$3\,R_2SnO + R_2SnX_2 + H_2O \longrightarrow 2XR_2Sn \cdot O \cdot SnR_2OH \tag{92}$$

Thirdly, these distannoxanes are also formed when the dialkyltin oxide is treated with the protic compound HX:

$$2\,R_2SnO + 2\,HX \longrightarrow H_2O + XR_2Sn \cdot O \cdot SnR_2X \tag{93}$$

A number of miscellaneous processes give rise to the same family of compounds, and they have been reported as adventitious by-products from a wide variety of reactions.

Much of the early work in this field was carried out by Harada in Tokyo between 1939 and 1947.

1. *Dialkyltin Hydroxides*, $R_2Sn(OH)X$

Relatively few hydroxides of the structure $R_2Sn(OH)X$ (Table 8) are known, dehydration usually readily giving the corresponding tetraalkyldistannoxanes, $XR_2Sn \cdot O \cdot SnR_2X$. Elemental analysis alone cannot readily distinguish between these two classes of compounds, and the data on diphenyltin chloride hydroxide do not preclude the distannoxane structure (*23*, *52*, *149*). Similarly, a series of dibutyltin hydroxide phenoxides, $Bu_2Sn(OH)OAr$, which were reported to be formed from the reaction between dibutyltin oxide and phenols in boiling dichloromethane (*344*), are, in fact, more likely to be the distannoxanes $(ArO)Bu_2Sn \cdot O \cdot SnBu_2(OAr)$ (*58*).

As with the trialkyltin hydroxides, bulky alkyl groups again hinder dehydration. Thus di(*o*-phenoxyphenyl)tin dichloride reacts with aqueous sodium hydroxide at 0°C to give the corresponding dihydroxide, $(o\text{-}PhO \cdot C_6H_4)_2Sn(OH)_2$, which shows an OH stretching vibration at 3610 cm^{-1} (*245*). With *t*-butyl and *t*-pentyl groups on the tin, both the hydroxide halides and dihydroxides appear to be formed, although di-*t*-butyltin dihydroxide is difficult to obtain pure (*26*, *168*).

With less bulky alkyl groups it is difficult to avoid dehydration to the distannoxane. Diphenyltin hydroxide cyanate, $Ph_2Sn(OH)CNO$, which was obtained from adventitious hydrolysis of diphenyltin dicyanate, readily loses water (*209*). Dibutyltin hydroxide chloride, $Bu_2Sn(OH)Cl$, ν_{max} 3509 (OH) cm^{-1}, is formed as a white solid when dibutyltin dichloride in methanol is added to a large volume of water, but it very readily reverts to water and the distannoxane (*16*, *105*), and is probably in equilibrium with these components in solution (*333*).

In contrast, the dialkyltin hydroxide nitrates, $R_2Sn(OH)NO_3$ (R = Me, Et, Pr, Bu), which can be prepared by treating the dialkyltin oxides with the appropriate amount of nitric acid, are stable and do not readily condense (*333*). They are insoluble in nonpolar solvents, and are probably coordinatively associated (*334*).

2. 1,1,3,3-*Tetraalkyldistannoxanes*, $XR_2Sn \cdot O \cdot SnR_2X$

The first publication on organotin chemistry, that by Löwig in 1852 (*186*), described the reaction between ethyl iodide with a tin-sodium alloy, which gave a compound thought to be the radical Et_2SnI. Shortly afterwards (*310*, *302*), Strecker showed that this compound had, in fact, the composition $IEt_2Sn \cdot O \cdot SnEt_2I$, and that it could also be prepared by heating diethyltin oxide with diethyltin diiodide.

As mentioned above, the three principal routes to these compounds are (i) the partial hydrolysis of R_2SnX_2, (ii) the telomerization of R_2SnO and R_2SnX_2, and (iii) the reaction of R_2SnO with the appropriate amount of the acid HX.

TABLE 8

DIALKYTIN HYDROXIDES

$R_2Sn(OH)X$ R	X	mp, °C	Reference
Me	NO_3	250	(*333, 334*)
Et	NO_3	214(decomp)	(*333, 334*)
Pr	NO_3	183(decomp)	(*333*)
Pr	NO_3, H_2O	183(decomp)	(*333*)
Bu	Cl	105–107	(*16, 105, 333*)
Bu	NO_3	92.5–95	(*333*)
Bu^t	OH	—	(*26, 168*)
Bu^t	Cl	170	(*168*)
Bu^t	Br	152	(*168*)
Pe^t	OH	200	(*168*)
Ph	Cl	—	(*23, 52, 149*)
Ph	NCO	99–100	(*166*)
cyclo-C_6H_{11}	OH	291(decomp)	(*258*
4-CH_2:$CH \cdot CH_2 \cdot C_6H_4$	OH	—	(*258*)
Ph(4-CH_2:$CH \cdot CH_2 \cdot C_6H_4$)$Sn(OH)_2$		—	(*258*)
2-$PhO \cdot C_6H_4$	OH	—	(*245*)

The partial hydrolysis can be carried out by adding the compound R_2SnX_2 to a solution of triethylamine (1 mole) in wet ethanol, or by adding the appropriate amount of aqueous alkali to R_2SnX_2 in ethanol (*17*):

$$2\,R_2SnX_2 + 2OH^- \longrightarrow XR_2Sn \cdot O \cdot SnR_2X + 2\,X^- + H_2O \qquad (94)$$

Largely because of the difficulty of analyzing (by difference) for oxygen in the distannoxanes ($XR_2Sn \cdot O \cdot SnR_2X$), some of these products of hydrolysis were at first believed to be the distannanes, $XR_2Sn \cdot SnR_2X$ (*142*, *143*), but this was later corrected (*12*, *17*, *141*); the authentic distannanes were prepared, and shown to react with oxygen to give the distannoxanes (*105*).

In the hydrolysis, the reactivity of various groups X follows the sequence F < Cl < Br < I, OAc. Thus Bu_2SnClF gives the fluoride $FBu_2Sn \cdot O \cdot SnBu_2F$, $Bu_2SnClBr$ gives the chloride $ClBu_2Sn \cdot O \cdot SnBu_2Cl$, and $Bu_2Sn(OAc)Br$ is hydrolyzed in air to acetic acid and the bromide $BrBu_2Sn \cdot O \cdot SnBu_2Br$. The di-iododistannoxane, $IBr_2Sn \cdot O \cdot SnBu_2I$, is readily hydrolyzed further to the iodide hydroxide $IBu_2Sn \cdot O \cdot SnBu_2OH$ (*17*).

The telomerization reaction between the compounds R_2SnX_2 and R_2SnO is usually carried out by heating equimolar amounts in an inert solvent such as benzene or toluene, whereupon the dialkyltin oxide dissolves as it reacts [Eq. (91)]. Progressively higher temperatures are needed as the group X is changed in the sequence F, Cl, Br, and I (*17*). It might be expected that mixed tetraalkyldistannoxanes, $XR_2Sn \cdot O \cdot SnR'_2X$ could be prepared from the appropriate compounds R_2SnX_2 and R'_2SnO, or R_2SnO and R'_2SnX_2. A few examples of such compounds have been reported, but the situation is complicated because they can apparently exist in isomeric dimeric forms; this is discussed below, and in Sec. II.C.5.

The nature of the products from the reaction of dialkyltin oxides with acids, HX, depends on the acid strength of HX and on the reaction conditions:

$$R_2SnO \xrightarrow{HX} XR_2Sn \cdot O \cdot SnR_2(OH) \xrightarrow{HX} XR_2Sn \cdot O \cdot SnR_2X \xrightarrow{HX} R_2SnX_2 \quad (95)$$

If HX is a strong acid, such as hydrochloric acid, it usually reacts completely, and the nature of the products depends only on the molar ratio of the reactants. If HX is a weaker acid, such as a phenol, progressively more severe conditions are needed to accomplish successive stages of the acidolysis, irrespective of the ratio of reactants. For example, if a dialkyltin oxide is heated with a phenol in boiling benzene (80°C), using a Dean and Stark trap to separate water, the reaction proceeds to the stage of the diphenoxydistannoxane, $(ArO)R_2Sn \cdot O \cdot SnR_2(OAr)$ (*58*), but, in boiling tetrahydronaphthalene (207°C) the reaction goes further to give the dialkyltin diphenoxide, $R_2Sn(OAr)_2$ (*265*). Primary alcohols such as allyl alcohol, butanol, and benzyl alcohol, react with dibutyltin oxide in boiling benzene to give the dialkoxydistannoxanes, $(R'O)Bu_2Sn \cdot O \cdot SnBu_2(OR')$, which dissociate on distillation into the corresponding dibutyltin dialkoxides and dibutyltin oxide; on the other hand, the secondary alcohols cyclohexanol, octan-2-ol, and diphenylmethanol showed no reaction with dibutyltin oxide in refluxing benzene or toluene (111°C) (*84*).

Miscellaneous reactions by which these distannoxanes have been prepared include the following.

(i) The replacement of one functional substituent in the distannoxane by another (*38*, *100*):

$$(AcO)Bu_2Sn \cdot O \cdot SnBu_2(OAc) + 2\ Bu^tO \cdot OH \longrightarrow (^tBuO \cdot O)Bu_2Sn \cdot O \cdot SnBu_2(O \cdot OBu^t) \quad (96)$$

$$ClBu_2Sn \cdot O \cdot SnBu_2Cl + 2\ Cl_2CH \cdot CO_2Na \longrightarrow (Cl_2CH \cdot CO_2)Bu_2Sn \cdot O \cdot SnBu_2(O_2C \cdot CHCl_2) \quad (97)$$

(ii) The cohydrolysis of a dialkyltin dichloride and a trialkylchlorosilane (*234*):

$$Et_2SnCl_2 + Me_3SiCl \xrightarrow{H_2O} (Me_3Si \cdot O)Et_2Sn \cdot O \cdot SnEt_2(O \cdot SiMe_3) \quad (98)$$

(iii) The autoxidation of *sym*-tetrabutyldichlorodistannane (*105*):

$$ClBu_2Sn \cdot SnBu_2Cl \xrightarrow{O_2} ClBu_2Sn \cdot O \cdot SnBu_2Cl \quad (99)$$

(iv) The hydrolysis of a dialkyltin benzoate alkoxide (*322*) or halide alkoxide (*72*):

$$Et_2Sn(OMe)Br \xrightarrow{H_2O} BrEt_2Sn\cdot O\cdot SnEt_2Br \qquad (100)$$

(v) The oxidation of triphenyltin hydroxide with peroxyacetic or peroxypropionic acid (*296*):

$$2\,Ph_3Sn\cdot OH + 2\,Me\cdot CO_3H \longrightarrow (MeCO_2)Ph_2Sn\cdot O\cdot SnPh_2(O_2C\cdot Me) + 2\,PhOH \qquad (101)$$

(vi) The reaction of a dialkylperfluorovinyltin chloride with aqueous potassium fluoride (*286*):

$$Et_2(CF_2{:}CF)SnCl + F^- \xrightarrow{H_2O} CF_2{:}CFH + FEt_2Sn\cdot O\cdot SnEt_2F \qquad (102)$$

The tetraalkyldihalogenodistannoxanes, $XR_2Sn\cdot O\cdot SnR_2X$, are usually highly crystalline, sometimes waxy solids which have sharp melting points, and are often very soluble in organic solvents. Compared with the bis(trialkyltin) oxides, the Sn—O bond is relatively unreactive. Thus whereas bis(tributyltin) oxide reacts rapidly with carbon dioxide to give bis(tributyltin) carbonate, and with hydrogen sulfide to give bis(tributyltin) sulfide, 1,1,3,3-tetrabutyl-1,3-dichlorodistannoxane is inert to both these reagents (*246*).

A considerable amount of work has been carried out on the structure of these compounds. Pfeiffer and Brock, in 1914 (*241*), assumed that they had the simple monomeric structure (**26**):

$$XR_2Sn\cdot O\cdot SnR_2X \qquad \left[\begin{array}{l} R_2Sn \\ \;\;| \\ \;\;O-SnR_2X_2 \end{array}\right]_n \qquad \left[\begin{array}{l} R_2SnX \\ \;\;| \\ \;\;O-SnR_2X \end{array}\right]_n$$

(**26**) (**27**) (**28**)

but in a long series of papers between 1939 and 1948 (*112–126*), Harada argued the case for a cyclic oligomeric structure (**27**), where the tin has coordination numbers of 4 and 5, and n has the value of 3. As halogen ligands are usually readily exchanged between different tin sites, the structure (**28**), in which the tin atoms again have the coordination numbers of 4 and 5, should also be considered.

The following evidence indicates that this structure (**28**), where $n = 2$, is correct (*15*, *229*).

(i) Early measurements of the molecular weight were confusing (*115*, *122*, *125*, *141–143*, *241*), but recent work, particularly by vapor pressure osmometry, shows that, for a wide variety of substituents X (e.g., F, Cl, Br, CNS, OAc, OAr, NO_3) these distannoxanes are dimeric in nonpolar solvents (*15*, *17*, *198*, *229*), although some dissociation into monomers may occur in more

TABLE 9

1,1,3,3-TETRAALKYLDISTANNOXANES

X	mp, °C	Reference
$XMe_2Sn\cdot O\cdot SnMe_2X$		
Cl	300	(*17, 45, 231, 233*)
Br	88(decomp)	(*121*)
I	94(decomp)	(*121*)
CNS		(*325*)
NO_3	250	(*333*)
$Me_3Si\cdot O$	167–168	(*56, 229, 234*)
CrO_4	360	(*274*)
HCO_2	185(decomp)	(*227*)
$MeCO_2$	240	(*197, 198, 231*)
$CH_2Cl\cdot CO_2$	226–227	(*227*)
$CHCl_2\cdot CO_2$	232–233	(*227*)
$CCl_3\cdot CO_2$	221–222	(*227*)
$NC\cdot CH_2\cdot CO_2$	300	(*227*)
$Br\cdot CH_2\cdot CO_2$	172–174	(*227*)
$Et\cdot CO_2$	185	(*227*)
$Pr\cdot CO_2$	161–162	(*227*)
$Ph\cdot CO_2$	234–235	(*262, 274*)
$4\text{-}MeO\cdot C_6H_4\cdot CO_2$	235–236	(*227*)
PhO	202	(*227*)
$2\text{-}MeO\cdot C_6H_4\cdot O$	235–236	(*227*)
O, NO	>230	(*274*)
$XEt_2Sn\cdot O\cdot SnEt_2X$		
F		(*286*)
Cl	175	(*17, 233*)
Br	173	(*72, 233*)
I		(*186*)
CNS		(*325*)
NO_3	214(decomp)	(*301, 333*)
Me_3SiO	126–130	(*234*)
$Bu^tO\cdot O$	135–143	(*38*)
$MeCO_2$	105–106	(*198*)
$CHCl_2\cdot CO_2$		(*100*)
$PhCO_2$	215–217	(*100, 262*)
$2\text{-}Cl\cdot C_6H_4\cdot CO_2$	90–91	(*100*)
$2\text{-}HO\cdot C_6H_4\cdot CO_2$	165–167	(*262*)
$2,4\text{-}(HO)_2C_6H_3\cdot CO_2$	>230	(*262*)
Camphorsulfonate	270–290(decomp)	(*17*)

TABLE 9 (*continued*)

X	mp, °C	Reference
$XEt_2Sn\cdot O\cdot SnEt_2X$ (*cont.*)		
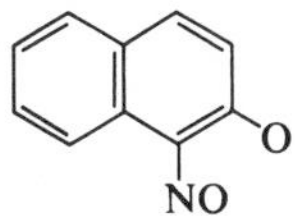 O NO	163	(*307*)
$XPr_2Sn\cdot O\cdot SnPr_2X$		
Cl	122	(*233, 241*)
Br	108	(*233, 241*)
CNS		(*325*)
NO_3	183(decomp)	(*333*)
$Me_3Si\cdot O$	107–108	(*234*)
$MeCO_2$	111–113	(*198*)
$X(NC\cdot CH_2\cdot CH_2)_2Sn\cdot O\cdot Sn(CH_2\cdot CH_2\cdot CN)_2X$		
OH	195(decomp)	(*268*)
$Oct^iO\cdot CO\cdot CH_2\cdot S$	Liquid	(*268*)
$Et_2CH\cdot CO\cdot O$	154–155	(*268*)
$XBu_2Sn\cdot O\cdot SnBu_2X$		
F	115–117	(*17*)
Cl	112	(*17*)
Br	108	(*17*)
NCO	100–102	(*209*)
NCS	84	(*17*)
NO_3	92.5–95	(*333*)
$Me_3Si\cdot O$	108–109	(*234*)
MeO	bp 176/0.1	(*309*)
$CH_2{=}CH{-}CH_2\cdot O$	oil	(*84*)
BuO	oil	(*84*)
$CF_3\cdot CH_2\cdot O$	oil	(*84*)
$PhCH_2\cdot O$	oil	(*84*)
$Bu^tO\cdot O$	78–79	(*38*)
PhO	137–139	(*56, 58*)
2,5-Cl,Br$\cdot C_6H_3\cdot O$	86–90	(*343*)
2,4-$Cl_2\cdot C_6H_3\cdot O$	104–108	(*343*)
2,4-$Br_2\cdot C_6H_3\cdot O$	127–128	(*343*)
2,4,5-$Cl_3\cdot C_6H_2\cdot O$	118–119	(*343*)
2,6-$Cl_2\cdot C_6H_3\cdot O$	109–111	(*343*)
2,5,4-$(NO_2)_2(C_6H_{11})C_6H_2\cdot O$	178–180	(*343*)
4,2-$Cl(C_6H_{11})C_6H_3\cdot O$	118–122	(*343*)
4-$MeO\cdot C_6H_4\cdot O$	86–89	(*58*)

TABLE 9 (*continued*)

X	mp, °C	Reference
$XBu_2Sn\cdot O\cdot SnBu_2X$ (*cont.*)		
$4\text{-}Bu^t\cdot C_6H_4\cdot O$	138–142	(*58*)
$4\text{-}NO_2\cdot C_6H_4\cdot O$	247–248	(*58*)
$2\text{-}C_{10}H_7\cdot O$	94–100	(*58*)
$CH_3\cdot CO_2$	58–60	(*17*)
$Cl_2CH\cdot CO_2$	192	(*100*)
$C_7H_{15}\cdot CO_2$	oil	(*17*)
$4\text{-}NO_2\cdot C_6H_4\cdot CO_2$	187	(*100*)
$3\text{-}NO_2\cdot C_6H_4\cdot CO_2$	148–149	(*100*)
$3,5\text{-}(NO_2)_2\cdot C_6H_3\cdot CO_2$	204–205	(*100*)
$2\text{-}Cl\cdot C_6H_4\cdot CO_2$	81–82	(*100*)
$4\text{-}(NH_2\cdot SO_2)C_6H_4\cdot CO_2$	250–252	(*100*)
$Ph\cdot CH(OH)\cdot CO_2$	180	(*100*)
$XPh_2Sn\cdot O\cdot SnPh_2X$		
Cl	187	(*143*)
NO_3	265–270	(*98*)
NCO	158–162	(*209, 210*)
PhO	~250	(*58*)
$4\text{-}Br\cdot C_6H_4\cdot O$	245–255	(*343*)
$2,5\text{-}Cl_2\cdot C_6H_3\cdot O$	210–225	(*343*)
$HO(2\text{-}PhO\cdot C_6H_4)_2Sn\cdot O\cdot Sn(2\text{-}PhO\cdot C_6H_4)_2OH$	164	(*245*)

polar solvents such as acetone [e.g., $Ph_4Sn_2(NO_3)_2O$] (*98*), camphor, or cyclopentadecanone [e.g., $Bu_4Sn_2(OPh)_2O$] (*58*).

(ii) In benzene or carbon tetrachloride at room temperature, the compounds of empirical formulas $Bu_4Sn_2Cl_2O$ and $Bu_4Sn_2Br_2O$, show two broad overlapping signals of approximately equal intensity in the ^{119}Sn nuclear magnetic resonance spectrum, implying the presence of nonequivalent tin atoms (*17*). The Mössbauer spectrum of $Bu_4Sn_2Cl_2O$, however, consists of only one doublet (*53*).

(iii) Similarly, in the proton magnetic resonance spectrum of the compound $Me_4Sn_2Cl_2O$, two signals are obtained indicating the presence of methyl groups in different environments (*17*). With the corresponding trimethylsiloxy compound (**25**) in carbon tetrachloride solution,

$$\begin{array}{c} Me_2Sn\cdot O\cdot SiMe_3 \\ | \\ O \\ Me_3Si\cdot O\cdot Me_2Sn \quad SnMe_2\cdot O\cdot SiMe_3 \\ O \\ | \\ Me_2Sn\cdot O\cdot SiMe_3 \\ \textbf{(29)} \end{array} \rightleftharpoons 2\ Me_3Si\cdot O\cdot SnMe_2\cdot O\cdot SnMe_2\cdot O\cdot SiMe_3 \qquad (103)$$

the relative intensity of the two signals for the different methyltin groups varies with temperature because the compound dissociates into monomeric units, showing that the average dissociation energy of each of the two SnO bonds is about 4.5 kcal/mole (*56*).

(iv) Harada reported (*125*), and Okawara confirmed (*228*), that the reactions between R_2SnO and R'_2SnX_2 on the one hand, and R'_2SnO and R_2SnX_2 on the other, gave isomeric compounds of the composition $R_2R'_2Sn_2X_2O$, which are assumed to have the structures:

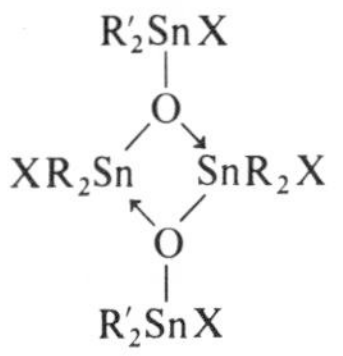

$$\begin{array}{ccc} & R'_2SnX & \\ & | & \\ & O & \\ \swarrow & & \searrow \\ XR_2Sn & & SnR_2X \\ \nwarrow & & \swarrow \\ & O & \\ & | & \\ & R'_2SnX & \end{array} \qquad \begin{array}{ccc} & R_2SnX & \\ & | & \\ & O & \\ \swarrow & & \searrow \\ XR'_2Sn & & SnR'_2X \\ \nwarrow & & \swarrow \\ & O & \\ & | & \\ & R_2SnX & \end{array}$$

(30) 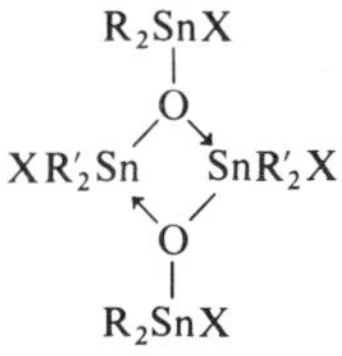 **(31)**

(v) An X-ray study of the crystalline silastannoxane (**29**) confirms that it has the structure shown. The peripheral tin atoms are arranged in a *trans* sense about the essentially square (2.2 × 2.8 Å) tin-oxygen ring (*230*).

3. 1,1,3,3-*Tetraalkylhydroxydistannoxanes*, $XR_2Sn{\cdot}O{\cdot}SnR_2(OH)$

Further hydrolysis of the compounds $XR_2Sn{\cdot}O{\cdot}SnR_2X$ gives the hydroxydistannoxanes $XR_2Sn{\cdot}O{\cdot}SnR_2(OH)$, which can also be prepared by dissolving the appropriate amount of the oxide R_2SnO in a hot solution of the compound R_2SnX_2 in a moist solvent (*16*). The same products are obtained when trialkyltin iodides are exposed to sunlight in the presence of oxygen and moisture (*116*, *120*). Examples are given in Table 10.

In carbon disulfide, they show the presence of an OH group at $\sim 3420\ cm^{-1}$ in the infrared spectra. They usually melt at about 200°C over a range of 5–10°C, but if they are "recrystallized" from alcohols, they are converted into the highly crystalline alkoxides, $XR_2Sn{\cdot}O{\cdot}SnR_2(OR')$, which, in the air, are rapidly hydrolyzed back to the hydroxides. They are dimeric in benzene, naphthalene, chloroform, and acetone (*16*, *115*), and presumably contain the same distannoxane ring structure as their progenitors $(R_4Sn_2X_2O)_2$. Okawara and Wada (*233*) suggested that the OH groups are bound to the tin atoms in the ring, and then coordinate to the peripheral tin atoms to render all four tins five coordinate:

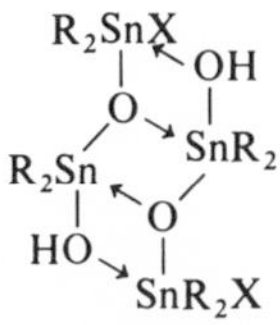

$$\begin{array}{cccc} & R_2SnX & \nwarrow & OH \\ & | & & | \\ & O & \searrow & SnR_2 \\ \swarrow & & & \swarrow \\ R_2Sn & \nwarrow & O & \\ | & & | & \\ HO & \searrow & SnR_2X & \end{array}$$

(32)

TABLE 10

1,1,3,3-TETRAALKYLHYDROXYDISTANNOXANES AND 1,1,3,3-TETRAALKYLALKOXYDISTANNOXANES

X	Y	mp, °C	Reference
$XMe_2Sn \cdot O \cdot SnMe_2Y$			
Cl	OH	>220	(*233*)
Cl	OMe	~200	(*233*)
Br	OH	>220	(*233*)
Br	OMe	~200	(*233*)
I	OH	>220	(*233*)
I	OMe	220	(*233*)
I	OEt	210	(*233*)
SCN	OH		(*325*)
$XEt_2Sn \cdot O \cdot SnEt_2Y$			
Cl	OH	218, 236	(*16, 233, 325*)
Cl	OMe	144	(*233*)
Br	OH	204–212	(*233*)
I	OH	195–208	(*233*)
I	OMe	148	(*233*)
I	OEt	142	(*233*)
CNS	OH		(*325*)
NO_3	OH		(*343*)
$Me \cdot CO_2$	OH	~200	(*198*)
$2\text{-}MeCO_2 \cdot C_6H_4 \cdot CO_2$	OH	70	(*127*)
$2\text{-}HO \cdot C_6H_4 \cdot CO_2$	OH	262–272	(*127*)
$XPr_2Sn \cdot O \cdot SnPr_2Y$			
Cl	OH	175–194	(*233*)
Cl	α-NO-β-naphthol	158–160	(*307*)
Br	OH	172–188	(*233*)
I	OH	178–182	(*233*)
CNS	OH		(*325*)
NO_3	OH	221–222.5	(*333*)
$Me \cdot CO_2$	OH	206–208	(*198*)
$XBu_2Sn \cdot O \cdot SnBu_2Y$			
Cl	OH	109–121	(*233*)
Cl	OEt	85–140	(*16*)
Cl	α-NO-β-naphthol	140–141	(*45*)
Br	OH	114–128	(*233*)
I	OH	134–140	(*233*)
I	OEt	71.5–85	(*233*)
NCO	OH	162–164	(*209*)
NO_3	OH	210–213	(*333*)
$CH_3 \cdot CO_2$	OH	~129	(*16*)
$CH_3 \cdot CO_2$	OMe	oil	(*337*)
$Cl_2CH \cdot CO_2$	OMe		(*337*)
$(ONC)Ph_2Sn \cdot O \cdot SnPh_2 \cdot OH$		300–301	(*209, 210*)
$HO(2\text{-}PhO \cdot C_6H_4)_2Sn \cdot O \cdot Sn(2\text{-}PhO \cdot C_6H_4)_2 \cdot OH$		164	(*245*)

4. *Oligomeric Dialkylstannoxanes*, $X(R_2Sn \cdot O)_nR_2SnX$

Further hydrolysis of a difunctional distannoxane can give some identifiable oligomeric dialkylstannoxanes before the ultimate polymeric dialkyltin oxide. Similar oligomers can be prepared by telomerization between the dialkyltin oxide and compound R_2SnX_2 or by the condensation of dialkyltin dialkoxides with dicarboxylates.

The most complete series which have been prepared are the chlorides $Cl(Bu_2Sn \cdot O)_xSnBu_2Cl$ ($x = 1$–6, 9, 12) which are formed when polymeric dibutyltin oxide is heated with the appropriate amount of dibutyltin dichloride in benzene (*77*). The lower members are crystalline, but the higher ones, which may be mixtures of different oligomers, are sometimes waxy solids.

$$x\, Bu_2SnO + Bu_2SnCl_2 \longrightarrow Cl(Bu_2Sn \cdot O)_xSnBu_2Cl \qquad (104)$$

All show a strong band in the infrared spectrum at 670–690 cm^{-1} which is ascribed to the asymmetric stretching of the Sn—O—Sn group.

Hexaethyldibenzoyloxytristannoxane, mp 220°C (and tetraethyldibenzoyloxydistannoxane, mp 165–168°C) has also been prepared by treating diethyltin oxide with diethyltin dibenzoate in hexane at room temperature (*259, 262*).

Diethyltin dichloride and dipropyltin dichloride react with sodium acetate in water at pH 7 at 50°C to give the mixtures of the oligomeric acetates, $AcO(R_2Sn \cdot O)_xSnR_2 \cdot OAc$, where x is 2–3; these condense further if they are heated with water (*339*). Similar mixtures, where $x = 2$–3 or 4–5 are formed by heating a mixture of a dialkyltin dialkoxide and dialkyltin diacetate at 180°C for 2 h (*337, 340*).

$$(x+1)\, Bu_2Sn(OBu)_2 + (x+1)\, Bu_2Sn(OAc)_2 \longrightarrow BuO(Bu_2SnO)_{2x+1}\, SnBu_2 \cdot OAc + (2x+1)BuOAc \qquad (105)$$

Oligomeric α,ω-diacetoxystannoxanes have also been prepared by hydrolyzing dibutyltin diacetate in acetone, and removing the acetic acid by methylating it with diazomethane (*338*).

$$Bu_2Sn(OAc)_2 + H_2O \xrightarrow{CH_2N_2} AcO(Bu_2SnO)_xSnBu_2OAc + MeOAc \qquad (106)$$
$$x = 1, 3, 7, 15$$

Examples of these oligomers are given in Table 11.

5. *Polymeric Dialkylstannoxanes*, $(R_2Sn \cdot O)_n$

The ultimate product of the hydrolysis of a dialkyltin dihalide is the polymeric dialkyltin oxide, $(R_2SnO)_n$. The reaction is often carried out by adding a solution of the dihalide in aqueous alcohol to an excess (3 mol.) of aqueous potassium hydroxide. The oxide separates immediately as a fine white powder, which is filtered off, thoroughly washed, and dried. The last traces of water can conveniently be removed by azeotropic distillation with benzene.

TABLE 11

OLIGOMERIC DIALKYLSTANNOXANES $X(R_2Sn\cdot O)_xSnR_2X$

x	mp, °C	Reference
$Cl(Bu_2Sn\cdot O)_xSnBu_2Cl$		
2	89–90	(*77*)
3	94–95	(*77*)
4	90–92	(*77*)
5	100–102	(*77*)
6	178–180	(*77*)
9	~140	(*77*)
12	~178	(*77*)
$AcO(Bu_2Sn\cdot O)_xSnBu_2OAc$		
3	148–149	(*337*)
	137–138	(*340*)
7	210 decomp.	(*340*)
15	330 decomp.	(*340*)
$(PhCO_2)(Et_2SnO)_xSnEt_2(O_2C\cdot Ph)$		
2	220	(*259, 262*)

Whereas the corresponding polydialkylsiloxanes are often soluble distillable liquids, the polydialkylstannoxanes are usually amorphous, insoluble solids which melt with decomposition in the range 200–400°C (*26*, *266*), the stability following the sequence p-R > s-R > t-R. Long-chain or branched-chain alkyl groups may confer solubility, and di-s-butyltin oxide has been found to be a 2-3-mer in chloroform, and dioctyltin oxide a 25-mer in xylene (*26*, *266*). No cyclic oligomers appear to have been isolated.

These properties suggest that the coordinative association which has been identified in the intermediate hydrolysis products $[XR_2Sn\cdot O\cdot SnR_2X]_2$, $[XR_2Sn\cdot O\cdot SnR_2(OH)]_2$, and $[(HO)R_2Sn\cdot O\cdot SnR_2(OH)]_2$, is preserved in the polydialkylstannoxanes, as in structure (**33**), conferring on these compounds the properties of cross-linked polymers. Other physical and chemical properties are in accord with this structure.

The Mössbauer spectra imply that the tin is 5-coordinate rather than 4-coordinate (*106*), with the degree of condensation decreasing with increasing size of the alkyl groups (*3*). In the infrared spectra, the Sn—O—Sn asymmetric stretching vibration has been identified at 560–580 cm^{-1}; whereas the corresponding frequency in bis(trialkyltin) oxides is about 775 cm^{-1}; this difference has been ascribed to the loss of p_π-d_π overlap when the p electrons are involved in coordination in the dialkylstannoxanes (*64*, *244*). The X-ray powder pattern confirms a fairly ordered structure (*266*):

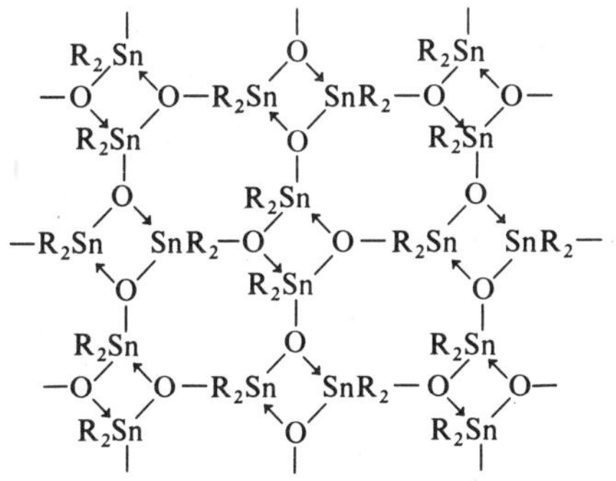

(33)

Harada reported that the reaction between diethyltin dibromide and dipropyltin oxide on the one hand, and between dipropyltin dibromide and diethyltin oxide on the other, gave a pair of isomers both with the composition $Et_2Pr_2Sn_2Br_2O$, with melting points 85–87° and 104–105°C respectively (*125*, *126*). This suggests that the distannoxane rings in the polymeric oxides retain their identity in the distannoxanes, so that the products have the structures:

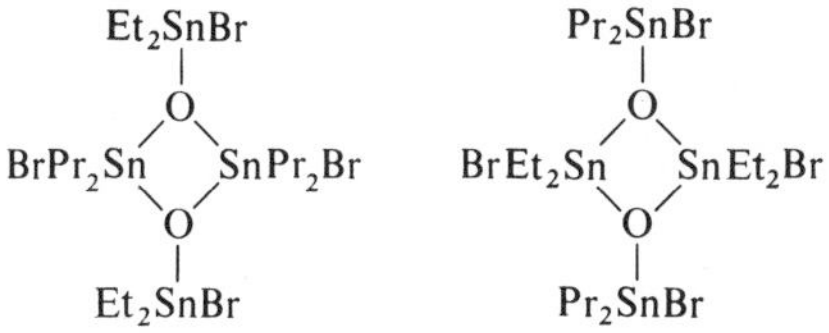

Et_2SnBr / $BrPr_2Sn$ $SnPr_2Br$ / Et_2SnBr

(34)

Pr_2SnBr / $BrEt_2Sn$ $SnEt_2Br$ / Pr_2SnBr

(35)

D. 1,1,3-TRIALKYLDISTANNOXANES $XR_2Sn{\cdot}O{\cdot}SnRX_2$ AND RELATED OLIGOMERS

If butyltin trichloride is added to dibutyltin oxide in benzene, the oxide rapidly dissolves to give 1,1,3-tributyl-1,3,3-trichlorodistannoxane as a crystalline solid, mp 34–35°C (*77*)

$$BuSnCl_3 + OSnBu_2 \longrightarrow Cl_2BuSn{\cdot}O{\cdot}SnBu_2Cl \qquad (107)$$

This distannoxane will react further with dibutyltin oxide, and a series of oligomers have been prepared by telomerization of the appropriate ratio of reactants (*77*). These compounds are listed in Table 13.

These compounds, like those derived from dibutyltin dichloride (Table 11) tend to become waxy as the molecular weight increases. They are usually soluble in benzene or toluene, and show the presence of a broad band in the infrared spectrum at 670–690 cm^{-1}, ascribed to the Sn—O—Sn asymmetric stretching vibration. All react rapidly with bipyridyl to give $Bu_2SnCl_2{\cdot}bipy$.

TABLE 12

POLYDIALKYLTIN OXIDES

R_2SnO	Reference
R	
Me	(*115, 137, 162, 266*)
$ClCH_2$	(*331*)
Et	(*115, 137, 341*)
$Me \cdot CHCl$	(*331*)
$CH_2{=}CH$	(*276, 287*)
Pr	(*126, 137*)
Pr^i	(*93*)
$CH_2{=}CH \cdot CH_2$	(*161*)
Bu	(*137, 266, 341*)
Bu^i	(*50*)
Pe^i	(*201*)
$Me_3Si \cdot CH_2$	(*284a*)
cyclo-C_6H_{11}	(*318*)
n-C_6H_{13}	(*318*)
$PhCH_2$	(*297*)
n-C_8H_{17}	(*318*)
$BuEtCH \cdot CH_2$	(*318*)
$Bu^t \cdot CH_2 \cdot CHMe \cdot CH_2 \cdot CH_2$	(*318*)
n-$C_{12}H_{25}$	(*298*)
Ph	(*137, 266*)
4-$F \cdot C_6H_4$	(*102*)
4-$Cl \cdot C_6H_4$	(*215, 256*)
4-$Br \cdot C_6H_4$	(*215, 256*)
4-$I \cdot C_6H_4$	(*256*)
4-$NO_2 \cdot C_6H_4$	(*215*)
4-$MeO \cdot C_6H_4$	(*305*)
4-$EtO \cdot C_6H_4$	(*307*)
2-$MeO \cdot CO \cdot C_6H_4$	(*215*)
2-$EtO \cdot CO \cdot C_6H_4$	(*257*)
3-$EtO \cdot CO \cdot C_6H_4$	(*257*)
4-$EtO \cdot CO \cdot C_6H_4$	(*95, 257*)
2-$Me \cdot C_6H_4$	(*215, 257*)
3-$Me \cdot C_6H_4$	(*151*)
4-$Me \cdot C_6H_4$	(*215, 256*)
4-$Pr^i \cdot C_6H_4$	(*66*)
4-$Ph \cdot C_6H_4$	(*305*)
1-$C_{10}H_7$	(*243*)
2-$C_{10}H_7$	(*215*)
MeEtSnO	(*46*)
EtPhSnO	(*177*)
Et(Ph_3Ge)SnO	(*62*)

TABLE 13

PRODUCTS OF THE REACTION

$$BuSnCl_3 + Bu_2SnO \longrightarrow BuSn\begin{cases}(OSnBu_2)_xCl\\(OSnBu_2)_yCl\\(OSnBu_2)_yCl\end{cases}$$

x	y	mp, °C
1	0	34–35
0	1	100–102
1	1	109–110
2	1	85–86
1	2	92–93
2	2	89–90
3	3	100–102
4	4	109–110

E. DIALKYLDISTANNOXANES AND RELATED COMPOUNDS

1. 1,1-*Dialkyldistannoxanes*, $XR_2Sn \cdot O \cdot SnX_3$

If a telomerization reaction is attempted between tin tetrachloride, or tetrabromide and a dialkyltin oxide, the dialkyltetrahalogenodistannoxane, which is presumably the initial product, can not be isolated, but decomposes further to dialkyltin dihalide:

$$SnX_4 + R_2SnO \longrightarrow [X_3Sn \cdot O \cdot SnR_2X] \longrightarrow X_2SnO + R_2SnX_2 \quad (108)$$

However, if the reaction is carried out with dialkyltin oxide which has not been completely dried, telomerization is accompanied by partial hydrolysis to give waxy solids which are apparently 1,1-dialkyl-1,3,3-trichloro-3-hydroxydistannoxanes, e.g., $ClBu_2Sn \cdot O \cdot Sn(OH)Cl_2$, mp 46–47°C, and $ClOct_2Sn \cdot O \cdot Sn(OH)Cl_2$, mp 42–44°C.

2. 1,3-*Dialkyldistannoxanes*, $X_2RSn \cdot O \cdot SnRX_2$, *Monoalkyltin Hydroxides*, $RSn(OH)X_2$, *and Alkanestannonic Acids*, $RSn(OH)O$

This family of compounds bears the same relation to the monoalkyltin trihalides, as do the tetraalkyldistannoxanes, which are discussed above, to the dialkyltin dihalides. Relatively little work however has been carried out in the present field.

If ethyl-, butyl-, or octyl-tin trichloride is exposed to moist air, the hydrated alkylhydroxytin dichlorides, $RSn(OH)Cl_2,H_2O$ are obtained as crystalline solids (mp: Et 94–96°, Bu 80–87°, and Oct 45–56°C) (*187*).

The trichlorides are hydrolyzed in water to give strongly acidic solutions, but the trichloride is recovered if the water is distilled off. If one or two equivalents of alkali are present, the alkyldihydroxytin chlorides, $RSn(OH)_2Cl$, are precipitated [mp: $EtSn(OH)_2Cl,H_2O > 255°$, $BuSn(OH)_2Cl$ 107–112°, $OctSn(OH)_2Cl$ 122–127°C]. These dihydroxytin chlorides condense when they are heated under reduced pressure to give glass-like polymers of the general composition —$[RSn(Cl)—O]_n$—; one specimen of the butyl compound showed $n = 6.6$ in methylene chloride (*187*).

If three equivalents of alkali are present, further hydrolysis occurs. Octyltin trichloride gives a precipitate of 1,3-dioctyl-3-chloro-1,1,3-trihydroxydistannoxane, $(HO)_2OctSn \cdot O \cdot Sn(OH)OctCl$, which undergoes complete hydrolysis to polymeric octanestannonic acid, $[OctSn(OH)O]_n$, only when it is boiled with aqueous alkali for several hours. Butyltin trichloride with three equivalents of alkali gives directly butanestannonic acid with no identifiable intermediate distannoxane. Ethyltin trichloride is similarly converted directly to ethanestannonic acid, but, whereas, butane- and octane-stannonic acids are insoluble in water, the ethane-stannonic acid remains in solution. It can be isolated as an infusible solid of the composition $HO[EtSn(OH)O]_nH$, by ion exchange on the trichloride followed by evaporation of the solution under reduced pressure. The freshly prepared material showed $n = 3.0$ (in acetone), but further condensation took place during several weeks to give a value of $n = 13.8$ (*187*).

A number of stannonic acids are reported in the literature (see Table 14). They are usually infusible solids which decompose without melting above 300°C. The higher homologues are usually soluble in organic solvents and insoluble in water, but soluble in both aqueous acid and alkali.

Methanestannonic acid reacts with carboxylic acids in excess to give the pentacarboxylatotristannoxanes (**36**) (*45*, *231*) (see Table 14). If a restricted amount of carboxylic acid is used, or if the above pentacarboxylates are hydrolyzed, a further family of carboxylates is formed, with the composition $[MeSn(O \cdot CO \cdot R)O]_n$. The acetate is insoluble in most organic solvents but the molecular weight was determined cryoscopically in phenol, giving a value of $n = 3$, and suggesting the probable structure (**37**) for the acetate, and (**38**) for the parent stannonic acid (*45*). The molecular weights of the higher carboxylates, however, could be determined cryoscopically in benzene, in which they showed a value of $n = 6$ (*231*):

$$(R \cdot CO_2)_2MeSn \cdot O \cdot SnMe(O_2C \cdot R)O \cdot SnMe(O_2C \cdot R)_2$$

(**36**)

```
        Me     OAc                 Me     OH
          \   /                      \   /
           Sn                         Sn
          /  \                       /  \
         O    O                     O    O
  Me     |    |     Me       Me     |    |     Me
    \    |    |    /           \    |    |    /
     Sn       Sn                Sn       Sn
    /  \     /  \              /  \     /  \
 OAc    O       OAc          HO    O       OH
         (37)                       (38)
```

TABLE 14

ALKANESTANNONIC ACIDS AND RELATED COMPOUNDS

RSn(OH)O	Reference
R	
Me	(*137*, *180*)
Cl_2CH	(*182*)
Et	(*92*, *187*)
$CH_2{=}CH$	(*299*)
$BrCH_2{\cdot}CH_2$	(*182*)
Pr^i	(*93*)
$CH_2{=}CH{\cdot}CH_2$	(*182*, *299*)
$CH_3{\cdot}CO{\cdot}CH_2$	(*182*)
Bu	(*137*, *187*)
$Ph{\cdot}CH_2$	(*182*)
C_8H_{17}	(*187*)
$C_{12}H_{25}$	(*299*)
PhCH:CPh	(*213*)
Ph	(*137*, *182*)
$4\text{-}Cl{\cdot}C_6H_4$	(*135*, *214*)
$4\text{-}Br{\cdot}C_6H_4$	(*135*, *214*)
$4\text{-}I{\cdot}C_6H_4$	(*135*, *214*)
$2\text{-}Me{\cdot}C_6H_4$	(*152*)
$3\text{-}Me{\cdot}C_6H_4$	(*151*)
$4\text{-}Me{\cdot}C_6H_4$	(*152*)
$1\text{-}C_{10}H_7$	(*182*, *243*)
$[CH_2{\cdot}Sn(OH)O]_2$	(*299*)
$CH_2[CH_2{\cdot}Sn(OH)O]_2$	(*130*)
$[CH_2{\cdot}CH_2{\cdot}Sn(OH)O]_2$	(*299*)

F. METALLOSTANNOXANES

Metallostannoxanes contain the structure M—O—Sn, where M is a metal or metalloid other than tin. The interest in these compounds stems partly from their structural relation to the polysiloxanes which are industrially important.

1. *Silastannoxanes*

The largest single group of metallostannoxanes are the silastannoxanes, containing the structural unit Si—O—Sn (*281*). The most common route to these compounds is to treat a tin halide with an alkali metal silanolate, which can be prepared from the silanol and alkali metal (*236*, *284*, *308*, *309*), or disiloxane and alkylmetallic compound (*236*, *285*):

$$Me_3SiOH + Na \longrightarrow Me_3Si{\cdot}ONa \quad (109)$$

$$2\,Me_3SiONa + Me_2SnCl_2 \longrightarrow Me_3Si{\cdot}O{\cdot}SnMe_2{\cdot}O{\cdot}SiMe_3 \quad (110)$$

$$(Me_3Si)_2O + MeLi \longrightarrow Me_4Si + Me_3Si{\cdot}OLi \quad (111)$$

$$Me_3Si{\cdot}OLi + Me_3SnBr \longrightarrow Me_3Si{\cdot}O{\cdot}SnMe_3 \quad (112)$$

The alternative method in which a silicon halide is treated with the alkali metal stannolate has not yet been widely exploited (*282*, *283*):

$$(Me_3Sn)_2O + MeLi \longrightarrow Me_4Sn + Me_3Sn{\cdot}OLi \quad (113)$$

$$Me_3Sn{\cdot}OLi + Me_2SiCl_2 \longrightarrow Me_3Sn{\cdot}O{\cdot}SiMe_2{\cdot}O{\cdot}SnMe_3 \quad (114)$$

The cohydrolysis of trimethylchlorosilane and trialkyltin chlorides or dialkyltin dichlorides probably follows a similar mechanism. Either the monostannoxane (*232*) or distannoxane (*234*) may be isolated, depending on the reaction conditions:

$$2\,Me_3SiCl + 2\,Et_2SnCl_2 \xrightarrow{H_2O} Me_3Si{\cdot}O{\cdot}SnEt_2{\cdot}O{\cdot}SnEt_2{\cdot}O{\cdot}SiMe_3 \quad (115)$$

The structures of the disiloxydistannoxanes have been discussed above.

An alternative versatile route to the heterostannoxanes involves the dehydration of a mixture of a silanol and a tin oxide or hydroxide (*78*, *159*). The water can often be removed conveniently by azeotropic distillation with a solvent such as benzene (*78*):

$$Ph_3Si{\cdot}OH + [BuSn(OH)O]_n \longrightarrow BuSn(O{\cdot}SiPh_3)_3 \quad (116)$$

$$Ph_3Si{\cdot}OH + [BuSn(OH)O]_n \longrightarrow BuSn(O{\cdot}SiPh_3)_3$$

$$Ph_2Si(OH)_2 + (Bu_2SnO)_n \longrightarrow \begin{array}{ccc} & O & \\ Ph_2Si & & SiPh_2 \\ | & & | \\ O & & O \\ & Bu_2Sn & \end{array} \quad (117)$$

A few silastannoxanes have also been prepared by heating together an ethoxysilane and a tin acetate, at, for example, 170°C for 20 h (*309*):

$$Ph_2Si(OEt)_2 + Ph_3Sn{\cdot}OAc \longrightarrow Ph_2Si(O{\cdot}SnPh_3)_2 + 2\,EtOAc \quad (118)$$

and also by treating an organotin oxide with a silicon hydride (*27*, *128*):

$$Bu_3SiH + (Bu_3Sn)_2O \longrightarrow Bu_3Si{\cdot}O{\cdot}SnBu_3 + Bu_3SnH \quad (119)$$

Reactions of these types using difunctional silicon and tin compounds, have been applied to the preparation of polymeric silastannoxanes (*21*). These reactions are considered in the Chapter on Organotin Polymers.

The known simple silastannoxanes are listed in Table 15.

TABLE 15

SILASTANNOXANES

	mp, °C	bp, °C/mm	Preparation[a]	References
$Me_3Sn{\cdot}O{\cdot}SiMe_3$	−59	141/720	A	(*282–285*)
$Me_3Sn{\cdot}O{\cdot}SiEt_3$		49/1	A	(*283*)
$Me_3Sn{\cdot}O{\cdot}SiPh_3$	82–84		D	(*86*)
$Et_3Sn{\cdot}O{\cdot}SiMe_3$		52/1	A	(*283*)
			E	(*234*)
$Et_3Sn{\cdot}O{\cdot}SiEt_3$		114–115/4	D	(*155*)
$Pr_3Sn{\cdot}O{\cdot}SiMe_3$		126/120	E	(*232*)
$Pr_3Sn{\cdot}O{\cdot}SiPh_3$		210–215/3	F	(*128*)
$Bu_3Sn{\cdot}O{\cdot}SiMe_3$		142/2	E	(*232*)
$Bu_3Sn{\cdot}O{\cdot}SiEt_3$		—	F	(*27*)
$Bu_3Sn{\cdot}O{\cdot}SiBu_3$		188–193/5	F	(*128*)
$Bu_3Sn{\cdot}O{\cdot}SiPh_3$		233–235/5	F	(*128*)
$Bu^i_3Sn{\cdot}O{\cdot}SiMeBu^tPh$		171–172.5/2.2	D	(*154*)
$Ph_3Sn{\cdot}O{\cdot}SiMe_3$		140/0.1	D	(*78*)
$Ph_3Sn{\cdot}O{\cdot}SiPh_3$	139–140		A	(*36, 309*)
			C	2 (*309*)
			D	(*78, 91*)
$(Me_3SnO)_2SiMe_2$		77/1	A	(*283*)
$(Bu_3SnO)_2SiPh_2$		oil	D	(*78*)
$(Bu^i_3SnO)_2SiEt_2$		202.5–203.5/3	D	(*154*)
$(Ph_3SiO)_2SiPh_2$	98–99		C	(*309*)
			D	(*78*)
$Me_2Sn(OSiPh_3)_2$	155–156		A	(*159, 309*)
			D	(*78*)
$Bu_2Sn(OSiMeBu^tPh)_2$		210–212/2.5	D	(*154*)
$Bu_2Sn(OSiPh_3)_2$	71		A	(*309*)
			C	(*309*)
			D	(*78*)
$Oct_2Sn(OSiMe_3)_2$		144/0.2	D	(*86*)
$Ph_2Sn(OSiPh_3)_2$	148.5–149.5		A	(*309*)
			D	(*78*)
$(PhCH_2)_2Sn(OSiPh_3)_2$	122–123		A	(*309*)
$Bu_2Sn{\cdot}O{\cdot}SiPh_2{\cdot}O{\cdot}SiPh_2{\cdot}O$ (cyclic)		190–200	D	(*78*)
$MeSn(OSiMe_3)_3$	34	49/1 (*283*)	A	(*284*)
		49/11 (*284*)	D	(*78*)
$BuSn(OSiPh_3)_3$	144–146			
$Sn(OSiMe_3)_2$	150–168		A	(*308*)
			E	(*234*)
$Sn(OSiMe_3)_4$	47–49	60/1 (*283*)	A	(*22, 283, 308*)
		83–85/1 (*22*)		
$Me_3Si{\cdot}O{\cdot}SnR_2{\cdot}O{\cdot}SnR_2{\cdot}OSiMe_3$			E	(*234*)
R = Me	167–168			
Et	126–130			
Pr	107–108			
Bu	108–109			

[a] Preparative methods; A. From metal silanolate and tin chloride; B. From metal stannolate and chlorosilane; C. From tin acetate and alkoxysilane; D. Azeotropic dehydration of silanol and tin oxide; E. Cohydrolysis of tin chloride and chlorosilane; F. From silicon hydride and tin oxide.

The silastannoxanes show an intense band in the infrared spectrum at 950–980 cm^{-1}, associated with the asymmetric stretching vibration of the Si—O—Sn group (*281*) (cf. Si—O—Si ~1050 cm^{-1}; Sn—O—Sn 735–785 cm^{-1}). In the proton magnetic resonance spectrum of $Me_3Si{\cdot}O{\cdot}SnMe_3$, the Me_3Si group is more shielded than in $Me_3Si{\cdot}O{\cdot}SiMe_3$, and the Me_3Sn group is less shielded than in $Me_3Sn{\cdot}O{\cdot}SnMe_3$, indicating increased polarity and reduced double bond character in the Sn—O bond, and reduced polarity and increased double bond character in the Si—O bond (*281*, *283*). Values for the chemical shifts (at 60 mcps) with respect to tetramethylsilane are as follows:

$Me_3Si{\cdot}O{\cdot}SnMe_3$: Me_3Si +2.5; Me_3Sn −20.5;
$Me_3Si{\cdot}O{\cdot}SiMe_3$: −3.5;
$Me_3Sn{\cdot}O{\cdot}SnMe_3$: −14.2 cps

The Sn—O bond shows a high reactivity towards nucleophilic reagents such as water, alcohols, and mercaptans (*283*, *285*), and the reactivity decreases in the sequence $Me_3Si{\cdot}O{\cdot}SnMe_3 \rightarrow (Me_3SiO)_4Sn$. The position of bond fission is indicated by the reaction of phenyl-lithium with hexamethylsilastannoxane (*285*):

$$Me_3Si{\cdot}O{\cdot}SnMe_3 + PhLi \longrightarrow PhSnMe_3 + Me_3SiOLi \qquad (120)$$

The compounds $Me_3Si{\cdot}O{\cdot}SnMe_3$ and $Me_3Si{\cdot}O{\cdot}SnEt_3$, are readily hydrolyzed in air (*232*, *285*), but larger alkyl groups confer stability. Thus $Me_3Si{\cdot}O{\cdot}SnPr_3$ and $Me_3Si{\cdot}O{\cdot}SnBu_3$ are fairly stable to air and are monomeric in benzene (*232*), and $Bu^i_3Sn{\cdot}O{\cdot}SiMeBu^iPh$ is stable to water but hydrolyzed by alkali (*155*).

Silastannoxanes with a functional substituent on the tin atom have been prepared by causing the appropriately substituted silane to react with a dialkyltin oxide in an inert solvent such as benzene (*75*). These reactions are parallel to those between, say, tin chlorides, and dialkyltin oxides, to give functionally substituted distannoxanes, which are discussed above. Examples of these silastannoxanes are shown in Table 16.

$$R_nSiX_{4-n} + R'_2SnO \longrightarrow X_{3-n}R_nSi{\cdot}O{\cdot}SnR'_2X \qquad (121)$$

2. *Other Metallostannoxanes*

Few attempts have yet been reported to extend the reactions described above to the preparation of metallostannoxanes involving elements other than silicon.

Lithium trimethylstannolate can be prepared by treating bis(trimethyltin) oxide with methyl-lithium in ether at room temperature:

$$Me_3Sn{\cdot}O{\cdot}SnMe_3 + MeLi \longrightarrow Me_4Sn + Me_3Sn{\cdot}OLi \qquad (122)$$

TABLE 16

FUNCTIONALLY SUBSTITUTED SILASTANNOXANES

	mp, °C
$ClBu_2Sn\cdot O\cdot SiMe_2Cl$	38.5–40.5
$(AcO)Bu_2Sn\cdot O\cdot SiMe_2(OAc)$	oil
$ClOct_2Sn\cdot O\cdot SiMePhCl$	81–84
$ClPh_2Sn\cdot O\cdot SiMePhCl$	oil
$ClMe_2Sn\cdot O\cdot SiMeCl_2$	180 (decomp)
$ClBu_2Sn\cdot O\cdot SiMePhCl$	38–39 (*76*)
$(PhCO_2)Bu_2Sn\cdot O\cdot SiMePh(O_2CPh)$	oil (*87*)

It is a colorless crystalline solid, very sensitive to moisture. It dissolves readily in ether, cyclohexane, and carbon tetrachloride, and exists as a hexamer in benzene (*282*).

Silastannoxanes and germastannoxanes can then be prepared from the lithium stannolate and appropriate metal halide (*282*). Hexamethylgermastannoxane is also accessible from lithium trimethylgermanolate and trimethyltin chloride (*277*):

$$Me_3Ge\cdot O\cdot GeMe_3 + MeLi \longrightarrow Me_4Ge + Me_3Ge\cdot OLi \quad (123)$$

$$Me_3Ge\cdot OLi + Me_3SnCl \longrightarrow Me_3Ge\cdot O\cdot SnMe_3 \quad (124)$$

and triphenylgermatriethylstannoxane is formed when a mixture of triphenylgermanol and triethyltin hydroxide is subjected to azeotropic dehydration (*78*).

Titanostannoxanes have been prepared by condensing tributyltin acetate or triphenyltin hydroxide with an alkoxytitanium compound (*54*):

$$Ph_3Sn\cdot OH + Pr^iO\cdot Ti(O\cdot CH_2\cdot CH_2)_3N \longrightarrow Ph_3Sn\cdot O\cdot Ti(O\cdot CH_2\cdot CH_2)_3N \quad (125)$$

$$Bu_3SnOAc + Ti(OBu)_4 \longrightarrow Bu_3Sn\cdot O\cdot Ti(OBu)_3 \xrightarrow{(HO\cdot CH_2\cdot CH_2)_3N} Bu_3Sn\cdot O\cdot Ti(O\cdot CH_2CH_2)_3N \quad (126)$$

Apart from these reactions a few tin-functional metallostannoxanes have been prepared by treating a dialkyltin oxide with a mercury, thallium, germanium, or lead halide (*75*). Examples of these compounds are given in Table 17.

G. ORGANOTIN PEROXIDES

The trialkyltin hydroperoxides, $R_3Sn\cdot O\cdot OH$, and bis(trialkyltin) peroxides, $R_3Sn\cdot O\cdot O\cdot SnR_3$, are structurally related to the trialkyltin hydroxides and

TABLE 17

METALLOSTANNOXANES

	mp, °C	bp, °C/mm	Reference
$Me_3Sn{\cdot}OLi$	200(decomp)		(*282*)
$Me_3Sn{\cdot}O{\cdot}GeMe_3$		51/12	(*277, 282*)
$Et_3Sn{\cdot}O{\cdot}GePh_3$		167–169/0.04	(*78*)
$Bu_3Sn{\cdot}O{\cdot}Ti(O{\cdot}C_2H_4)_3N$		74/0.05	(*54*)
$Ph_3Sn{\cdot}O{\cdot}Ti(OC_2H_4)_3N$		205–208(decomp)	(*54*)
$ClBu_2Sn{\cdot}O{\cdot}HgPh$	206–208		(*75*)
$BrBu_2Sn{\cdot}O{\cdot}Tl(C_6F_5)_2$	160(decomp)		(*75*)
$ClBu_2Sn{\cdot}O{\cdot}GeMe_2Cl$	~25		(*75*)
$ClBu_2Sn{\cdot}O{\cdot}PbBu_2Cl$	120–130(decomp)		(*75*)

oxides respectively (*300*). Only a few of these types of peroxide are known, and they are much less stable than the trialkyltin alkyl peroxides, $R_3Sn{\cdot}O{\cdot}OR$, which are discussed in Sec. I.B.

Both the tin peroxides and the tin hydroperoxides can be prepared by treating a trialkyltin oxide or hydroxide with the appropriate amount of hydrogen peroxide, with provision for separating the water which is liberated:

$$R_3Sn{\cdot}O{\cdot}SnR_3 \xrightarrow{HO\cdot OH} R_3Sn{\cdot}O{\cdot}O{\cdot}SnR_3 \xrightarrow{HO\cdot OH} 2\,R_3Sn{\cdot}O{\cdot}OH \quad (127)$$

By this method, trimethyltin hydroperoxide (*67*), triphenyltin hydroperoxide (*67*), and bis(triethyltin) peroxide (*1, 7*) have been isolated, but bis(triethyltin) oxide reacted with 1–3 moles of anhydrous hydrogen peroxide in ether to give the perhydrate $[Et_3Sn{\cdot}O{\cdot}OH]_2H_2O_2$ as a crystalline solid (*8*). Bis(tripropyltin) peroxide has also been prepared by treating the methoxide with hydrogen peroxide, and bis(triphenyltin) peroxide by treating the chloride with hydrogen peroxide and ammonia (*271–273*). Characteristics of these peroxides are given in Table 18.

TABLE 18

ORGANOTIN HYDROPEROXIDES AND PEROXIDES

	mp, °C	Reference
$Me_3Sn{\cdot}O{\cdot}OH$	97–98	(*67*)
$[Et_3Sn{\cdot}O{\cdot}OH]_2H_2O_2$	35–36	(*7*)
$Et_2Sn(OH)O{\cdot}OH$	150–180	(*8*)
$Ph_3Sn{\cdot}O{\cdot}OH$	75 (expl.)	(*67*)
$Et_3Sn{\cdot}O{\cdot}O{\cdot}SnEt_3$	liquid	(*1*)
$Pr_3Sn{\cdot}O{\cdot}O{\cdot}SnPr_3$	solid	(*271–273*)
$Ph_3Sn{\cdot}O{\cdot}O{\cdot}SnPh_3$	solid	(*271–273*)

Trimethyltin hydroperoxide is stable for a few hours at room temperature, but triphenyltin hydroxide decomposes much more rapidly. Bis(triethyltin) peroxide breaks down completely in 24 h at room temperature to give diethyltin hydroxyhydroperoxide, $Et_2Sn(OH)O \cdot OH$, as a crystalline solid. Bis(tripropyltin) peroxide, however, appears to be more stable, and can be distilled without decomposition (*273*).

Bis(triethyltin peroxide) reacts quickly with hexaethylditin at room temperature:

$$Et_3Sn \cdot O \cdot O \cdot SnEt_3 + Et_3Sn \cdot SnEt_3 \longrightarrow 2\, Et_3Sn \cdot O \cdot SnEt_3 \qquad (128)$$

The decomposition in nonane gives diethyltin oxide (93%) and triethyltin ethoxide (98%). The reaction is of the first order in peroxide, but the rate increases with increasing peroxide concentration, indicating induced decomposition. The presence of free radicals is confirmed by the fact that the peroxide initiates the polymerization of acrylonitrile and methyl methacrylate. The following mechanism for the decomposition was therefore proposed (*8*):

$$Et_3Sn \cdot O \cdot O \cdot SnEt_3 \longrightarrow 2\, Et_3SnO\cdot \qquad (129)$$

$$Et_3SnO\cdot + Et_3Sn \cdot O \cdot O \cdot SnEt_3 \longrightarrow Et_3Sn \cdot OEt + Et_2\dot{S}n \cdot O \cdot O \cdot SnEt \qquad (130)$$

$$Et_2\dot{S}n \cdot O \cdot O \cdot SnEt_3 \longrightarrow Et_2SnO + Et_3SnO\cdot \qquad (131)$$

Bis(triethyltin) peroxide has also been detected in the products of the autoxidation of hexaethylditin, although the main products are bis(triethyltin) oxide and diethyltin oxide. A complex radical chain mechanism was proposed, in which the initiating step was (*5, 6, 44*):

$$Et_3Sn \cdot SnEt_3 + O_2 \longrightarrow Et_3Sn \cdot O \cdot O\cdot + Et_3Sn\cdot \qquad (132)$$

III. Reactions of the Sn—O Bond

The two principal basic reactions by which the Sn—O bond may be broken are those of substitution and of addition:

$$\text{Substitution} \quad Sn{-}O + A{-}B \longrightarrow Sn{-}A + B{-}O \qquad (133)$$

$$\text{Addition} \quad Sn{-}O + A{=}B \longrightarrow Sn{-}A{-}B{-}O \qquad (134)$$

The substitution reactions are as old as organotin chemistry itself, but the addition reactions involving an Sn—O bond have been recognized only in the past few years.

Many important reactions of the organotin oxides and alkoxides are brought about by sequential combinations of two or more reactions of these two basic types. In this Section, we discuss, firstly, the simple substitution reactions, secondly, the simple addition and related elimination reactions, and, thirdly, combinations of these processes.

A. Substitution Reactions

Very little work has been reported on the mechanisms of these substitutions, but the majority are clearly heterolytic processes. The reagent A—B is usually a polar molecule, in which A nucleophilically attacks the tin, and B electrophilically attacks the oxygen. These reagents can conveniently be divided into the following four categories: (a) protic reagents (where B = H); (b) polar organic compounds; (c) organotin compounds (where B = Sn); (d) derivatives of other metals.

1. *Protic Reagents*

Many reactions of organotin oxides and alkoxides with protic reagents have already been discussed in the earlier sections of this chapter.

Two of the important routes to trialkyltin alkoxides involve treating the corresponding methoxide or oxide with the appropriate alcohol, with provision to remove the methanol or water which is formed, (Sec. I.A.1.*b* and Sec. I.A.1.*c*):

$$Et_3Sn{\cdot}OMe + HO{\cdot}CHMe{\cdot}C{\vdots}CH \longrightarrow Et_3Sn{\cdot}O{\cdot}CHMe{\cdot}C{\vdots}CH + MeOH \quad (135)$$

$$Bu_3Sn{\cdot}O{\cdot}SnBu_3 + 2\,HO{\cdot}CH_2{\cdot}Ph \longrightarrow 2\,Bu_3Sn{\cdot}O{\cdot}CH_2{\cdot}Ph + H_2O \quad (136)$$

Phenols and alkyl hydroperoxides react in a similar manner. With the same reagents, dialkyltin oxides often undergo only partial substitution to give the difunctional distannoxanes, e.g., $PhO{\cdot}Bu_2Sn{\cdot}O{\cdot}SnBu_2{\cdot}OPh$.

Conversely, water will hydrolyze most alkoxides and some oxides. Simple alkoxides are very readily hydrolyzed, but the cyclic dialkyltin dialkoxides derived from 1,2-diols are surprisingly stable. Stability also appears to increase with the increasing acidity of the parent alcohol, and the phenoxides are relatively stable.

As discussed in Sec. II.A.2, some bis(trialkyltin) oxides are readily hydrolyzed to hydroxides. Triphenyltin hydroxide in organic solvents appears to be in equilibrium with the oxide and water, and bis(trimethyltin) oxide fumes in air as it is hydrolyzed to the hydroxide.

Other protic compounds have a reactivity depending on their acidity. Amides undergo *N*-stannylation by both oxides (*83*) and alkoxides (*224*); again the reaction is reversible, and the water or methanol must be separated, for example by distillation, to displace the equilibrium:

$$(Bu_3Sn)_2O + PhNH{\cdot}CO{\cdot}NHPh \xrightarrow{-H_2O} Bu_3Sn{\cdot}NPh{\cdot}CO{\cdot}NPh{\cdot}SnBu_3 \quad (137)$$

$$Et_3Sn{\cdot}OMe + PhNH{\cdot}CHO \xrightarrow{-MeOH} Et_3Sn{\cdot}NPh{\cdot}CHO \quad (138)$$

Thioamides, on the other hand, appear to undergo *S*-stannylation:

$$Bu_3Sn{\cdot}O{\cdot}SnBu_3 + PhNH{\cdot}CS{\cdot}OMe \longrightarrow PhN{:}C(OMe)S{\cdot}SnBu_3 \quad (139)$$

The Sn—S bond is also formed when organotin oxides or alkoxides are treated with hydrogen sulfide or alkane thiols (*303*):

$$Bu_2Sn(OMe)Cl + H_2S \longrightarrow ClBu_2Sn{\cdot}S{\cdot}SnBu_2Cl + 2MeOH \quad (140)$$

$$3Bu_2SnO + 3H_2S \longrightarrow (Bu_2Sn{\cdot}S)_3 + 3H_2O \quad (141)$$

Acetylenes are stannylated by oxides, hydroxides, and alkoxides (*293*, *294*):

$$Et_3Sn{\cdot}O{\cdot}SnEt_3 + HC{\vdots}C{\cdot}CH_2{\cdot}OMe \longrightarrow Et_3Sn{\cdot}C{\vdots}C{\cdot}CH_2{\cdot}OMe \quad (142)$$

Stronger acids, such as the mineral acids and carboxylic acids, react very readily with the organotin oxides, hydroxides, and alkoxides. For example, oxalic acid has been used for protodestannylation of the organotin alkoxides resulting from Reformatsky-like reactions (*225*).

In their reactions with organotin oxides or alkoxides, the organotin hydrides may be regarded as a special class of protic compounds. Water or an alcohol is liberated, and a tin-tin bond is formed, as shown in the following examples (*221*, *222*, *280*):

$$2\,Bu_3SnH + Bu_3Sn{\cdot}O{\cdot}SnBu_3 \xrightarrow[\text{several days}]{100^\circ} H_2O + 2\,Bu_3Sn{\cdot}SnBu_3 \quad (143)$$

$$Bu_2SnH_2 + Bu_2Sn(OMe)_2 \xrightarrow{\text{Room temperature}} 2\,MeOH + (Bu_2Sn)_n \quad (144)$$

$$Bu_2SnH_2 + (Bu_3Sn)_2O \xrightarrow[8h]{100^\circ} H_2O + Bu_3Sn{\cdot}SnBu_2{\cdot}SnBu_3 \quad (145)$$

Radical generators or inhibitors have no effect on the rates, and the reactions are faster in more polar solvents. Electron-releasing substituents

$$\begin{matrix} \gtrdot Sn{-}OR \\ \gtrdot Sn{-}H \end{matrix} \longrightarrow \begin{matrix} \diagdown | \diagup \\ Sn \\ | \\ Sn \\ \diagup | \diagdown \end{matrix} + \begin{matrix} OR \\ | \\ H \end{matrix} \quad (146)$$

accelerate the reactions of phenoxides, and triaryltin hydrides are more reactive than trialkyltin hydrides. It was concluded that the reactions involve nucleophilic attack of the oxygen on the hydrogen atom (*61*).

2. *Polar Organic Compounds*

The Sn—O bond in organotin oxides, alkoxides, and related compounds will react readily with a variety of carboxyl derivatives.

The reaction of bis(trialkyltin) oxides with pyrocarbonates and with carbonates has already been discussed (Sec. I.A.1.*c*) as a means of preparing the alkoxides (*85*):

$$(Bu_3Sn)_2O + EtO{\cdot}CO{\cdot}O{\cdot}CO{\cdot}OEt \longrightarrow 2\,Bu_3Sn{\cdot}O{\cdot}CO{\cdot}OEt \quad (147)$$

$$(C_8H_{17})_3Sn{\cdot}O{\cdot}Sn(C_8H_{17})_3 + MeO{\cdot}CO{\cdot}OMe \longrightarrow (C_8H_{17})_3Sn{\cdot}OMe + (C_8H_{17})_3Sn{\cdot}O{\cdot}CO{\cdot}OMe \quad (148)$$

A similar reaction with carbonyl chloride provides a route to organotin chlorides (*90*).

Trialkyltin alkoxides react with other acid chlorides to give a mixture of an ester and a trialkyltin chloride (*119*):

$$Bu_3Sn\cdot OMe + Me\cdot CO\cdot Cl \longrightarrow Bu_3SnCl + Me\cdot CO\cdot OMe \qquad (149)$$

Compounds with Sn—O bonds also react readily with phthalic anhydride or 3-nitrophthalic anhydride to give tin phthalates which, when solid, can be used for characterizing the initial organotin compounds (*38*):

$$Bu_3Sn\cdot O\cdot OBu^t + C_6H_4(CO)_2O \longrightarrow C_6H_4(C(=O)\cdot O\cdot OBu^t)(C(=O)\text{—}O\cdot SnBu_3) \qquad (150)$$

Benzoyl peroxide reacts exothermically with triethyltin alkoxides; the triethyltin peroxybenzoate, which presumably is the initial product, is unstable, and undergoes a nucleophilic 1,2-rearrangement to give diethyltin ethoxide benzoate (*324*).

$$Ph\cdot CO\cdot O\cdot O\cdot CO\cdot Ph + Et_3Sn\cdot OMe \longrightarrow [Ph\cdot CO\cdot OMe + Et_2Sn(Et)\text{—}O\text{—}O\cdot CO\cdot Ph] \longrightarrow Et_2Sn(OEt)(O\cdot CO\cdot Ph) \qquad (151)$$

β-Propionolactone reacts with trimethyltin methoxide predominantly by acyl-oxygen fission to give methyl *β*-trimethylstannyloxypropionate, but as the polarity of the solvent increases, an increasing proportion of trimethyltin *β*-methoxypropionate is formed by alkyl-oxygen fission (*138*) (see Sec. I.A.2.*b*):

$$H_2C\text{—}CH_2\text{—}C(=O)\text{—}O\ (\text{ring}) + Me_3Sn\cdot OMe \longrightarrow \begin{cases} Me_3Sn\cdot O\cdot CH_2\cdot CH_2\cdot CO\cdot O\cdot Me & \text{Acyl-oxygen fission} \\ MeO\cdot CH_2\cdot CH_2\cdot CO\cdot O\cdot SnMe_3 & \text{Alkyl-oxygen fission} \end{cases} \qquad (152)$$

An interesting reaction has been reported to take place at room temperature, in which triethyltin ethoxide exothermically catalyzes the fragmentation of diethyl azodicarboxylate into nitrogen, carbon monoxide, and diethyl carbonate (*191*). The mechanism has not been fully established, but the authors suggested that it might involve a seven-center cyclic process:

$$\begin{array}{l} O{=}C(OEt)\text{–}N{=}N\text{–}C(OEt){=}O \\ \quad + \; EtO\text{–}SnEt_3 \end{array} \longrightarrow \begin{array}{c} O{=}C(OEt)_2 \\ N{\equiv}N \quad SnEt\text{–}OEt \\ C{=}O \end{array} \qquad (153)$$

Only the more reactive organic halides have been shown to react with organotin alkoxides. Dibutyltin dimethoxide reacts with allyl bromide or 1,2-dibromoethane to give dibutyltin bromide methoxide (*249*), and methyl iodide (*290*) and methoxymethyl chloride (*207*) react similarly with trialkyltin alkoxides.

$$Et_3Sn{\cdot}O{\cdot}CH_2{\cdot}CH_2{\equiv}CH + MeO{\cdot}CH_2Cl \longrightarrow Et_3SnCl + MeO{\cdot}CH_2{\cdot}O{\cdot}CH_2{\cdot}CH_2{\cdot}C{\equiv}CH \qquad (154)$$

3. *Organotin Compounds*

The simplest reaction of this type is the symmetrical exchange of the oxygen ligand between two identical organotin sites. This exchange can be seen in the proton magnetic resonance spectra of organotin alkoxides. For example, at low temperatures, tributyltin methoxide in dilute solution shows sidebands to the main singlet of the methoxy group, which are due to coupling between the ^{117}Sn and ^{119}Sn nucleii, with a spin of one half, and the methoxyl protons, where the value of $J_{^{119}Sn—H}$ is about 35 cps. If the solution is allowed to warm to about $-10°C$, these sidebands disappear because the rate of exchange of methoxy groups between different tributyltin sites approaches this frequency.

The value of the coupling constant is approximately the same in a variety of alkoxides, and the temperature.at which the coupling is broken thus gives a measure of the relative rate of exchange of the alkoxy groups. Typical results are shown in Table 19, from which it appears that the increasing size of the alkoxy group slows down the rate of exchange (*81*). Parallel results have been reported for dialkylaminotin compounds (*345*).

The exchange of dissimilar groups between organotin centers has been described in Sec. I.A.1*a*, where reactions of the following type are reported:

$$R_2Sn(OMe)_2 + R_2SnX_2 \rightleftharpoons 2\,R_2Sn(OMe)X \qquad (155)$$

These reactions take place rapidly and often exothermically at room temperature, and the products (X = e.g., halide, CNS, carboxylate) can be isolated, usually as crystalline solids. The reactions are reversible, and, if bipyridyl is added, the complex R_2SnX_2,bipy is precipitated, regenerating the dialkyltin dimethoxide (*72*).

A similar exchange of the ligands O and X with a polymeric dialkyltin

TABLE 19

TEMPERATURE DEPENDENCE OF TIN-ALKOXYL COUPLING IN TRIBUTYLTIN ALKOXIDES

Tributyltin alkoxide [a]	Temperature, °C [b]
$Bu_3Sn \cdot O \cdot CH_3$	−20
$Bu_3Sn \cdot O \cdot CH_2 \cdot Me$	+40
$Bu_3Sn \cdot O \cdot CH_2 \cdot Ph$	+90
$Bu_3Sn \cdot O \cdot CHMe_2$	>160
$Bu_3Sn \cdot O \cdot CH_2 \cdot CMe_3$	>160
$Bu_3Sn \cdot O \cdot CHMePh$	>160
$Bu_3Sn \cdot O \cdot CHPh_2$	>160

[a] 1 *M* in benzene or chlorobenzene.
[b] Temperature at which coupling breaks between ^{119}Sn and OCH.

oxide, leads to the formation of functionally substituted distannoxanes by the general equation:

$$\geq Sn-X + -(O-SnR_2)_n- \longrightarrow \geq Sn-O-SnR_2X \qquad (156)$$

Examples of such reactions are known where the product is a di-, tri-, tetra-, or penta-alkyldistannoxane (*74*, *75*), and, by similar reactions, oligomeric dialkylstannoxanes can be prepared (*77*). Typical examples are shown in the following equations; these reactions have already been discussed in Sec. II.B,C,D, as routes to organotin oxides:

$$Et_3SnCl + OSnBu_2 \longrightarrow Et_3Sn \cdot O \cdot SnBu_2Cl \qquad (157)$$

$$Bu_2Sn(CNS)_2 + OSnBu_2 \longrightarrow (SNC)Bu_2Sn \cdot O \cdot SnBu_2(CNS) \qquad (158)$$

$$BuSnCl_3 + OSnOct_2 \longrightarrow Cl_2BuSn \cdot O \cdot SnOct_2Cl \qquad (159)$$

$$SnCl_4 + O\,SnBu_2 \xrightarrow{H_2O} (HO)Cl_2Sn \cdot O \cdot SnBu_2Cl \qquad (160)$$

$$BuSnCl_3 + 12\,OSnBu_2 \longrightarrow BuSn[(OSnBu_2)_4Cl]_3 \qquad (161)$$

4. *Derivatives of Other Metals*

These reactions follow the general equation:

$$Sn—O + X—M \longrightarrow Sn—X + O—M$$

where the nucleophile X is carried by the element M which is a metal other than tin. Most of the reactions in this category have already been mentioned in this chapter (Sec. II.F).

An example where X is an alkyl group, is given by the reaction of methyllithium with bis(trimethyltin) oxide, to give lithium trimethylstannolate and tetramethyltin (*282*). The equivalent reaction of dimethyltin oxide, which would give the stannolate as the only product, does not appear to have been reported.

$$MeLi + Me_3Sn{\cdot}O{\cdot}SnMe_3 \longrightarrow Me_4Sn + LiO{\cdot}SnMe_3 \quad (162)$$

The reaction of the nucleophilic hydrogen of silicon hydrides with organotin oxides and alkoxides has been recommended as a route to the organotin hydrides (*27*, *128*):

$$Et_3SiH + Bu_3Sn{\cdot}O{\cdot}SnBu_3 \longrightarrow Et_3Si{\cdot}O{\cdot}SnBu_3 + Bu_3SnH \quad (163)$$

A similar exchange reaction takes place between chlorosilanes and tin alkoxides (*247*):

$$Et_3SiCl + Bu_3SnOMe \longrightarrow Et_3Si{\cdot}OMe + Bu_3SnCl \quad (164)$$

The same reaction with dialkyltin oxides gives functionally substituted silastannoxanes, and other metal halides will react similarly to give halogeno-metallostannoxanes. Two examples are given in the following equations; these reactions have been discussed fully in Sec. II.F:

$$PhMeSiCl_2 + Bu_2SnO \longrightarrow PhMeClSi{\cdot}O{\cdot}SnBu_2Cl \quad (165)$$

$$(C_6F_5)_2TlCl + Bu_2SnO \longrightarrow (C_6F_5)_2Tl{\cdot}O{\cdot}SnBu_2Cl \quad (166)$$

Some nonfunctional metallostannoxanes have also been prepared by azeotropic dehydration of a mixture of the appropriate metal hydroxide and organotin hydroxide:

$$Me_3SiOH + Ph_3SnOH \longrightarrow Me_3Si{\cdot}O{\cdot}SnPh_3 + H_2O \quad (167)$$

Again, these reactions have been discussed in Sec. II.F.

B. Addition Reactions

1. *Trialkyltin Compounds*

Trialkyltin oxides (**39**; $X = SnR_3$) and alkoxides (**39**; $X = R'$) add, often reversibly, to a variety of multiply-bonded acceptors (A=B) to give 1:1 adducts:

$$\underset{\mathbf{(39)}}{R_3Sn{-}OX} + A{=}B \underset{b}{\overset{a}{\rightleftharpoons}} R_3Sn{-}A{-}B{-}OX \quad (168)$$

Trialkyltin alkyl peroxides (**39**; $X = OR'$) behave similarly but they are weaker addenda.

The most widely studied compounds are tributyltin oxide and methoxide, which serve as model compounds for their respective classes since variations of alkyl groups on tin (R = Me, Et, Pr, Bu, and Ph) have little effect upon the reaction. A change of alkoxy group can have a more significant effect; for example phenoxides (**39**; X = Ph) are considerably less powerful as addenda than methoxides (**39**; X = Me).

Additions to aldehydes and ketones have been described [Eq. (49)] since they provide routes to new, substituted alkoxides. Tributyltin oxide and methoxide also add to carbon dioxide (*33, 39*), carbodiimides (*33, 39*), isocyanates (*33, 39*) sulfur dioxide (*33, 39, 79*) sulfodiimides (RN:S:NR) (*79*), sulfinylamines (RN:S:O) (*79*), carbon disulfide (*33, 39*), isothiocyanates (*33, 39*), imines (*80*), nitriles (*33, 39*), ketene (*71, 190*), and acetylenedicarboxylic esters (*191*) to give 1:1 adducts, examples of which are shown in Table 20.

It will be seen that, with the exception of the acetylene dicarboxylic ester, the multiple bond to which the oxide or methoxide adds is polar, the tributyltin group becoming attached to the negative end (A) of the dipole, $\overset{\delta-}{A}=\overset{\delta+}{B}$. The principal requirement in an acceptor molecule appears to be that it should be susceptible to nucleophilic attack, and structural alterations within a given class of acceptors which increase the electrophilicity of B, increase their reactivity. This may be achieved by inductive withdrawal of electrons from B, e.g., trichloroacetonitrile reacts, whereas simple alkylnitriles do not, or by mesomeric stabilization of positive charge on B, e.g., aryl carbodiimides are much more reactive than alkylcarbodiimides, presumably because of contribution from canonical forms of the type (**40**):

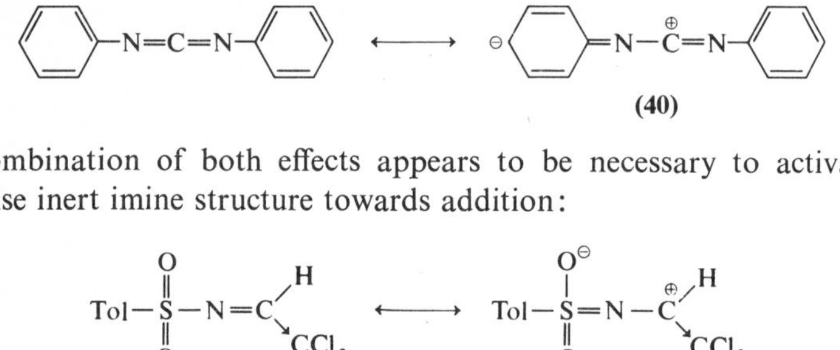

$$C_6H_5-N=C=N-C_6H_5 \longleftrightarrow {}^{\ominus}C_6H_5{=}N-\overset{\oplus}{C}=N-C_6H_5$$

(40)

A combination of both effects appears to be necessary to activate the otherwise inert imine structure towards addition:

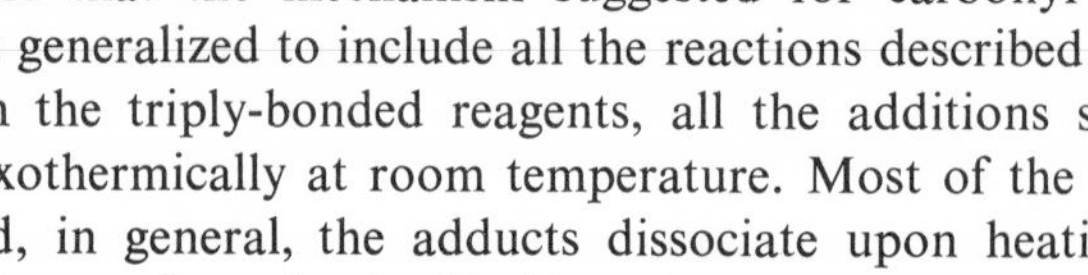

$$\text{Tol}-\overset{O}{\underset{O}{\overset{\|}{\underset{\|}{S}}}}-N=C\begin{matrix}H\\CCl_3\end{matrix} \longleftrightarrow \text{Tol}-\overset{O^{\ominus}}{\underset{O}{\overset{|}{\underset{\|}{S}}}}=N-\overset{\oplus}{C}\begin{matrix}H\\CCl_3\end{matrix}$$

The reactivity sequence in addenda, $R_3Sn{\cdot}OMe > R_3Sn{\cdot}OPh$, again suggests that the nucleophilicity of the oxygen bonded to tin is important, and it appears that the mechanism suggested for carbonyl additions in Sec. I may be generalized to include all the reactions described here.

Apart from the triply-bonded reagents, all the additions shown in the table occur exothermically at room temperature. Most of the reactions are reversible and, in general, the adducts dissociate upon heating, although alkoxide adducts of arylcarbodiimides, isocyanates, arylisothiocyanates and acetylene dicarboxylic esters can be purified by distillation. Many of the other products may be obtained pure by treating the organotin compound

TABLE 20

PRODUCTS OF THE REACTION $Bu_3Sn\cdot OX + A{:}B \longrightarrow Bu_3Sn\cdot A\cdot B\cdot OX$

A=B	$Bu_3Sn\cdot A\cdot B\cdot OX$	
	X = Me	X = $SnBu_3$
$O{=}CH\cdot CCl_3$	$Bu_3Sn\cdot OCH(CCl_3)OMe$	$Bu_3Sn\cdot OCH(CCl_3)OSnBu_3$
$O{=}C(CCl_3)_2$	$[Bu_3Sn\cdot OC(CCl_3)_2\cdot OMe]$	$[Bu_3Sn\cdot OC(CCl_3)_2\cdot OSnBu_3]$
$O{=}C{:}O$	$Bu_3Sn\cdot O\cdot CO\cdot OMe$	$Bu_3Sn\cdot O\cdot CO\cdot OSnBu_3$
$NpN{=}C{:}NNp$	$Bu_3Sn\cdot NNp\cdot C({:}NNp)OMe$	$Bu_3Sn\cdot NNp\cdot C({:}NNp)OSnBu_3$
$MeN{=}C{:}O$	$Bu_3Sn\cdot NMe\cdot CO\cdot OMe$	$Bu_3Sn\cdot NMe\cdot CO\cdot OSnBu_3$
$O{=}S{:}O$	$Bu_3Sn\cdot O\cdot SO\cdot OMe$	$Bu_3Sn\cdot O\cdot SO\cdot OSnBu_3$
$Tol\cdot N{=}S{:}N\cdot Tol$	no reaction	$[Bu_3Sn\cdot NTol\cdot SO\cdot N(Tol)SnBu_3]$
$ArN{=}S{:}O$	$[Bu_3Sn\cdot NAr\cdot SO\cdot OMe]$[a]	[b]
$S{=}C{:}S$	$[Bu_3Sn\cdot S\cdot CS\cdot OMe]$	[c]
$S{=}C{:}NPh$	$Bu_3Sn\cdot S\cdot C({:}NPh)OMe$	[d]
$Tol\cdot SO_2\cdot N{=}CH\cdot CCl_3$	$[Bu_3Sn\cdot N(SO_2\cdot Tol)CH(CCl_3)OMe]$	$[Bu_3Sn\cdot N(SO_2\cdot Tol)CH(CCl_3)OSnBu_3]$
$N{\equiv}C\cdot CCl_3$	$[Bu_3Sn\cdot N{:}C(CCl_3)OMe]$	$Bu_3Sn\cdot N{:}C(CCl_3)OSnBu_3$
$H_2C{=}C{:}O$	$[Bu_3Sn\cdot CH_2\cdot CO\cdot OMe]$	$[Bu_3Sn\cdot CH_2\cdot CO\cdot OSnBu_3]$
$MeO\cdot CO\cdot C{\equiv}C\cdot CO\cdot OMe$	$Et_3Sn\cdot C(CO\cdot OMe){:}C(CO\cdot OMe)OMe$[e]	not attempted

Np = 1-naphthyl; Tol = *p*-tolyl; Ar = *p*-$NO_2\cdot C_6H_4$.
Although there is ample evidence for their existence, compounds enclosed in parentheses have not been obtained analytically pure.
[a] The 1:1 adduct only exists in equilibrium with its precursors.
[b] A 1:1 mixture of $Bu_3Sn\cdot O\cdot SO\cdot OSnBu_3$ and $Bu_3Sn\cdot NAr\cdot SO\cdot NAr\cdot SnBu_3$ is obtained.
[c] Bis(tributyltin sulphide) and $Bu_3Sn\cdot O\cdot CO\cdot OSnBu_3$ are obtained.
[d] Oxygen-sulfur exchange occurs, giving bis(tributyltin) sulfide and the oxide-phenyl isocyanate adduct.
[e] From triethyltin methoxide.

with a slight excess of the acceptor then removing the excess at room temperature *in vacuo*.

Decomposition can proceed by a route other than simple retrogression [Eq. (168b)], particularly where oxide-adducts are concerned. Clearly the product formed by addition of bis(trialkyltin) oxide to A=B (**41**), could equally arise from addition of $(R_3Sn)_2A$ to O=B:

$$R_3Sn{\cdot}O{\cdot}SnR_3 + A{=}B \rightleftharpoons \underset{(\mathbf{41})}{R_3Sn{\cdot}A{\cdot}B{\cdot}O{\cdot}SnR_3} \rightleftharpoons R_3Sn{\cdot}A{\cdot}SnR_3 + O{=}B \qquad (169)$$

It follows that there are two competitive routes for the dissociation of (**41**) and the products which are isolated appear to be determined by thermodynamic control. Thus the formation of the strong Sn—S bond is very favorable and carbon disulfide or isothiocyanate adducts cannot be isolated even at room temperature:

$$(Bu_3Sn)_2O + CS_2 \longrightarrow [Bu_3Sn{\cdot}S{\cdot}CS{\cdot}OSnBu_3] \longrightarrow (Bu_3Sn)_2S + [COS] \qquad (170)$$

$$(Bu_3Sn)_2O + PhNCS \longrightarrow [Bu_3Sn{\cdot}S{\cdot}C(:NPh)OSnBu_3] \longrightarrow (Bu_3Sn)_2S + [PhNCO] \qquad (171)$$

Equation (169), in principle, provides the basis of a preparative route from bis(trialkyltin) oxides to other distannyl derivatives. In practice this has been exploited by preparing bis(tributyltin) derivatives of primary aromatic amines from the reaction of bis(tributyltin) oxide with sulfodiimide (*79*) [e.g. Eq. (172)]; these compounds are not, as yet, accessible by any other method.

$$(Bu_3Sn)_2O + PhN{:}S{:}NPh \longrightarrow Bu_3Sn{\cdot}NPh{\cdot}S(:NPh)OSnBu_3 \xrightarrow{\Delta} (Bu_3Sn)_2NPh + PhNSO \qquad (172)$$

An exchange of groups between the organotin compound and acceptor can also occur as a result of structural modification in the acceptor. One such modification is that which is often employed to activate the molecule towards addition, namely the introduction of a trihalogenomethyl group onto B. The product from this type of acceptor often decomposes by elimination of the trihalogenomethyltrialkyltin [Eq. (173)], and we have previously encountered this in Sec. I.A.2b for aldehydes and ketones [e.g., Eq. (50)

and Eq. (51)]:

$$R_3Sn{-}OX + A{=}B{\diagup}^{CHal_3} \rightleftharpoons R_3Sn{-}A{-}B\begin{matrix}\diagup CHal_3 \\ \diagdown OX\end{matrix} \tag{173}$$

$$\downarrow$$

$$(X = Me, SnR_3) \qquad R_3Sn{\cdot}CHal_3 + A{=}B{\diagdown}_{OX}$$

Equations (173) and (169) then represent alternative routes to simple retrogression [Eq. (168b)] by which certain types of adduct may decompose, and these reactions can be of preparative value in organotin chemistry.

Most of the acceptors are heterocumulenes, where two similar or dissimilar multiple bonds compete for the organotin compound, and with ketenes, isocyanates, sulfinylamines, and isothiocyanates, two isomeric adducts are formally possible:

$$R_3Sn{-}OX + A{=}B{=}C \begin{cases} \nearrow R_3Sn{-}A{-}\overset{\overset{C}{\|}}{B}{-}OX \\ \quad \text{or} \\ \searrow R_3Sn{-}C{-}\overset{\overset{A}{\|}}{B}{-}OX \end{cases} \tag{174}$$

It has been suggested (*33*) that, in some cases, association by intermolecular coordination to the trialkyltin group might provide an easy route for interconversion of the two structures:

$$\begin{matrix} \rightarrow R_3Sn{-}A{-}\underset{\underset{OX}{|}}{B}{=}C \rightarrow R_3Sn{-}A{-}\underset{\underset{OX}{|}}{B}{=}C \rightarrow R_3Sn{-}A{-}\underset{\underset{OX}{|}}{B}{=}C \rightarrow \\ \upharpoonleft\downharpoonright \\ {-}R_3Sn \leftarrow A{=}\underset{\underset{OX}{|}}{B}{-}C{-}R_3Sn \leftarrow A{=}\underset{\underset{OX}{|}}{B}{-}C{-}R_3Sn \leftarrow A{=}\underset{\underset{OX}{|}}{B}{-}C{-} \end{matrix} \tag{175}$$

If such a mechanism operates, it follows that when $X = SnR_3$ (oxide adducts) and A and $C \neq O$, a third isomeric structure is possible, namely $R_3Sn{\cdot}A{\cdot}B({:}O){\cdot}CSnR_3$.

These structural problems are considered in the following brief review of the reactions of each class of acceptor molecule.

a. Carbon Dioxide. Carbon dioxide is one of the weakest acceptors in the heterocumulene class. It does not react with tributyltin phenoxide (*39*) or tributyltin *t*-butyl peroxide (*31*), and it is readily displaced from adducts by more powerful acceptors such as isocyanates (see Sec. III.B.3). Both oxide and methoxide adducts dissociate at 60–100°C, a property which has been used in a new route to trialkyltin alkoxides [cf. Eq. (24)] (*85*). Bis(trialkyltin) carbonates are formed when readily hydrolyzable trialkyltin compounds are exposed to the atmosphere; they exhibit a broad carbonyl stretching band in the infrared at about 1515 cm^{-1}.

b. Carbodiimides. Diarylcarbodiimides are more powerful acceptors than the corresponding dialkyl compounds. For example, tributyltin methoxide reacted exothermically with di-1-naphthylcarbodiimide in solution to give *O*-methyl-*N*,*N'*-di-1-naphthyl-*N*-tributylstannylisourea [Eq. 176)], but it did not add to di-isopropylcarbodiimide even after 11 h at 80°C (*39*):

$$Bu_3Sn{\cdot}OMe + NpN{:}C{:}NNp \rightleftharpoons Bu_3Sn{\cdot}NNp{\cdot}C({:}NNp)OMe \quad (176)$$

However even the diarylcarbodiimides are not very strong acceptors. They do not react with tributyltin phenoxide at room temperature, and the methoxide adducts dissociate during distillation; a slow reaction takes place with tributyltin *t*-butyl peroxide but it appears to be complicated by subsequent decomposition. The *O*-alkyl-*N*-stannylisoureas are readily hydrolyzed to liberate the parent *O*-alkylisoureas.

Bis(trialkyltin) oxides are more reactive than trialkyltin alkoxides in adding to carbodiimides. Bis(tributyltin) oxide added to di-isopropyl- and dicyclohexyl-carbodiimides when the reagents were heated at 80–100° C for 20 h:

$$\underset{(R = Pr^i \text{ or cyclohexyl})}{(Bu_3Sn)_2O + RN{:}C{:}NR} \rightleftharpoons \underset{\mathbf{(42)}}{Bu_3Sn{\cdot}NR{\cdot}C({:}NR)OSnBu_3} \quad (177)$$

These adducts dissociated to some extent upon distillation at 150°C, but distannylureas derived from diarylcarbodiimides are thermally stable.

The same *N*,*N'*-dialkyldistannylureas have also been prepared by stannylation of the parent urea (*83*):

$$(Bu_3Sn)_2O + RNH{\cdot}CO{\cdot}NHR \xrightarrow{-H_2O} \underset{\mathbf{(43)}}{Bu_3Sn{\cdot}NR{\cdot}CO{\cdot}NR{\cdot}SnBu_3} \quad (178)$$

or by heating bis(trialkyltin) oxide with two moles of isocyanate (see below). The properties of these distannylureas appear to be compatible with either the *N*,*O*- or the *N*,*N'*-bonded structure [(**42**) and (**43**) respectively]. These compounds are indeed probably cationotropic and the apparent difference between the extreme forms (**42**) and (**43**) may be reduced by intermolecular association.

c. Isocyanates. Isocyanates are probably the most powerful acceptors which have been studied. Both alkyl and aryl isocyanates react exothermically with trialkyltin alkoxides and oxides to give 1 : 1 adducts at room temperature. The alkyl *N*-alkyl-*N*-trialkylstannylcarbamates from alkoxide addition may be distilled without decomposition, but phenoxide adducts, although formed exothermically, dissociate upon heating (*33*). The only alkoxide so far found to be inert to addition to isocyanates is tributyltin pentafluorophenoxide, which did not react with phenyl isocyanate after 5 h at 100°C (*90*). This inactivity can be ascribed to the weak nucleophilicity of the oxygen of the pentafluorophenoxy group.

The trialkyltin *N*-alkyl-*N*-trialkylstannylcarbamates from oxide-isocyanate addition are not distillable. Decarboxylation occurs if distillation is attempted and a 1:1 mixture of the oxide and bis(trialkyltin) urea is obtained (*34*):

$$\underset{(\mathbf{44})}{2\,Bu_3Sn\cdot NPh\cdot CO\cdot OSnBu_3} \xrightarrow{\Delta} CO_2 + (Bu_3Sn)_2O + Bu_3Sn\cdot NPh\cdot CO\cdot NPh\cdot SnBu_3 \qquad (179)$$

Compound (**44**) can also be regarded as an adduct between carbon dioxide and bis(tributylstannyl) aniline, so that the mechanism of the reaction described by Eq. (179) could initially involve a unimolecular decarboxylation [Eq. (180)], thereby bringing about nitrogen-oxygen exchange between acceptor and addendum (cf. Eq. (169); A = NPh, B = CO):

$$Bu_3Sn\cdot NPh\cdot CO\cdot OSnBu_3 \longrightarrow (Bu_3Sn)_2NPh + CO_2 \qquad (180)$$

However in the presence of isocyanates, stannylcarbamates can decarboxylate under conditions where a unimolecular mechanism is extremely unlikely. Thus tributyltin *N,N*-diethylcarbamate is stable to distillation at 120°C/0.1 mm, but, in the presence of ethyl isocyanate it loses carbon dioxide at 60–80°C to give *N,N,N'*-triethyl-*N'*-tributylstannylurea. It appears, then, that the isocyanate displaces the weaker acceptor, carbon dioxide, and a six-center process has been suggested as a possible mechanism (*40*):

$$\begin{matrix} & O-CO & \\ Bu_3Sn & & NEt_2 \\ & EtN{=}CO & \end{matrix} \longrightarrow \begin{matrix} Bu_3Sn & & NEt_2 + CO_2 \\ & EtN-CO & \end{matrix} \qquad (181)$$

A parallel mechanism might operate for the reaction where compound (**44**) decarboxylates at room temperature when treated with 1-naphthyl isocyanate to yield bistributylstannyl-*N*-phenyl-*N'*-1-naphthylurea. When this reaction was originally reported, an addition-elimination sequence was suggested as a possible mechanism, intramolecular coordination assisting unimolecular decarboxylation of the intermediate allophanate (*37*):

$$Bu_3Sn\cdot NPh\cdot CO\cdot OSnBu_3 + Np\cdot NCO \longrightarrow \left[\begin{matrix} & NPh & \\ Bu_3Sn\cdot NNp-C & & C{=}O \\ \| & & | \\ O & & O \\ & SnBu_3 & \end{matrix}\right] \qquad (182)$$

$$\xrightarrow{-CO_2} Bu_3Sn\cdot NNp\cdot C(:NPh)OSnBu_3$$

Distannylallophanates had not then been isolated, but subsequently, bis(triphenylstannyl)allophanates have been prepared and they do indeed decarboxylate upon heating to yield distannylureas (*84*).

It appears then that three feasible mechanisms may be proposed for the thermal decomposition of oxide-isocyanate adducts. All three require a partial dissociation of the adduct into isocyanate and oxide, analogous to that observed, without further complications, for phenoxide adducts. The liberated isocyanate then, either decarboxylates unchanged carbamate by an addition-elimination sequence [cf. Eq. (182)] or by a six-center process [cf. Eq. (181)], or it adds to the distannylamine formed by competitive unimolecular decarboxylation of the carbamate [cf. Eq. (180)].

By sequential reaction with appropriate combinations of isocyanates, bis(trialkyltin) derivatives of ureas and biurets may be obtained under very mild conditions from bis(trialkyltin) oxides (*37*):

$$(Bu_3Sn)_2O \xrightarrow{EtNCO} Bu_3Sn\cdot NEt\cdot CO\cdot OSnBu_3$$

$$\Big\downarrow PhNCO, \; -CO_2 \qquad (183)$$

$$Bu_3Sn\cdot NPh\cdot CO\cdot NEt\cdot CO\cdot NPh\cdot SnBu_3 \underset{PhNCO}{\overset{-PhNCO}{\rightleftharpoons}} Bu_3Sn\cdot NPh\cdot CO\cdot NEt\cdot SnBu_3$$

A detailed discussion of these Sn—N bonded compounds does not come within the scope of this chapter.

Isocyanate adducts are generally formulated as products of addition across the N=C group (carbamates), rather than across the C=O group (carbimidates), although evidence as to which is correct is by no means conclusive. The alkoxide adducts absorb strongly in the infrared at 1635–1690 cm^{-1} and the oxide adducts show three strong bands at about 1550, 1590 and 1610 cm^{-1}. These bands could reasonably be ascribed to either the C=O or the N=C group, since the stretching frequency in model compounds overlaps, and little is known about the effect that the tin atoms would have on these frequencies. The *N*-methyl group of methyl isocyanate adducts shows $^{119/117}$Sn—H coupling of about 30 cps. Although this is perhaps more consistent with the carbamate structure, it is again not conclusive.

In their reactions, the adducts behave as *N*-stannylcarbamates but chemical reactions may here be a particularly poor criterion of structure. There is evidence that trialkyltin carboxylates, under some conditions, can associate through the carbonyl group giving the system a threefold axis of symmetry about the trialkyltin group. If a similar association obtained with the stannylcarbamates it could provide an easy route for the interconversion of the alternative structures [Eq. (175); A = NR′, B = C, C = O].

Alkyl triphenyltin peroxides yield crystalline adducts with alkyl isocyanates (*30*):

$$Ph_3Sn{\cdot}O{\cdot}OBu^t + MeNCO \longrightarrow Ph_3Sn{\cdot}NMe{\cdot}CO{\cdot}O{\cdot}OBu^t \quad (184)$$

These adducts show a single broad band at 1690 cm^{-1} in the solid, but in solution show an additional band at 1740 cm^{-1}. This could be interpreted as some evidence in support of intermolecular association, the aggregates breaking down to some extent in solution.

The *N*-triphenylstannylperoxycarbamates slowly decompose at room temperature with the loss of peroxide. Adducts between alkyl tributyltin peroxide and alkyl isocyanates could not be isolated as they decompose readily by homolysis of the peroxide linkage to yield products derived from radical interaction with the tributyltin group.

Adducts could not be isolated by addition of either tributyltin or triphenyltin alkyl peroxides to phenyl isocyanate; in contrast to the alkyl isocyanate reactions, the decomposition involved the formation of carbon dioxide.

d. Sulfur Dioxide, Sulfodiimides, and Sulfinylamines (*79*). These lesser known heterocumulenes are related to carbon dioxide, carbodiimides and isocyanates by replacement of the central carbon atom by sulfur. Like their carbon analogues these acceptors, in general, react exothermically with trialkyltin alkoxides and oxides to give 1:1 adducts. Furthermore, an exchange of moieties, which is largely determined by relative acceptor strengths, frequently occurs during the addition or during attempted distillation. In contrast to their carbon analogues, however, the weakest acceptors here appear to be the sulfinylamines (RNSO).

Sulfur dioxide behaves as a rather weak acceptor. The sulfites which are formed ($R_3Sn{\cdot}O{\cdot}SO{\cdot}OR'$ and $R_3Sn{\cdot}O{\cdot}SO{\cdot}OSnR_3$) dissociate on heating, and the sulfur dioxide may be displaced by a more powerful acceptor such as an isocyanate:

$$SO_2 + Bu_3Sn{\cdot}OEt \rightleftharpoons Bu_3Sn{\cdot}O{\cdot}SO{\cdot}OEt \xrightarrow{PhNCO} Bu_3Sn{\cdot}NPh{\cdot}CO{\cdot}OEt + SO_2 \quad (185)$$

Bis(trialkyltin) oxides (but not trialkyltin alkoxides) react exothermically with diaryl sulfodiimides. The bis(*N*-trialkylstannyl-*N*-arylamino)sulfoxides ($R_3Sn{\cdot}NAr{\cdot}SO{\cdot}NAr{\cdot}SnR_3$) which are formed, behave in subsequent reactions as adducts between bis(trialkylstannyl) amines ($R_3Sn{\cdot}NAr{\cdot}SnBu_3$) and sulfinylamines, and these components can be separated by distillation [e.g., Eq. (172)]. Acceptor molecules such as isocyanates, sulfur dioxide, and chloral, displace the sulfinylamine and give an adduct between the distannylamine and the new acceptor [e.g., Eq. (186)]:

$$PhNSNPh + (Bu_3Sn)_2O \longrightarrow \underset{(\mathbf{45})}{Bu_3Sn{\cdot}NPh{\cdot}SO{\cdot}NPh{\cdot}SnBu_3} \xrightarrow{Np{\cdot}NCO} Bu_3Sn{\cdot}NPh{\cdot}CO{\cdot}NNp{\cdot}SnBu_3 + PhNSO \qquad (186)$$

The alternative structure for the adduct, $Bu_3Sn{\cdot}NPh{\cdot}S(:NPh)O{\cdot}SnBu_3$ (**45a**) cannot be ruled out, but it seems less likely. Firstly, the NMR spectrum of a similar product from di-*p*-tolylsulfodiimide shows only one pattern for a *para*-substituted aromatic ring, whereas this alternative structure (**45a**) might be expected to show the presence of two nonequivalent rings. Secondly, the possibility of an energetically favorable cationotropic rearrangement (**45a** → **45**) may help to account for the fact that no reaction could be observed between trialkyltin alkoxides and sulfodiimides. The situation, however, is similar to that which is discussed above for the adducts of the carbodiimides: the two possible isomeric products are probably readily interconvertible cationotropically, and the apparent difference between them may be reduced or removed by intermolecular association through the trialkyltin groups.

Sulfinylamines react with bis(trialkyltin) oxides to give sulfimates which appear to be in equilibrium with the corresponding diaminosulfoxide and sulfite:

$$2\,(Bu_3Sn)_2O + 2\,PhNSO \rightleftharpoons 2\,Bu_3Sn{\cdot}NPh{\cdot}SO{\cdot}OSnBu_3 \rightleftharpoons Bu_3Sn{\cdot}NPh{\cdot}SO{\cdot}NPh{\cdot}SnBu_3 + Bu_3Sn{\cdot}O{\cdot}SO{\cdot}O{\cdot}SnBu_3 \qquad (187)$$

However, more powerful acceptors still displace sulfinylamine from the adduct.

Sulfimates from trialkyltin alkoxides and aryl sulfinylamines are only formed in equilibrium with their precursors, but the acceptor power of the sulfinylamine can be greatly enhanced by replacing the aryl group by an arylsulfonyl group (*80*):

$$Tol{\cdot}SO_2{\cdot}NSO + Bu_3Sn{\cdot}OPh \longrightarrow Bu_3Sn{\cdot}N(SO_2{\cdot}Tol){\cdot}SO{\cdot}OPh \qquad (188)$$

e. Carbon Disulfide and Isothiocyanates. These acceptors exchange their sulfur for the oxygen of bis(trialkyltin) oxides by addition-elimination sequences. The product obtained at room temperature is a mixture of bis(tributyltin) sulfide and the adduct between oxide and the new heterocumulene [cf. Eq. (170) and (171)]:

$$3(Bu_3Sn)_2O + CS_2 \longrightarrow 2\,(Bu_3Sn)_2S + Bu_3Sn{\cdot}O{\cdot}CO{\cdot}OSnBu_3 \quad (189)$$

$$2\,(Bu_3Sn)_2O + PhNCS \longrightarrow (Bu_3Sn)_2S + Bu_3Sn{\cdot}NPh{\cdot}CO{\cdot}OSnBu_3 \quad (190)$$

Tributyltin methoxide and carbon disulfide appear to give *O*-methyl-*S*-tributyltin dithiocarbamate, but it has not been isolated analytically pure; the same product was obtained from the reaction of tributyltin chloride and sodium methyl xanthate (*90*):

$$Bu_3Sn{\cdot}OMe + CS_2 \rightleftharpoons Bu_3Sn{\cdot}S{\cdot}CS{\cdot}OMe \xleftarrow{-NaCl} Bu_3SnCl + Na^+ \;\; {}^-S{\cdot}CS{\cdot}OMe \quad (191)$$

Phenyl isothiocyanate and allyl isothiocyanate reacted exothermically with tributyltin methoxide to give the corresponding formimidates [Eq. (192)], but the adduct from allyl isothiocyanate dissociated during distillation, and tributyltin phenoxide did not react with phenyl isothiocyanate during 8 days at room temperature. The isothiocyanates are, then, less powerful acceptors than the isocyanates.

$$\underset{(R = \text{phenyl or allyl})}{Bu_3Sn{\cdot}OMe + RNCS} \rightleftharpoons \underset{\mathbf{(46)}}{Bu_3Sn{\cdot}S{\cdot}C(:NR)OMe} \quad (192)$$

The evidence is in favor of the formimidate (**46**) rather than the thiocarbamate ($Bu_3Sn{\cdot}NR{\cdot}CS{\cdot}OMe$) structure. The products are hydrolytically stable and, indeed, they can be prepared by simply mixing the appropriate thiocarbamate and bis(trialkyltin) oxide. This stability would be surprising for a compound in which nitrogen is bonded to 4-coordinate tin, but would be expected for a compound containing the Sn—S bond. The adducts show a strong infrared absorption at 1625 cm^{-1} which may be assigned to the stretching frequency of the C=N bond in structure (**46**). Noltes assumed a similar structure for the product of triethylstannylation of *O*-ethyl *N*-phenylthiocarbamate (*224*).

The evidence, however, must be interpreted with caution, as both hydrolytic stability and infrared spectra would be modified if inter- or intramolecular association were to occur.

f. Imines and Nitriles. Both imines and nitriles are very weak acceptors and the basic structures require considerable activation before adducts can be prepared from either tributyltin methoxide or oxide. Thus, simple nitriles such as propionitrile do not react, but trichloroacetonitrile slowly gives 1:1 adducts (*39*):

$$Bu_3Sn{\cdot}OMe + CCl_3{\cdot}C{:}N \longrightarrow Bu_3Sn{\cdot}N{:}C(CCl_3)OMe \quad (193)$$

Even the introduction of α-halogeno-substituents is insufficient activation for imines. Thus compounds of the type $RN{:}CH{\cdot}CCl_3$, where R = Et, Pr,

Ph, Tol, do not react with either oxide or methoxide and indeed they can be isolated from mixtures of bis(tributylstannyl)amines and chloral, presumably by an addition-elimination sequence (*80*):

$$(Bu_3Sn)_2NR + CCl_3\cdot CHO \underset{b}{\overset{a}{\rightleftharpoons}} [Bu_3Sn\cdot NR\cdot CH(CCl_3)OSnBu_3] \xrightarrow{c} (Bu_3Sn)_2O + RN{:}CH\cdot CCl_3 \quad (194)$$

Only by the further introduction of an activating arylsulfonyl group onto the nitrogen has an imine capable of adding to methoxide and oxide been obtained:

$$Bu_3Sn\cdot OMe + Tol\cdot SO_2\cdot N{:}CH\cdot CCl_3 \longrightarrow Bu_3Sn\cdot N(SO_2\cdot Tol)CH(CCl_3)OMe \quad (195)$$

but even this acceptor remains inert to tributyltin phenoxide. The oxide adduct decomposed on heating to give a complex mixture of products which included chloral, presumably formed via the route (194b), where $R = Tol\cdot SO_2$—.

Tributylstannyl-*N*-tributylstannyl-trichloroacetimidate (from oxide and trichloroacetonitrile) decomposed when it was heated, to give a mixture of tributyltin chloride and tributyltin isocyanate [Eq. (196)]. This appears to be an example of the ligand-transfer reaction generalized in Eq. (173), which is parallel to the corresponding reactions with chloral and hexachloroacetone described in Sec. I.A.2b.

$$(Bu_3Sn)_2O + CCl_3\cdot C{\vdots}N \longrightarrow Bu_3Sn\cdot N{:}C(CCl_3)OSnBu_3 \longrightarrow Bu_3Sn\cdot CCl_3 + Bu_3Sn\cdot N{:}C{:}O$$
$$Bu_3Sn\cdot CCl_3 \longrightarrow [{:}CCl_2] + Bu_3SnCl \quad (196)$$

g. Ketene and Acetylenedicarboxylic Esters. The reaction of trialkyltin alkoxides with ketene [Eq. (197)] was the first example to be recognized of an addition reaction of the Sn—O bond [Eq. (168)]:

$$R_3Sn\cdot OR' + CH_2{:}C{:}O \longrightarrow \underset{\mathbf{(47)}}{R_3Sn\cdot CH_2\cdot CO\cdot OR'} \quad (197)$$

An exothermic reaction occurs in ether solution, and 70–80% yield of α-stannylesters, where R = Et, Pr, and R′ = Me, Et and Pr, were isolated by distillation (*190*); the reaction was later extended to include tributyltin methoxide and oxide (*71*).

The evidence favors structure (**47**) over the alternative $CH_2{:}C(OR')OSnR_3$,

(**47a**). The products all show an intense absorption at 1706–1718 cm^{-1} which can be confidently assigned to the carbonyl stretching vibration. Saturated esters show a carbonyl band at 1735–1750 cm^{-1}, but the C=C vibration, required for structure (**47a**), is expected in the range 1600–1625 cm^{-1}. The infrared spectra will, of course, be modified by any intermolecular association of the type discussed for products with other acceptors [Eq. (175)], and indeed, the lowering of the carbonyl stretching vibration by about 30 cm^{-1} may arise partly in this way.

Trialkyltin alkoxides can also be added to acetylene dicarboxylic esters, although the reagents have to be heated to 100°C for 1 h; the products show a C=C stretching vibration at about 1612 cm^{-1} (*191*).

$$Et_3Sn{\cdot}OMe + MeO{\cdot}CO{\cdot}C{\vdots}C{\cdot}CO{\cdot}OMe \longrightarrow MeO{\cdot}CO{\cdot}C(SnEt_3){:}C(OMe)CO{\cdot}OMe \quad (198)$$

No reaction was observed with ordinary alkenes (e.g., styrene) under similar conditions.

2. *Dialkyltin Compounds*

All the foregoing discussion of addition reactions has been concerned with trialkyltin compounds. Dialkyltin dialkoxides undergo similar reactions [Eq. (199a) and Eq. (199b)], but they have been examined less extensively. Dibutyltin dimethoxide adds to aldehydes, carbon dioxide, carbodiimides, isocyanates, sulfur dioxide, carbon disulfide, isothiocyanates, and trichloroacetonitrile (*73*), and diethyltin and dipropyltin alkoxides have been added to ketene (*250*).

$$Bu_2Sn(OMe)_2 \xrightarrow[a]{A=B} Bu_2Sn(OMe)A{\cdot}B{\cdot}OMe \xrightarrow[b]{A=B} Bu_2Sn(A{\cdot}B{\cdot}OMe)_2$$

$$Bu_2Sn(OMe)_2 \xrightarrow[c]{Bu_2SnX_2} Bu_2Sn(X)OMe \xrightarrow[d]{A=B} Bu_2Sn(X)A{\cdot}B{\cdot}OMe \xleftarrow[e]{Bu_2SnX_2} Bu_2Sn(A{\cdot}B{\cdot}OMe)_2 \quad (199)$$

In most cases both 1:1 [$Bu_2Sn(OMe)A{\cdot}B{\cdot}OMe$] and 1:2 adducts [$Bu_2Sn(A{\cdot}B{\cdot}OMe)_2$] can be prepared by using the appropriate molar ratios of reactants. Examples are given in Table 21.

The 1:2 adducts react with compounds Bu_2SnX_2, where X=Cl, Br, I, CNS, or O·CO·R, to give the products of disproportionation, $Bu_2Sn(X)A{\cdot}B{\cdot}OMe$ [Eq. (199e)], which can also be obtained by adding the monomethoxides, $Bu_2Sn(X)OMe$ to the acceptors, A=B [Eq. (199d)]. The limited and qualitative observations on the reactivity of various methoxides, $R_2Sn(X)OMe$, are again consistent with a rate-determining nucleophilic attack of alkoxy group on the acceptor.

TABLE 21

ADDUCTS FORMED FROM $Bu_2Sn(OMe)_2$

A=B	$Bu_2Sn(OMe)A \cdot B \cdot OMe$	$Bu_2Sn(A \cdot B \cdot OMe)_2$
O=CHMe	$Bu_2Sn(OMe)O \cdot CHMe \cdot OMe$	$Bu_2Sn(O \cdot CHMe \cdot OMe)_2$
O=C:O		$Bu_2Sn(O \cdot CO \cdot OMe)_2$
NpN=C:NNp	$Bu_2Sn(OMe)NNp \cdot C(:NNp)OMe$	$Bu_2Sn[NNp \cdot C(:NNp)OMe]_2$
EtN=C:O	$Bu_2Sn(OMe)NEt \cdot CO \cdot OMe$	$Bu_2Sn(NEt \cdot CO \cdot OMe)_2$
O=S:O	$Bu_2Sn(OMe)O \cdot SO \cdot OMe$	
S=C:S	$Bu_2Sn(OMe)S \cdot CS \cdot OMe$	$Bu_2Sn(S \cdot CS \cdot OMe)_2$
S=C:NPh	$Bu_2Sn(OMe)S \cdot C(:NPh)OMe$	$Bu_2Sn[S \cdot C(:NPh)OMe]_2$
$N \equiv C \cdot CCl_3$	$Bu_2Sn(OMe)N:C(CCl_3)OMe$	
H_2C=C:O		$Et_2Sn(CH_2 \cdot CO \cdot OMe)_2$[a]

Np = 1-naphthyl.
[a] From diethyltin dimethoxide.

Dialkyltin oxides and alkyltin trialkoxides can be expected to undergo addition reactions, but the consequences of introducing these further structural complications into the addendum molecule have not yet been investigated.

C. Combinations of Substitution and/or Addition Reactions

The Sn—A bond of an adduct formed by addition of an Sn—O bonded compound to A=B [Eq. (200a)] is, in general, also susceptible to the simple processes of substitution [Eq. (200b)] or addition [Eq. (200c)]:

$$\text{Sn—O} + \text{A=B} \xrightarrow{a} \text{Sn—A—B—O} \begin{cases} \xrightarrow[b]{\text{A}'\text{—B}'} \text{Sn—A}' + \text{B}'\text{—A—B—O} \\ \xrightarrow[c]{\text{A}'\text{=B}'} \text{Sn—A}'\text{—B}'\text{—A—B—O} \end{cases} \qquad (200)$$

These subsequent reactions only fall strictly within the context of this chapter if A = O, but the combinations represent a variety of reactions of Sn—O bonded compounds which are useful in organic synthesis.

1. *Addition Followed by Substitution*

The majority of adducts (Sn—A—B—O) are susceptible to protolysis (B′ = H) thus providing a route to compounds of the type (H—A—B—O). An important example of this sequence involves an isocyanate as the acceptor molecule. A small amount of an organotin compound will catalyze the addition of an alcohol to an isocyanate [Eq. 203)] (*33*), a reaction which is industrially important in the preparation of polyurethanes. Since both the addition of a tin alkoxide to an isocyanate [Eq. (201)] and the subsequent alcoholysis of the Sn·N bond [Eq. (202)] are very fast, in combination these reactions represent a very likely mechanism for the catalysis.

$$Bu_3Sn{\cdot}OEt + PhN{:}C{:}O \longrightarrow Bu_3Sn{\cdot}NPh{\cdot}CO{\cdot}OEt \quad (201)$$

$$Bu_3Sn{\cdot}NPh{\cdot}CO{\cdot}OEt + EtOH \longrightarrow Bu_3Sn{\cdot}OEt + PhNH{\cdot}CO{\cdot}OEt \quad (202)$$

$$PhN{:}C{:}O + EtOH \xrightarrow{Sn(IV)} PhNH{\cdot}CO{\cdot}OEt \quad (203)$$

Tin-catalysis of additions of protic reagents to heterocumulenes is probably general; catalysis in the formation of peroxycarbamates [Eq. (204)] (*30*), ureas [Eq. (205)] (*39*), and isourea ethers [Eq. (206)] (*39*) has been demonstrated:

$$RN{:}C{:}O + R'O{\cdot}OH \xrightarrow{Sn(IV)} RNH{\cdot}CO{\cdot}O{\cdot}OR' \quad (204)$$

$$RN{:}C{:}NR + H_2O \xrightarrow{Sn(IV)} RNH{\cdot}CO{\cdot}NHR \quad (205)$$

$$RN{:}C{:}NR + R'OH \xrightarrow{Sn(IV)} RNH{\cdot}C({:}NR)OR' \quad (206)$$

In a closely related reaction, tributyltin methoxide catalyzes the methanolysis of hexachloroacetone to give chloroform and methyl trichloroacetate [Eq. (209)]. The initial step is again addition [Eq. (207)] but the substitution stage is modified and probably involves a six-center process [Eq. (208)] (*89*):

$$Bu_3Sn{\cdot}OMe + (CCl_3)_2CO \longrightarrow Bu_3Sn{\cdot}O{\cdot}C(CCl_3)_2OMe \quad (207)$$

$$\begin{array}{l} \quad O-C(CCl_3)OMe \\ Bu_3Sn \qquad CCl_3 \\ \quad MeO-H \end{array} \longrightarrow Bu_3Sn{\cdot}OMe + CCl_3{\cdot}CO{\cdot}OMe + H{\cdot}CCl_3 \quad (208)$$

$$MeOH + (CCl_3)_2CO \xrightarrow{Sn(IV)} CCl_3{\cdot}CO{\cdot}OMe + H{\cdot}CCl_3 \quad (209)$$

The combination of additions with substitution by aprotic reagents permits the introduction of other groups into an organic structure (*33*):

$$Bu_3Sn{\cdot}OMe + PhNCO \longrightarrow Bu_3Sn{\cdot}NPh{\cdot}CO{\cdot}OMe \xrightarrow{AcO{\cdot}Ac} Bu_3Sn{\cdot}OAc + AcNPh{\cdot}CO{\cdot}OMe \quad (210)$$

2. *Addition Followed by Addition*

If the Sn—A bond of the 1:1 adduct (Sn—A—B—O) will undergo further addition to A=B, a 1:2 adduct is formed and the Sn—A bond is regenerated. Repeated insertion of A=B into the Sn—A bond [Eq. (211)] then provides a mechanism for the polymerization of A=B which is equivalent to Ziegler's growth reaction, in which an aluminium-carbon bond regeneratively, and repeatedly, adds to a carbon-carbon double bond:

$$\text{Sn—O} \xrightarrow{\text{A=B}} \text{Sn—A—B—O} \xrightarrow{\text{A=B}} \text{Sn—A—B—A—B—O} \xrightarrow{n\,\text{A=B}} \text{Sn—(A—B)}_{n+2}\text{—O} \quad (211)$$

Chloral has been successfully polymerized in this way by treatment with tributyltin methoxide or oxide (*89*):

$$Bu_3Sn{\cdot}OMe + n\,CCl_3{\cdot}CHO \longrightarrow Bu_3Sn[O{\cdot}CH(CCl_3)]_n{\cdot}OMe \quad (212)$$

The oxide was the more efficient catalyst, perhaps because the polymer can grow at both ends whereas the polymer derived from the methoxide can grow from only one end. Formation of the intermediate 1:2 adduct is supported by spectroscopic evidence.

With isocyanates, the degree of polymerization is limited to the formation of cyclic trimers, the isocyanurates. Different isocyanates can be interbred to give mixed isocyanurates (*35*):

$$Bu_3Sn{\cdot}OMe + PhNCO \longrightarrow Bu_3Sn{\cdot}NPh{\cdot}CO{\cdot}OMe$$

$$\xrightarrow{\text{EtNCO}} [Bu_3Sn{\cdot}NEt{\cdot}CO{\cdot}NPh{\cdot}CO{\cdot}OMe]$$

$$\xrightarrow{\text{EtNCO}} [Bu_3Sn{\cdot}NEt{\cdot}CO{\cdot}NEt{\cdot}CO{\cdot}NPh{\cdot}CO{\cdot}OMe] \longrightarrow [Bu_3Sn{\cdot}OMe] + \text{(1,3-diethyl-5-phenyl isocyanurate: ring of Et—N, C=O, N—Et, C=O, N—Ph, C=O)} \quad (213)$$

Attempts to copolymerize aldehydes or ketones were not successful. An alternative reaction takes place whereby the second acceptor displaces the first, probably by a six-center type of process, and an equilibrium containing four species is set up (*89*):

$$Bu_3Sn(O{-}CHMe{\cdot}OMe) + O{=}CH{\cdot}CCl_3 \rightleftharpoons O{=}CHMe + Bu_3Sn{-}O{-}CH(OMe){\cdot}CCl_3 \quad (214)$$

Equations (213) and (214) represent examples of the alternative routes [which are generalized in Eq. (200c) and Eq. (215)], which the reaction of a 1:1 adduct with a second, dissimilar acceptor can take:

$$\text{Sn—A—B—O} + \text{A}'{=}\text{B}' \longrightarrow \text{Sn—A}'\text{—B}'\text{—A—B—O} \quad (200c)$$

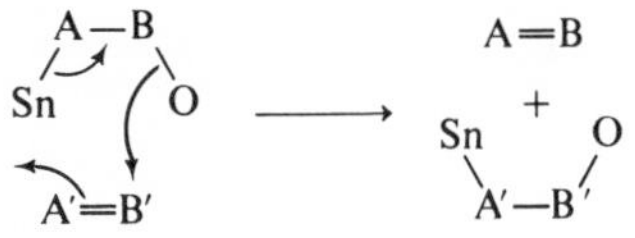

$$\text{Sn}(\text{O—B=A})\ldots\text{A}'\text{=B}' \longrightarrow \text{A=B} + \text{Sn—A}'\text{—B}'\text{—O} \tag{215}$$

The decarboxylation (A=B = CO_2) of oxide-isocyanate adducts by a second isocyanate (A′=B′ = RNCO), which we have previously encountered [Eq. (183)], is another example of reaction (215). If the decarboxylation is effected with an isothiocyanate (A′=B′ = RNCS), the new adduct is unstable and dissociates into bis(trialkyltin) sulfide and carbodiimide. This reaction provides a route to simple and mixed carbodiimides (*40*):

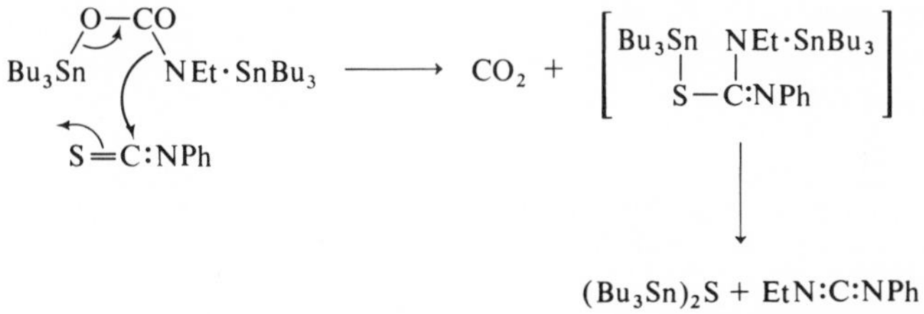

$$\text{Bu}_3\text{Sn—O—CO—NEt·SnBu}_3 + \text{S=C:NPh} \longrightarrow CO_2 + [\text{Bu}_3\text{Sn—S—C(:NPh)—NEt·SnBu}_3] \longrightarrow (\text{Bu}_3\text{Sn})_2\text{S} + \text{EtN:C:NPh} \tag{216}$$

The treatment of tributyltin methoxide with chloral and isocyanate yields the same organotin alkoxide (**48**), regardless of the order of addition [Eq. (217)]. It appears therefore that both processes of displacement [Eq. (217a)]

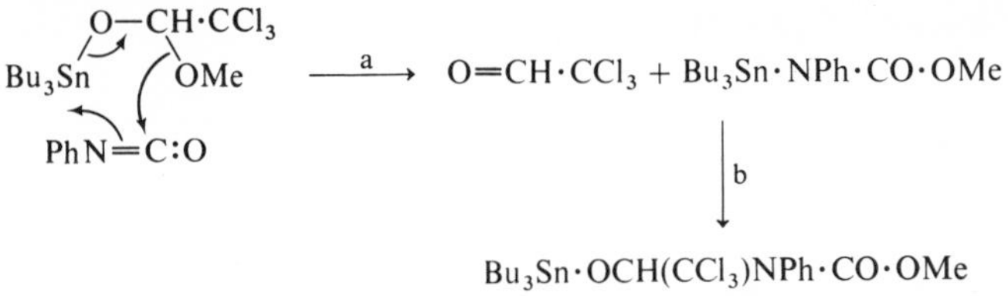

$$\text{Bu}_3\text{Sn—O—CH·CCl}_3\text{—OMe} + \text{PhN=C:O} \xrightarrow{a} \text{O=CH·CCl}_3 + \text{Bu}_3\text{Sn·NPh·CO·OMe} \xrightarrow{b} \underset{(\mathbf{48})}{\text{Bu}_3\text{Sn·OCH(CCl}_3\text{)NPh·CO·OMe}} \tag{217}$$

and addition [Eq. (217b)] must be involved in the reaction of the methoxide-chloral adduct with isocyanate (*36*, *84*):

REFERENCES

1. Yu. A. Aleksandrov, *Trudy Khim. Khim. Tekhnol.*, **3**, 642 (1960); through *CA*, **55**, 25570 (1961).

2. Yu. A. Aleksandrov, *Trudy Khim. Khim. Tekhnol.*, **5**, 485 (1962); through *CA*, **59**, 8777 (1963).

3. Yu. A. Aleksandrov, K. P. Mitrofanov, O. Yu. Okhlobystin, L. S. Polak, and V. A. Shpinel, *Proc. Acad. Sci. U.S.S.R., Sect. Phys. Chem.*, **153**, 974 (1963).

4. Yu. A. Aleksandrov and B. A. Radbil', *J. Gen. Chem. U.S.S.R.*, **36**, 562 (1966); and preceding publications.
5. Yu. A. Aleksandrov, B. A. Radbil', and V. A. Shushunov, *Trudy Khim. Khim. Tekhnol.*, **3**, 388 (1960); through *CA*, **55**, 27023 (1961).
6. Yu. A. Aleksandrov, B. A. Radbil', and V. A. Shushunov, *Trudy Khim. Khim. Tekhnol.*, **4**, 3 (1961); through *CA*, **56**, 592 (1962).
7. Yu. A. Aleksandrov and V. A. Shushunov, *Proc. Acad. Sci. U.S.S.R., Sect. Chem.*, **140**, 922 (1961).
8. Yu. A. Aleksandrov and V. A. Shushunov, *J. Gen. Chem. U.S.S.R.*, **35**, 113 (1965).
9. Yu. A. Aleksandrov and B. V. Sul'din, *Zhur. Obschei Khim.*, **37**, 2350 (1967).
10. Yu. A. Aleksandrov, B. V. Sul'din, and S. N. Kokurina, *Trudy Khim. Khim. Tekhnol.*, 228 (1965); through *CA*, **67**, 3128 (1967).
11. Yu. A. Aleksandrov, B. V. Sul'din, and S. N. Kokurina, *J. Gen. Chem. U.S.S.R.*, **36**, 2192 (1966).
12. D. L. Alleston and A. G. Davies, *Chem. Ind. London*, 949 (1961).
13. D. L. Alleston and A. G. Davies, *J. Chem. Soc.*, 2050 (1962).
14. D. L. Alleston and A. G. Davies, *J. Chem. Soc.*, 2465 (1962).
15. D. L. Alleston, A. G. Davies, and B. N. Figgis, *Proc. Chem. Soc.*, 457 (1961).
16. D. L. Alleston, A. G. Davies, and M. Hancock, *J. Chem. Soc.*, 5744 (1964).
17. D. L. Alleston, A. G. Davies, M. Hancock, and R. F. M. White, *J. Chem. Soc.*, 5469 (1963).
18. E. Amberger and M. Kula, *Chem. Ber.*, **96**, 2562 (1963).
19. E. Amberger, M. Kula, and J. Lorberth, *Angew. Chem., Int. Ed.*, **3**, 138 (1964).
20. H. H. Anderson, *J. Org. Chem.*, **19**, 1766 (1954); **22**, 147 (1957).
21. K. A. Andrianov, *Metalorganic Polymers*, Interscience, New York, 1965, pp. 311–314.
22. K. A. Andrianov and M. N. Emakova, *Vysokomolekul. Soedin*, **5**, 217 (1963).
23. B. Aronheim, *Annalen*, **194**, 145 (1878).
24. A. S. Atavin, R. I. Dubova, and N. P. Vasil'ev, *J. Gen. Chem. U.S.S.R.*, **36**, 1511 (1966).
25. G. Bähr and G. Zoche, *Chem. Ber.*, **88**, 1450 (1955).
26. G. A. Baum and W. J. Considine, *J. Polymer Sci. B, Polymer Letters*, **1**, 517 (1963).
27. B. Bellegarde, M. Pereyre, and J. Valade, *Bull. soc. chim. France*, 3082 (1967).
28. V. A. Belyaev and M. S. Nemtsov, *J. Gen. Chem. U.S.S.R.*, **31**, 3599 (1961).
29. D. Blake, G. E. Coates, and J. M. Tate, *J. Chem. Soc.*, 756 (1961).
30. A. J. Bloodworth, *J. Chem. Soc. C*, 2380 (1968).
31. A. J. Bloodworth, unpublished work.
32. A. J. Bloodworth and A. G. Davies, *Proc. Chem. Soc.*, 315 (1963).
33. A. J. Bloodworth and A. G. Davies, *J. Chem. Soc.*, 5238 (1965).
34. A. J. Bloodworth and A. G. Davies, *J. Chem. Soc.*, 6245 (1965).
35. A. J. Bloodworth and A. G. Davies, *J. Chem. Soc.*, 6858 (1965).
36. A. J. Bloodworth and A. G. Davies, unpublished work; A. J. Bloodworth, Ph.D. Thesis, London, 1965.
37. A. J. Bloodworth and A. G. Davies, *J. Chem. Soc. C*, 299 (1966).
38. A. J. Bloodworth, A. G. Davies, and I. F. Graham, *J. Organometal. Chem.*, **13**, 351 (1968).
39. A. J. Bloodworth, A. G. Davies, and S. C. Vasishtha, *J. Chem. Soc. C*, 1309 (1967).
40. A. J. Bloodworth, A. G. Davies, and S. C. Vasishtha, *J. Chem. Soc. C*, 2640 (1968).
41. J. Bornstein, B. R. Laliberte, T. M. Andrews, and J. C. Montermoso, *J. Org. Chem.*, **24**, 886 (1959).

42. T. G. Brilkina, M. K. Safonova, and V. A. Shushunov, *Trudy Khim. Khim. Tekhnol.*, 67, 74 (1965); through *CA*, **66**, 54769, 54770 (1967).
43. T. G. Brilkina, M. K. Safonova, and N. A. Sokolov, *J. Gen. Chem. U.S.S.R.*, **36**, 2196 (1966).
44. T. G. Brilkina and V. A. Shushunov, *Reactions of Organometallic Compounds with Oxygen and Peroxides*, Nauka, Moskow, 1966, pp. 175–183.
45. M. P. Brown, R. Okawara, and E. G. Rochow, *Spectrochim. Acta*, **16**, 595 (1960).
46. R. H. Bullard and F. R. Holden, *J. Am. Chem. Soc.*, **53**, 3150 (1931).
47. R. H. Bullard and R. A. Vingee, *J. Am. Chem. Soc.*, **51**, 892 (1929).
48. F. K. Butcher, W. Gerrard, E. F. Mooney, R. G. Rees, and H. A. Willis, *Spectrochim. Acta*, **20**, 51 (1964).
49. A. Cahours and E. Demarçay, *Compt. rend.*, **88**, 1112 (1879).
50. A. Cahours and E. Demarçay, *Compt. rend.*, **89**, 68 (1878).
51. R. Calas, J. Valade, and J. C. Pommier, *Compt. rend.*, **255**, 1450 (1962).
52. R. F. Chambers and P. C. Scherer, *J. Am. Chem. Soc.*, **48**, 1054 (1926).
53. A. C. Chapman and A. G. Davies, unpublished.
54. H. J. Cohen, *J. Organometal. Chem.*, **9**, 177 (1967).
55. W. J. Considine, *J. Organometal. Chem.*, **5**, 263 (1966).
56. W. J. Considine, G. A. Baum, and R. C. Jones, *J. Organometal. Chem.*, **3**, 308 (1965).
57. W. J. Considine and J. J. Ventura, *J. Org. Chem.*, **28**, 221 (1963).
58. W. J. Considine, J. J. Ventura, A. J. Gibbons, and A. Ross, *Can. J. Chem.*, **41**, 1239 (1963).
59. W. J. Considine, J. J. Ventura, B. G. Kushlefsky, and A. Ross, *J. Organometal. Chem.*, **1**, 299 (1964).
60. D. Cleverdon, J. J. P. Staudinger, D. Faulkner, and J. N. Milne, U.S. Pat. 2,623,892, through *CA*, **47**, 3036 (1953).
61. H. M. J. C. Creemers and J. G. Noltes, *Rec. trav. chim.*, **84**, 1589 (1965).
62. H. M. J. C. Creemers and J. G. Noltes, *J. Organometal. Chem.*, **7**, 237 (1967).
63. W. R. Cullen and G. E. Styan, *Inorgan. Chem.*, **4**, 1437 (1965).
64. R. A. Cummins, *Australian J. Chem.*, **18**, 98 (1965).
65. R. A. Cummins and J. V. Evans, *Spectrochim. Acta*, **21**, 1016 (1965).
66. O. Danek, *Coll. Czech. Chem. Commun.*, **26**, 2035 (1961).
67. R. L. Dannley and W. A. Aue, *J. Org. Chem.*, **30**, 3845 (1965).
68. J. D'ans and H. Gold, *Chem. Ber.*, **92**, 3076 (1959).
69. A. G. Davies, *Organic Peroxides*, Butterworths, London, 1961, p. 144.
70. A. G. Davies, *Chemistry in Britain*, **4**, 403 (1968).
71. A. G. Davies and I. F. Graham, unpublished work; I. F. Graham, Ph.D. Thesis, London, 1964.
72. A. G. Davies and P. G. Harrison, *J. Chem. Soc. C*, 298 (1967).
73. A. G. Davies and P. G. Harrison, *J. Chem. Soc. C*, 1313 (1967).
74. A. G. Davies and P. G. Harrison, *J. Organometal. Chem.*, **7**, P13 (1967).
75. A. G. Davies and P. G. Harrison, *J. Organometal. Chem.*, **10**, P31 (1967).
76. A. G. Davies and P. G. Harrison, unpublished work; P. G. Harrison, Ph.D. Thesis, London, 1968.
77. A. G. Davies, P. G. Harrison, and P. R. Palan, *J. Organometal. Chem.*, **10**, P33 (1967).
78. A. G. Davies, P. G. Harrison, and T. A. G. Silk, *Chem. Ind. London*, 949 (1968).
79. A. G. Davies and J. D. Kennedy, *J. Chem. Soc. C*, 2630 (1968).
80. A. G. Davies and J. D. Kennedy, unpublished work; J. D. Kennedy, Ph.D. Thesis, London, 1968.

81. A. G. Davies and D. C. Kleinschmidt, unpublished work; quoted by A. G. Davies, in ref. 70.
82. A. G. Davies and T. N. Mitchell, unpublished work; T. N. Mitchell, Ph.D. Thesis, London, 1967.
83. A. G. Davies, T. N. Mitchell, and W. R. Symes, *J. Chem. Soc. C*, 1311 (1966).
84. A. G. Davies and P. R. Palan, unpublished work; P. R. Palan, Ph.D. Thesis, London, 1967.
85. A. G. Davies, P. R. Palan, and S. C. Vasishtha, *Chem. Ind. London*, 229 (1967).
86. A. G. Davies and T. A. G. Silk, unpublished work.
87. A. G. Davies and P. J. Smith, unpublished work.
88. A. G. Davies and W. R. Symes, *J. Organometal. Chem.*, **5**, 394 (1966).
89. A. G. Davies and W. R. Symes, *J. Chem. Soc. C*, 1009 (1967).
90. A. G. Davies and S. C. Vasishtha, unpublished work; S. C. Vasishtha, Ph.D. Thesis, London, 1967.
91. A. D. Delman, A. A. Stein, B. B. Simms, and R. J. Katzenstein, *J. Polymer Sci. AI*, **4**, 2307 (1966).
92. J. G. F. Druce, *J. Chem. Soc.* **119**, 758 (1921).
93. J. G. F. Druce, *J. Chem. Soc.*, **121**, 1859 (1922).
94. H. J. Emeléus and J. J. Zuckerman, *J. Organometal. Chem.*, **1**, 328 (1964).
95. I. T. Eskin, A. N. Nesmeyanov, and K. A. Kocheshkov, *J. Gen. Chem. U.S.S.R.*, **8**, 35 (1938); through *CA*, **32**, 5386 (1938).
96. G. Faraglia, L. Roncucci, and R. Barbieri, *Ricerche Sci. Rend. Serz. A.*, **8**, 205 (1965); through *CA*, **63**, 12654 (1965).
97. Farbenfabriken Bayer A.G., British Pat. 945,068; through *CA*, **60**, 12051 (1964).
98. A. N. Fenster and E. I. Becker, *J. Organometal. Chem.*, **11**, 549 (1968).
99. I. Foldesi and G. Straner, *Acta Chim. Acad. Sci. Hung.*, **45**, 313 (1965); through *CA*, **64**, 3591 (1966).
100. M. Frankel, D. Gertner, D. Wagner, and A. Zilkha, *J. Organometal. Chem.*, **9**, 83 (1967).
101. E. Friebe and H. Kelker, *Z. anal. Chem.*, **192**, 267 (1963).
102. O. Fuchs and H. W. Post, *Rec. trav. chim.*, **78**, 566 (1959).
103. W. Gerrard, E. F. Mooney, and W. G. Peterson, unpublished work; W. G. Peterson, Ph.D. Thesis, London, 1966.
104. W. Gerrard, E. F. Mooney, and R. G. Rees, *J. Chem. Soc.*, 740 (1964).
105. A. J. Gibbons, A. K. Sawyer, and A. Ross, *J. Org. Chem.*, **26**, 2304 (1961).
106. V. I. Goldanskii, E. F. Makarov, R. A. Stukan, V. A. Trukhtanov, and V. V. Khrapov, *Proc. Acad. Sci. U.S.S.R., Sect. Phys. Chem.*, **151**, 598 (1963).
107. I. P. Goldshtein, N. N. Zemlyanski, O. P. Shamagina, E. N. Gur'yanova, E. M. Panov, N. A. Slovokhotova, and K. A. Kocheshkov, *Proc. Acad. Sci. U.S.S.R.*, **163**, 715 (1965).
108. R. D. Gorsich, *J. Am. Chem. Soc.*, **84**, 2486 (1962).
109. R. D. Gorsich, *J. Organometal. Chem.*, **5**, 105 (1966).
110. V. S. Griffiths and G. A. W. Derwish, *J. Mol. Spectry.*, **7**, 233 (1961); **9**, 83 (1962).
111. I. M. Gverdtsiteli and S. V. Adamiya, *Soobshcheniya Akad. Nauk Gruzin. S.S.R.* **47**, 55 (1967); through *CA*, **68**, 29817 (1968).
112. T. Harada, *Bull. Chem. Soc. Japan*, **2**, 105 (1927).
113. T. Harada, *Bull. Chem. Soc. Japan*, **4**, 226 (1929).
114. T. Harada, *Bull. Chem. Soc. Japan*, **6**, 240 (1931).
115. T. Harada, *Sci. Papers Inst. Phys. Chem. Res. Tokyo*, **35**, 290 (1939).
116. T. Harada, *Sci. Papers Inst. Phys. Chem. Res. Tokyo*, **36**, 497 (1939).

117. T. Harada, *Sci. Papers Inst. Phys. Chem. Res. Tokyo*, **36**, 501 (1939).
118. T. Harada, *Sci. Papers Inst. Phys. Chem. Res. Tokyo*, **36**, 504 (1939).
119. T. Harada, *Sci. Papers Inst. Phys. Chem. Res. Tokyo*, **38**, 115 (1940).
120. T. Harada, *Sci. Papers Inst. Phys. Chem. Res. Tokyo*, **38**, 146 (1940).
121. T. Harada, *Bull. Chem. Soc. Japan*, **15**, 455 (1940).
122. T. Harada, *Sci. Papers Inst. Phys. Chem. Res. Tokyo*, **39**, 419 (1942).
123. T. Harada, *Sci. Papers Inst. Phys. Chem. Res. Tokyo*, **42**, 57 (1947).
124. T. Harada, *Sci. Papers Inst. Phys. Chem. Res. Tokyo*, **42**, 62 (1947).
125. T. Harada, *Sci. Papers Inst. Phys. Chem. Res. Tokyo*, **42**, 64 (1947).
126. T. Harada, *Rep. Sci. Res. Inst. Japan*, **24**, 177 (1948).
127. T. Harada, *Sci. Papers Inst. Phys. Chem. Res. Tokyo*, **57**, 25 (1963).
128. K. Hayashi, J. Iyoda, and I. Shiihara, *J. Organometal. Chem.*, **10**, 81 (1967).
129. A. Heymons and H. Croon, German Pat. 1,006,664; through *CA*, **54**, 12472 (1960).
130. W. Hieber and R. Bren, *Angew. Chem.*, **68**, 679 (1956).
131. J. R. Holmes and H. D. Kaesz, *J. Am. Chem. Soc.*, **83**, 3903 (1961).
132. J. M. Holmes, R. D. Peacock, and J. C. Tatlow, *J. Chem. Soc. A*, 150 (1966).
133. J. R. Horder and M. F. Lappert, *Chem. Commun.*, 485 (1967).
134. F. Huber and R. Kaiser, *J. Organometal. Chem.*, **6**, 126 (1966).
135. R. Hulse and H. J. Twitchett, British Pat. 899,948; through *CA*, **58**, 9137 (1963).
136. R. Hulse and H. J. Twitchett, British Pat. 957,841; through *CA*, **61**, 9526 (1964).
137. R. K. Ingham, S. D. Rosenberg, and H. Gilman, *Chem. Rev.*, **60**, 459 (1960).
138. K. Itoh, S. Kobayashi, S. Sakai, and Y. Ishii, *J. Organometal. Chem.*, **10**, 451 (1967).
139. E. Jacobi, S. Lust, O. Zima, and A. Van Schoor, German Pat. 1,045,716; through *CA*, **54**, 25546 (1960).
140. M. J. Janssen and J. G. A. Luijten, *Rec. trav. chim.*, **82**, 1008 (1963).
141. O. H. Johnson, *J. Org. Chem.*, **25**, 2262 (1960).
142. O. H. Johnson and H. E. Fritz, *J. Org. Chem.*, **19**, 74 (1954).
143. O. H. Johnson, H. E. Fritz, D. O. Halvorson, and R. L. Evans, *J. Am. Chem. Soc.*, **77**, 5857 (1955).
144. W. J. Jones, D. P. Evans, T. Gulwell, and D. C. Griffiths, *J. Chem. Soc.*, 39 (1935).
145. K. Jones and M. F. Lappert, *J. Organometal. Chem.*, **3**, 295 (1965).
146. T. Katsumura, *Nippon Kagaku Zasshi*, **83**, 729 (1962); through *CA*, **59**, 5185 (1963).
147. K. Kawakami and R. Okawara, *J. Organometal. Chem.*, **6**, 249 (1966).
148. Y. Kawasaki, T. Tanaka, and R. Okawara, *Bull. Chem. Soc. Japan*, **37**, 903 (1964).
149. F. B. Kipping, *J. Chem. Soc.*, 2365 (1928).
150. W. Kitching, *J. Organometal. Chem.*, **6**, 586 (1966).
151. K. A. Kocheshkov and M. M. Nad, *J. Gen. Chem. U.S.S.R.*, **4**, 1434 (1934); through *CA*, **29**, 3660 (1935).
152. K. A. Kocheshkov and M. M. Nad, *J. Gen. Chem. U.S.S.R.*, **5**, 1158 (1935); through *CA*, **30**, 1036 (1936).
153. K. A. Kocheshkov and A. N. Nesmeyanov, *Ber.*, **64**, 628 (1931).
154. D. A. Kochkin and Y. N. Chirgadze, *J. Gen. Chem. U.S.S.R.*, **32**, 3932 (1962).
155. D. A. Kochkin, L. V. Luk'yanova, and E. B. Reznikova, *J. Gen. Chem. U.S.S.R.*, **33**, 1892 (1963).
156. D. A. Kochkin, V. N. Kotrelev, M. F. Shostakovskii, S. P. Kalinina, G. I. Kuznetsova, and V. V. Borisenko, *Vysokomoleculyanye Soedieniya*, **1**, 482 (1959); through *CA*, **54**, 5150 (1960).
157. D. A. Kochkin, V. I. Vashkov, and V. P. Dremova, *J. Gen. Chem. U.S.S.R.*, **34**, 321 (1964).

158. P. E. Koenig and R. D. Crain, WADC TR 58-44, Part II, June, 1959, quoted in Ref. (*90*).
159. P. E. Koenig and J. M. Hutchinson, WADC TR 58-44, Part I, May, 1958, quoted in Ref. (*90*).
160. K. König and W. P. Neumann, *Tetrahedron Letters*, 495 (1967).
161. M. M. Koton and T. M. Kiseleva, *Zhur. Obshchei Khim.*, **27**, 2553 (1957); through *CA*, **52**, 7136 (1958).
162. C. A. Kraus and R. H. Bullard, *J. Am. Chem. Soc.*, **51**, 3605 (1929).
163. C. A. Kraus and R. H. Bullard, *J. Am. Chem. Soc.*, **52**, 4056 (1930).
164. C. A. Kraus and T. Harada, *J. Am. Chem. Soc.*, **47**, 2416 (1925).
165. C. A. Kraus and A. M. Neal, *J. Am. Chem. Soc.*, **51**, 2403 (1929).
166. E. Krause and R. Pohland, *Ber.*, **57**, 532 (1924).
167. E. Krause and R. Pohland, *Ber.*, **57**, 540 (1924).
168. E. Krause and K. Weinberg, *Ber.*, **63**, 381 (1930).
169. H. Kreigsmann and H. Geissler, *Z. Anorg. Allgem. Chem.*, **323**, 170 (1963).
170. H. Kriegsmann and H. Hoffmann, *Z. Anorg. Allgem. Chem.*, **321**, 224 (1963).
171. H. Kriegsmann, H. Hoffmann, and H. Geissler, *Z. Anorg. Allgem. Chem.*, **341**, 24 (1965).
172. H. Kriegsmann, H. Hoffmann, and S. Pischtchau, *Z. Anorg. Allgem. Chem.*, **315**, 283 (1962).
173. H. G. Kuivila and O. F. Beumel, *J. Am. Chem. Soc.*, **80**, 3250 (1958).
174. H. G. Kuivila and O. F. Beumel, *J. Am. Chem. Soc.*, **80**, 3793 (1958); **83**, 1246 (1961).
175. B. G. Kushlefsky and A. Ross, *Anal. Chem.*, **34**, 1666 (1962).
176. B. G. Kushlefsky, I. Simmons, and A. Ross, *Inorg. Chem.*, **2**, 187 (1963).
177. A. Ladenburg, *Ber.*, **4**, 17 (1871).
178. A. Ladenburg, *Ann. Suppl.*, **8**, 55 (1872).
179. B. R. Laliberte, W. Davidson, and M. C. Henry, *J. Organometal. Chem.*, **5**, 526 (1966).
180. H. Lambourne, *J. Chem. Soc.*, **121**, 2533 (1922).
181. H. Lambourne, *J. Chem. Soc.*, **125**, 2013 (1924).
182. M. Lesbre and G. Glotz, *Compt. rend.*, **198**, 1426 (1934).
183. A. J. Leusink, *Hydrostannation*, Schotanus and Jens, Utrecht, 1966, p. 98.
184. A. J. Leusink and J. G. Noltes, *Tetrahedron Letters*, 2221 (1966).
185. J. Lorberth and M. R. Kula, *Chem. Ber.*, **97**, 3444 (1964).
186. G. Löwig, *Annalen*, **84**, 308 (1852).
187. J. G. A. Luijten, *Rec. trav. chim.*, **85**, 873 (1966).
188. J. G. A. Luijten and G. J. M. van der Kerk, *Investigations in the Field of Organotin Chemistry*, Tin Research Institute, Greenford, England, 1959.
189. J. G. A. Luijten and G. J. M. van der Kerk, *J. Appl. Chem.*, **11**, 35 (1961).
190. I. F. Lutsenko and S. V. Ponomarev, *J. Gen. Chem. U.S.S.R.*, **31**, 1894 (1961).
191. I. F. Lutsenko, S. V. Ponomarev and O. P. Petrii, *J. Gen. Chem. U.S.S.R.*, **32**, 886 (1962).
192. W. McFarlane, unpublished work.
193. M. M. McGrady and R. S. Tobias, *J. Am. Chem. Soc.*, **87**, 1909 (1965).
194. G. P. Mack, U.S. Pat. 2,745,820; through *CA*, **51**, 6219 (1957).
195. G. P. Mack and E. Parker, U.S. Pat. 2,700,675; through *CA*, **50**, 397 (1956).
196. G. P. Mack and E. Parker, U.S. Pat. 2,727,917; through *CA*, **50**, 10761 (1956).
197. Y. Maeda, C. R. Dillard, and R. Okawara, *Inorg. Nucl. Chem. Letters*, **2**, 197 (1966).
198. Y. Maeda, and R. Okawara, *J. Organometal. Chem.*, **10**, 247 (1967).
199. O. L. Mageli and J. B. Harrison, U.S. Pat. 3,152,156; through *CA*, **61**, 16903 (1964).

200. Z. M. Manulkin, F. A. Yukubova, A. B. Kuchkarev, and A. M. Rashkes, *Uzbeksk. Khim. Zhur.*, **6**, 52 (1962); through *CA*, **59**, 3942 (1963).
201. S. Matsuda and H. Matsuda, *Kogyo Kagaku Zasshi*, **63**, 114 (1960).
202. N. A. Matwiyoff and R. S. Drago, *J. Organometal. Chem.*, **3**, 393 (1965).
203. R. C. Mehrotra and V. D. Gupta, *J. Organometal. Chem.*, **4**, 145 (1965).
204. R. C. Mehrotra and V. D. Gupta, *J. Organometal. Chem.*, **4**, 237 (1965).
205. J. Mendelsohn, A. Marchand, and J. Valade, *J. Organometal. Chem.*, **6**, 25 (1966).
206. J. Mendelsohn, J. Pommier, and J. Valade, *Compt. rend.*, **263C**, 921 (1966).
207. R. G. Mirskov and V. M. Vlasov, *J. Gen. Chem. U.S.S.R.*, **36**, 176 (1966).
208. R. G. Mirskov and V. M. Vlasov, *J. Gen. Chem. U.S.S.R.*, **36**, 581 (1966).
209. A. S. Mufti and R. C. Poller, *J. Chem. Soc.*, 5055 (1965).
210. A. S. Mufti and R. C. Poller, *J. Organometal. Chem.*, **3**, 99 (1965).
211. W. H. Nelson and D. F. Martin, *J. Inorg. Nuclear Chem.*, **27**, 89 (1965).
212. W. H. Nelson and D. F. Martin, *J. Organometal. Chem.*, **4**, 67 (1965).
213. A. N. Nesmeyanov, A. E. Borisov, and L. G. Makarova, *Izv. Akad. Nauk. S.S.S.R., Otdel Khim. Nauk.*, 380 (1954); through *CA*, **49**, 5346 (1955).
214. A. N. Nesmeyanov and K. A. Kocheshkov, *J. Gen. Chem. U.S.S.R.*, **1**, 219 (1931); through *CA*, **26**, 2182 (1932).
215. A. N. Nesmeyanov and L. G. Makarova, *Dokl. Akad. Nauk S.S.S.R.*, **87**, 421 (1952); through *CA*, **48**, 623 (1954).
216. A. N. Nesmeyanov, I. F. Lutsenko, and S. V. Ponomarev, *Dokl. Akad. Nauk S.S.S.R.*, **124**, 1073 (1959); through *CA*, **53**, 14984 (1959).
217. W. P. Neumann and E. Heymann, *Angew. Chem. Int. Ed.*, **2**, 100 (1963).
218. W. P. Neumann and E. Heymann, *Annalen*, **683**, 11 (1965).
219. W. P. Neumann and H. Lind, *Angew. Chem., Int. Ed.*, **6**, 76 (1967).
220. W. P. Neumann and J. A. Pedain, German Pat. 1,214,237; through *CA*, **65**, 5490 (1966).
221. W. P. Neumann and B. Schneider, *Angew. Chem. Int. Ed.*, **3**, 751 (1964).
222. W. P. Neumann, B. Schneider, and R. Sommer, *Annalen*, **692**, 1 (1966).
223. W. P. Neumann, R. Sommer, and E. Müller, *Angew. Chem. Int. Ed.*, **5**, 514 (1966).
224. J. G. Noltes, *Rec. trav. chim.*, **84**, 799 (1965).
225. J. G. Noltes, H. M. J. C. Creemers, and G. J. M. van der Kerk, *J. Organometal. Chem.*, **11**, P21 (1968).
226. J. Nosek, *Coll. Czech. Chem. Commun.*, **29**, 597 (1964).
227. M. O'Hara, R. Okawara, and Y. Nakamura, *Bull. Chem. Soc. Japan*, **38**, 1379 (1965).
228. R. Okawara, personal communication, quoted in Ref. (*30*).
229. R. Okawara, *Proc. Chem. Soc.*, 383 (1961).
230. R. Okawara, *Adv. Organometal. Chem.*, **5**, 137 (1967).
231. R. Okawara and E. G. Rochow, *J. Am. Chem. Soc.*, **82**, 3285 (1960).
232. R. Okawara and K. Sugita, *J. Am. Chem. Soc.*, **83**, 4480 (1961).
233. R. Okawara and M. Wada, *J. Organometal. Chem.*, **1**, 81 (1963).
234. R. Okawara, D. G. White, K. Fujitani, and H. Sato, *J. Am. Chem. Soc.*, **83**, 1342 (1961).
235. R. Okawara and K. Yasuda, *J. Organometal. Chem.*, **1**, 356 (1964).
236. S. Papetti and H. W. Post, *J. Org. Chem.*, **22**, 527 (1957).
237. M. Pereyre, B. Bellegarde, J. Mendelsohn, and J. Valade, *J. Organometal. Chem.*, **11**, 97 (1968).
238. M. Pereyre, B. Bellegarde, and J. Valade, *Compt. rend.*, **265C**, 939 (1967).
239. M. Pereyre and J. Valade, *Compt. rend.*, **258**, 4785 (1964).
240. M. Pereyre and J. Valade, *Compt. rend.*, **260**, 581 (1965).

241. P. Pfeiffer and O. Brack, *Z. Anorg. Allgem. Chem.*, **87**, 229 (1914).
242. R. Piekós and A. Radecki, *Roczniki Chem.*, **39**, 767 (1965).
243. E. I. Pikina, T. V. Talalaeva, and K. A. Kocheshkov, *J. Gen. Chem. U.S.S.R.*, **8**, 1844 (1938); through *CA*, **33**, 5839 (1939).
244. R. C. Poller, *J. Inorg. Nucl. Chem.*, **24**, 593 (1962).
245. R. C. Poller, *J. Chem. Soc.*, 706 (1963).
246. R. C. Poller and J. A. Spillman, *J. Organometal. Chem.*, **7**, 259 (1967).
247. J. C. Pommier, M. Pereyre, and J. Valade, *Compt. rend.*, **260**, 6397 (1965).
248. J. C. Pommier and J. Valade, *Bull. soc. chim. France*, 975 (1965).
249. J. C. Pommier and J. Valade, *Compt. rend.*, **260**, 4549 (1965).
249a. J. C. Pommier and J. Valade, *J. Organometal. Chem.*, **12**, 433 (1968).
250. S. V. Ponomarev, Yu. I. Baukov, and I. F. Lutsenko, *J. Gen. Chem. U.S.S.R.*, **34**, 1951 (1964).
251. S. V. Ponomarev, Z. M. Lisina, and I. F. Lutsenko, *J. Gen. Chem. U.S.S.R.*, **36**, 1810 (1966).
252. S. V. Ponomarev and I. F. Lutsenko, *J. Gen. Chem. U.S.S.R.*, **34**, 3492 (1964).
253. S. V. Ponomarev, E. V. Machigin, and I. F. Lutsenko, *J. Gen. Chem. U.S.S.R.*, **36**, 566 (1966).
254. S. V. Ponomarev, B. G. Rogachev, and I. F. Lutsenko, *J. Gen. Chem. U.S.S.R.*, **36**, 1363 (1966).
255. M. V. Proskurnina, Z. S. Novikova, and I. F. Lutsenko, *Proc. Acad. Sci. U.S.S.R.*, **159**, 1240 (1964).
256. O. A. Ptitsyna, O. A. Reutov, and M. F. Turchinskii, *Dokl. Akad. Nauk. S.S.S.R.*, **114**, 110 (1957); through *CA*, **52**, 1090 (1958).
257. O. A. Ptitsyna, O. A. Reutov, and M. F. Turchinskii, *Nauk Dokl. Vysshei Shkoly. Khim. Khim. Technol.*, 138 (1959); through *CA*, **53**, 17030 (1959).
258. E. A. Puchinyan and M. Z. Manulkin, *Dokl. Akad. Nauk. Uz.S.S.R.*, **19**, 47 (1962); through *CA*, **57**, 13788 (1962).
259. B. A. Radbil, V. N. Glushakova, and Yu. A. Aleksandrov, *J. Gen. Chem. U.S.S.R.*, **37**, 195 (1967).
260. K. Ramaiah and D. F. Martin, *Chem. Commun.*, 130 (1965).
261. H. E. Ramsden and C. K. Banks, U.S. Pat. 2,789,994; through *CA*, **51**, 14786 (1957).
262. G. A. Razuvaev, O. A. Schchepetkova, and N. S. Vyazankin, *J. Gen. Chem. U.S.S.R.*, **19**, 2121 (1949).
263. G. A. Razuvaev, N. S. Vyazankin and O. S. D'Yachkovskaya, *J. Gen. Chem. U.S.S.R.*, **32**, 2129 (1962).
264. G. A. Razuvaev, N. S. Vyazankin and O. A. Shchepetkova, *Zhur. Obshei. Khim.*, **30**, 2498 (1960); through *CA*, **55**, 14290 (1961).
265. R. G. Rees and A. F. Webb, *J. Organometal. Chem.*, **12**, 239 (1968).
266. W. T. Reichle, *J. Polymer Sci.*, **49**, 521 (1961).
267. W. T. Reichle, *Inorg. Chem.*, **5**, 87 (1966).
268. G. H. Reifenberg and W. J. Considine, *J. Organometal. Chem.*, **10**, 279 (1967).
269. A. Rieche and T. Bertz, *Angew. Chem.*, **70**, 507 (1958).
270. A. Rieche and T. Bertz, German Pat. 1,081,891; through *CA*, **55**, 13315 (1961).
271. A. Rieche and J. Dahlmann, *Angew. Chem.*, **71**, 194 (1959).
272. A. Rieche and J. Dahlmann, *Monatsber. Deut. Akad. Wiss. Berlin*, **1**, 491 (1959).
273. A. Rieche and J. Dahlmann, *Annalen*, **675**, 19 (1964).
274. E. G. Rochow, D. Seyferth, and A. C. Smith, *J. Am. Chem. Soc.*, **75**, 3099 (1953).
275. L. Roncucci, G. Faraglia, and R. Barbieri, *J. Organometal. Chem.*, **1**, 427 (1964).
276. S. D. Rosenberg and A. J. Gibbons, *J. Am. Chem. Soc.*, **79**, 2138 (1957).

277. I. Ruidisch and M. Schmidt, *Angew. Chem. Int. Ed.*, **2**, 328 (1963).
278. G. S. Sasin, *J. Org. Chem.*, **18**, 1142 (1953).
279. R. Sasin and G. S. Sasin, *J. Org. Chem.*, **20**, 770 (1955).
280. A. K. Sawyer, *J. Am. Chem. Soc.*, **87**, 537 (1965).
281. H. Schmidbaur, *Angew. Chem. Int. Ed.*, **4**, 201 (1965).
282. H. Schmidbaur and H. Husseh, *Angew. Chem. Int. Ed.*, **2**, 328 (1963).
283. H. Schmidbaur and H. Husseh, *J. Organometal. Chem.*, **1**, 244 (1964).
284. H. Schmidbaur and M. Schmidt, *J. Am. Chem. Soc.*, **83**, 2963 (1961).
284a. D. Seyferth, *J. Am. Chem. Soc.*, **79**, 5881 (1957).
285. D. Seyferth and D. L. Alleston, *Inorg. Chem.*, **2**, 418 (1963).
286. D. Seyferth, G. Raab, and K. A. Brändle, *J. Org. Chem.*, **26**, 2934 (1961).
287. D. Seyferth and F. G. A. Stone, *J. Am. Chem. Soc.*, **27**, 907 (1955).
288. M. F. Shostakovskii, N. V. Komarev, I. S. Guseva, V. K. Misyunas, A. M. Sklyanova, and T. D. Burnashova, *Proc. Acad. Sci. U.S.S.R.*, **163**, 678 (1965).
289. M. F. Shostakovskii, V. N. Kotrelev, D. A. Kochkin, C. I. Kuznetsova, S. P. Kalinina, and V. V. Borisenko, *Zhur. prikl. Khim.*, **31**, 1434 (1958); through *CA*, **53**, 3040 (1959).
290. M. F. Shostakovskii, R. G. Mirskov, V. M. Vlasov, and Sh. I. Tarpishchev, *Zhur. Obschei Khim.*, **37**, 1738 (1967); through *CA*, **68**, 13903 (1968).
291. M. F. Shostakovskii, V. M. Vlasov, and R. G. Mirskov, *J. Gen. Chem. U.S.S.R.*, **33**, 320 (1963).
292. M. F. Shostakovskii, V. M. Vlasov, and R. G. Mirskov, *J. Gen. Chem. U.S.S.R.*, **34**, 1355 (1964).
293. M. F. Shostakovskii, V. M. Vlasov, and R. G. Mirskov, *Proc. Acad. Sci. U.S.S.R. Chem. Sect.*, **159**, 1296 (1964).
294. M. F. Shostakovskii, V. M. Vlasov, R. G. Mirskov, and I. M. Korotaeva, *J. Gen. Chem. U.S.S.R.*, **35**, 403 (1965).
295. M. F. Shostakovskii, V. M. Vlasov, R. G. Mirskov, and V. N. Petrova, *J. Gen. Chem. U.S.S.R.*, **35**, 43 (1965).
296. V. A. Shushunov and T. G. Brilkina, *Proc. Acad. Sci. U.S.S.R.*, **141**, 1310 (1961).
297. T. A. Smith and F. S. Kipping, *J. Chem. Soc.*, **103**, 2034 (1913).
298. A. Solerio, *Gazzetta*, **81**, 664 (1951).
299. A. Solerio, *Gazzetta*, **85**, 61 (1955).
300. G. Sosnovsky and J. H. Brown, *Chem. Rev.*, **66**, 529 (1966).
301. A. Strecker, *Annalen*, **105**, 306 (1858).
302. A. Strecker, *Annalen*, **123**, 365 (1862).
303. D. Sukhani, V. D. Gupta, and R. C. Mehrotra, *J. Organometal. Chem.*, **7**, 85 (1967).
304. Y. Takeda, Y. Hayakawa, T. Fueno, and J. Furukawa, *Makromol. Chem.*, **83**, 234 (1965).
305. T. V. Talalaeva, N. A. Zaitseva, and K. A. Kocheshkov, *J. Gen. Chem. U.S.S.R.*, **16**, 901 (1946); through *CA*, **41**, 2014 (1947).
306. T. Tanaka, M. Komura, Y. Kawasaki, and R. Okawara, *J. Organometal. Chem.*, **1**, 484 (1964).
307. T. Tanaka, R. Ueeda, M. Wada, and R. Okawara, *Bull. Chem. Soc. Japan*, **37**, 1554 (1964).
308. W. S. Tatlow and E. G. Rochow, *J. Org. Chem.*, **17**, 1555 (1952).
309. C. Thies and J. B. Kinsinger, *Inorg. Chem.*, **3**, 551 (1964).
310. I. M. Thomas, *Can. J. Chem.*, **39**, 1386 (1961).
311. R. S. Tobias, *Organometal. Chem. Rev.*, **1**, 93 (1966).
312. A. Tzschach and E. Reiss, *J. Organometal. Chem.*, **8**, 255 (1967).

313. R. Ueeda, Y. Kawasaki, T. Tanaka, and R. Okawara, *J. Organometal. Chem.*, **5**, 194 (1966).
314. J. Valade and M. Pereyre, *Compt. rend.*, **254**, 3693 (1962).
315. J. Valade, M. Pereyre, and R. Calas, *Compt. rend.*, **253**, 1216 (1961).
316. G. J. M. van der Kerk and J. G. A. Luijten, *J. Appl. Chem.*, **6**, 49 (1956).
317. G. J. M. van der Kerk and J. G. A. Luijten, *J. Appl. Chem.*, **6**, 56 (1956).
318. G. J. M. van der Kerk and J. G. A. Luijten, *J. Appl. Chem.*, **7**, 369 (1957).
319. V. M. Vlasov, R. G. Mirskov, and V. N. Petrova, *J. Gen. Chem. U.S.S.R.*, **37**, 902 (1967).
320. N. S. Vyazankin, G. A. Razuvaev, and T. N. Brevnova, *J. Gen. Chem. U.S.S.R.*, **35**, 2024 (1965).
321. N. S. Vyazankin, G. A. Razuvaev, and T. N. Brevnova, *Proc. Acad. Sci. U.S.S.R.*, **163**, 801 (1965).
322. N. S. Vyazankin, G. A. Razuvaev, O. S. D'Yachkovskaya, and O. A. Shchepetkova, *Proc. Acad. Sci. U.S.S.R.*, **143**, 343 (1962).
323. N. S. Vyazankin, G. A. Razuvaev, and O. A. Kruglaya, *Izvest. Akad. Nauk S.S.S.R. Otdel. Khim. Nauk*, 2008 (1962); through *CA*, **58**, 9114 (1963).
324. N. S. Vyazankin, G. A. Razuvaev, O. A. Kruglaya-Shchepetkova, and O. S. D'Yachkovskaya, *Khim. Perekisnykh Soedin. Akad. Nauk S.S.S.R.*, 298 (1963); through *CA*, **60**, 12040 (1964).
325. M. Wada, M. Nishiho, and R. Okawara, *J. Organometal. Chem.*, **3**, 70 (1965).
326. C. Walling and L. Heaton, *J. Am. Chem. Soc.*, **87**, 48 (1965).
327. R. West and R. H. Baney, *J. Phys. Chem.*, **64**, 822 (1960).
328. R. West, R. H. Baney, and D. L. Powell, *J. Am. Chem. Soc.*, **82**, 6269 (1960).
329. A. H. Westlake and D. F. Martin, *J. Inorg. Nucl. Chem.*, **27**, 1579 (1965).
330. P. J. Wheatley, *J. Chem. Soc.*, 5027 (1961).
331. A. Ya. Yakubovich, S. P. Makarov, and G. I. Gavrilov, *Zhur. Obshchei. Khim.*, **22**, 1534 (1952).
332. Y. A. Yakubovich, S. P. Makarov, and V. A. Ginsburg, *Zhur. Obschei Khim.*, **28**, 1036 (1958); through *CA*, **52**, 17094 (1958).
333. K. Yasuda, H. Matsumoto, and R. Okawara, *J. Organometal. Chem.*, **6**, 528 (1966).
334. K. Yasuda and R. Okawara, *J. Organometal. Chem.*, **3**, 76 (1965).
335. L. I. Zakharkin and O. Yu. Okhlobystin, *Dokl. Akad. Nauk S.S.S.R.*, **116**, 236 (1957); through *CA*, **52**, 6167 (1958).
336. L. I. Zakharkin and O. Yu. Okhlobystin, *Izv. Akad. Nauk S.S.S.R.*, 1942 (1959); through *CA*, **54**, 9738 (1960).
337. N. N. Zemlyanski, E. M. Panov, O. P. Shamagina, and K. A. Kocheshkov, *J. Gen. Chem. U.S.S.R.*, **35**, 1034 (1965).
338. N. N. Zemlyanski, E. M. Panov, N. A. Slovokhotova, O. P. Shamagina, and K. A. Kocheshkov, *Proc. Acad. Sci. U.S.S.R.*, **149**, 205 (1963).
339. S. M. Zhivukhin, E. D. Dudikova, and V. V. Kireev, *J. Gen. Chem. U.S.S.R.*, **31**, 2895 (1961).
340. S. M. Zhivukhin, E. D. Dudikova, and E. M. Ter-Sarkisyan, *J. Gen. Chem. U.S.S.R.*, **32**, 3010 (1962).
341. J. R. Zietz, S. M. Blitzer, H. E. Redman, and G. C. Robinson, *J. Org. Chem.*, **22**, 60 (1957).
342. H. Zimmer, I. Hechenbleikner, O. A. Honberg, and M. Danzik, *J. Org. Chem.*, **29**, 2632 (1964).
343. M. L. Zweigle, U.S. Pat. 3,113,144; through *CA*, **60**, 5550 (1964).
344. M. L. Zweigle and H. Tolkmith, U.S. Pat. 3,099,668; through *CA*, **60**, 550 (1964).
345. J. Zuckermann, E. W. Randall, and C. H. Yoder, *J. Am. Chem. Soc.*, **89**, 3438 (1967).

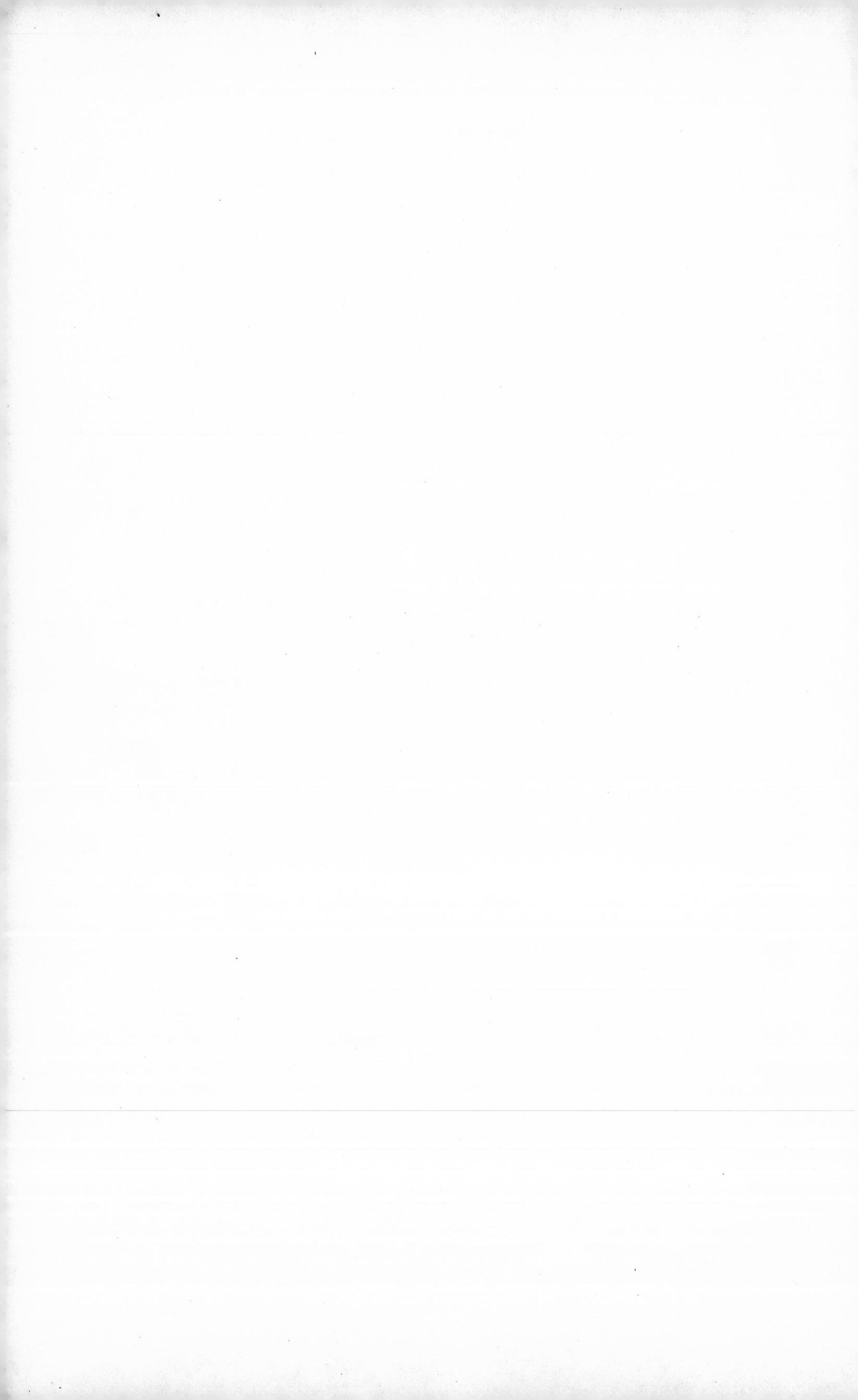